Restoring North America's Birds

Restoring North America's Birds

Lessons from Landscape Ecology

Second Edition

Robert A. Askins

Illustrations by Julie Zickefoose

Yale University Press   New Haven & London

Frontispiece: Detail of fig 3.8.

Designed by James Johnson and set in Quadraat Roman type by
The Composing Room of Michigan, Inc., Grand Rapids, Michigan.

Printed in the United States of America by R. R. Donnelley & Sons Company,
Harrisonburg, Virginia.

Library of Congress control number: 2001099750

ISBN 0-300-09316-0 (pbk. : alk. paper)

A catalogue record for this book is available from the British Library.

The paper in this book meets the guidelines for permanence
and durability of the Committee on Production Guidelines
for Book Longevity of the Council on Library Resources.

10 9 8 7 6 5 4 3 2 1

To my wife, Karen, with whom I have shared most of the birds and natural areas described in this book

Contents

Preface   ix

CHAPTER 1. Grassland Birds of the East Coast:
Pleistocene Parkland to Hay Meadow   1

CHAPTER 2. Another Quiet Decline: Birds of the
Eastern Thickets   27

CHAPTER 3. The Great Plains: Birds of the Shifting Mosaic   55

CHAPTER 4. Lost Birds of the Eastern Forest   75

CHAPTER 5. Deep-forest Birds and Hostile Edges   99

CHAPTER 6. Industrial Forestry and the Prospects for
Northern Birds   131

CHAPTER 7. Birds of the Western Mountain Slopes   155

CHAPTER 8. Declining Birds of Southwestern Floodplains   185

CHAPTER 9. Red-cockaded Woodpeckers and the
Longleaf Pine Woodland   209

CHAPTER 10. Landscape Ecology: The Key to
Bird Conservation   229

Afterword   246
Appendixes   259
Notes   269
References   283
Index   319

Color plates follow page 144.

Preface

I was inspired to write this book after a long series of public presentations about conservation of birds in eastern North America. I spoke at many different types of meetings; the audience might consist of foresters, serious birders, wildlife managers, people with a general interest in natural history, members of land trusts and other conservation organizations, private forest owners, or faculty and students in ecology and conservation biology programs. I discovered that scientific research on ecology and conservation of birds had become confusing to both amateurs and professionals. Thousands of excellent scientific studies have a direct bearing on how habitats can be managed to sustain bird populations, but most address very specific and often very local issues. Also, different papers sometimes present apparently contradictory results.

I found that people who had attempted to understand the basic issues by piecemeal reading of scientific papers were often confused by both the apparent contradictions and the amount of detailed information. This is even more true for people who occasionally read reports on bird conservation in newspapers and magazines. For example, in recent years the *New York Times* printed two stories on bird populations in eastern North America: the first (on April 4, 1995) was headlined "Songbird Population Losses Tied to Forest Fragmentation" while the second (on June 10, 1997) was entitled "Something To Sing About: Songbirds Aren't in Decline." The first headline leaves the impression that forest birds are in serious trouble, while the second implies that there may not be a problem after all. Had the situation changed dramatically in the two years between these articles? Actually, the articles are not really contradictory, and both are generally thorough accounts of the issues, but people who had read them were confused about the status of migratory birds. Are migratory birds in trouble or not? Has the widespread concern about them been a false alarm?

Many of the issues that confront conservationists are surrounded by confusion and debate. Should we be concerned by the decline of old-field birds in eastern North America? If so, why? Are we trying to preserve all biological diversity, or just some types of biological diversity? Were the massive fires in Yellowstone National Park in 1988 a disaster precipitated by poor management of the park, or an inevitable part of a natural cycle? Would we lose biological diversity (including some species of birds) without such fires? If the effects of deforestation are so severe in the tropics, why did so few species of woodland birds become extinct in eastern North America in the eighteenth and nineteenth centuries, when most of the eastern deciduous forest was cleared? Can we sustain a diversity of birds and other organisms on land that is used for timber harvesting, ranching, farming, and other economic activities, or is it more effective to concentrate on establishing and protecting strict nature preserves?

All of these questions can be addressed separately, but when they are addressed together, the answers begin to converge on a few common themes. To some extent I expected this to be the case when I began writing this book, but I was surprised at how strongly the same basic ecological principles applied to habitats and birds that seemed to be distinctly different. For example, birds in southwestern streamside woodlands, northeastern thickets, and southern pine savannas all depend on periodic disturbances—floods, severe storms and beavers, and fires, respectively—to generate the habitat they need. Protecting an area of good habitat will be fruitless if natural disturbances are suppressed. Other species of birds depend on stable systems where natural disturbances occur infrequently, so their habitats must be managed differently. Spotted Owls, Marbled Murrelets, and Ivory-billed Woodpeckers are especially vulnerable because they depend on old-growth forests, and their populations melt away as ancient trees are harvested. Species may also be in trouble because they require large blocs of their preferred habitat. The Cerulean Warbler, Upland Sandpiper, Golden-winged Warbler, and Spotted Owl have probably suffered severe population declines primarily because of this requirement—even though they use distinctly different habitats. I also identify a group of species that are potentially vulnerable because of their nomadic behavior. These birds need to move to different nesting areas in different years. This requirement may have doomed the Passenger Pigeon, and it may eventually threaten the Red Crossbill, Pinyon Jay, Cape May Warbler, and Clark's Nutcracker.

Although this is overtly a book about bird conservation, at a deeper level it is an introduction to some of the concepts of landscape ecology. Instead of abstract chapters on theory, however, I have let the general principles emerge from case studies of particular ecosystems. Even this approach can be too distant from the real world of birds and their habitats, so I include profiles of the ecology and behavior of representative or particularly intriguing species, such as Dickcissel, Yellow-breasted Chat, Bachman's Warbler, Kirtland's Warbler,

Pinyon Jay, Clark's Nutcracker, Red Crossbill, Phainopepla, Mountain Plover, Marbled Murrelet, and Red-cockaded Woodpecker.

The principles of landscape ecology are made explicit in the last chapter, but, if the book works, the reader should already know them well by this point. Again, I try to avoid an overly abstract approach in the final chapter by showing how several of these principles can be applied to particular conservation problems in the Texas Hill Country, were two species of endangered songbirds live.

This book covers many of the major habitats of the temperate zone in the United States and Canada, but I have not attempted to survey all of the major habitats on the continent. I could have used other important habitats—such as upland deserts, California coastal chaparral, and tidal marshes—to illustrate the same principles, but my goal was not to provide an overview of all of the habitats of North America. Instead, I focus on a few major ecosystems to explain how they work and what their bird populations require to survive.

Although I hope that this synthesis will be interesting to researchers and students of ecology, conservation biology, and ornithology, I wrote this book primarily for the land managers and conservationists who are largely responsible for protecting the rich natural diversity of North America, and for the birders and other naturalists who have an interest in the ecology and behavior of birds and are looking for ways to tie together what they know. One of my primary goals is to distill the voluminous research on bird conservation and present it in a form that is accessible and useful for foresters, wildlife managers, nature preserve managers, biologists with The Nature Conservancy and other conservation organizations, and the numerous citizens who donate their time and money for conservation. These people are interested not only in protecting natural areas but also in doing it the right way, with the best information available from scientific research.

I have avoided scientific jargon as much as possible, and I have explained the few technical terms that are used. Scientific names are relegated to an appendix, and I have avoided cluttering the text with "author, year" citations by using endnotes. Citations are critically important in a book like this, however, because many readers may want to study the original sources to improve their understanding of a particular ecosystem or bird species. Although I believe in the advantages of the metric system of measurement, I realize that English measurements (acres, miles, feet) are more familiar to most land mangers and naturalists in the United States. I therefore used the English system with the metric equivalent in parentheses.

My own field work has taken place in only a few regions: southern New England, Minnesota, the Virgin Islands, Guatemala, and Japan. Although I have visited virtually all of the habitats discussed in this book, walking through the terrain and identifying the birds, I do not have research experience in most of these habitats. I gathered information from the scientific literature, but ultimately I depended on researchers who had worked in these habitats to check my interpretations. The following researchers reviewed early drafts of this book and greatly improved it with their corrections and suggestions: Peter Vickery, Massachusetts Audubon Society (eastern grasslands); John Confer, Ithaca College, and David Ewert, The Nature Conservancy (eastern shrublands); Fritz Knopf, National Biological Service, and John Zimmerman, Kansas State University (prairies); Stuart Pimm, University of Tennessee (extinction of eastern forest birds); Edmund Telfer, Canadian Wildlife Service (boreal forests); Richard Hutto, University of Montana (western mountains); Kenneth Rosenberg, Cornell University (southwestern streamside woodlands); and Todd Engstrom, Tall Timbers Research Center, and Ralph Costa of the Red-cockaded Woodpecker Field Office, U.S. Fish and Wildlife Service (longleaf pine woodlands). John Kricher of Wheaton College read two of the chapters on western birds and gave me considerable general advice. George Willauer of Connecticut College and Alan Brush of the University of Connecticut also provided helpful comments on some chapters. Jean-Pierre Savard of the Canadian Wildlife Service and Ralph Costa sent me copies of important references and helped me locate other references on birds of western coniferous forests and Red-cockaded Woodpeckers, respectively. David Krueper of the Bureau of Land Management and Robert Ohmart of Arizona State University provided me with information and illustrations of riparian restoration sites in Arizona and California. In addition, Richard De-Graaf of the U.S. Forest Service provided invaluable advice and assistance when I was planning this book, and he carefully reviewed most of the chapters. Finally, Thomas Sherry of Tulane University and an anonymous reviewer provided invaluable advice on improving the organization and clarity of this book.

It was a great pleasure working with Julie Zickefoose on the paintings that introduce each chapter. She immediately comprehended my goal—to show a species in the context of its habitat—and, with her deep knowledge of North American birds, she was able to depict species beautifully, dynamically, and accurately.

I appreciate the efforts of my editor at Yale University Press, Jean Thomson Black, who patiently guided me through every step in the preparation of this book, and of Heidi Downey, who carefully edited the manuscript.

Much of the research for this book was completed when I was on sabbatical leave from Connecticut College and held an appointment as visiting professor in the School of Forestry and Environmental Studies at Yale University. I greatly benefited from discussions with Steven Beissinger when I worked in his lab that year, and I was able to search the literature broadly and deeply because of the excellent library system at Yale. I especially depended on the superb Ornithology Library in the Peabody Museum of Natural History, where Celia Lewis and Fred Sibley assisted me.

The research on eastern ecosystems was supported by funds provided by the U.S. Department of Agriculture, Forest Service, Northeastern Forest Experiment Station. Funding for research on field projects related to the subject of this book was received from the following organizations: Connecticut Department of Environmental Protection, Wildlife Division; Connecticut College Arboretum; Connecticut Forest and Parks Association; Institute of Tropical Forestry, U.S. Forest Service; U.S. National Biological Service; U.S. National Park Service; The Nature Conservancy (Connecticut and Virgin Islands chapters); National Geographic Society; and World Nature Association.

Restoring North America's Birds

CHAPTER 1

Grassland Birds of the East Coast

Pleistocene Parkland to Hay Meadow

The Bobolink is gone—
The Rowdy of the Meadow—
And no one swaggers now but me—
The Presbyterian Birds
Can now resume the Meeting
He boldly interrupted that overflowing Day
When supplicating mercy
In a portentous way
He swung upon the Decalogue
And shouted let us pray—

—EMILY DICKINSON, c. 1883

EMILY Dickinson's surroundings, in Amherst, Massachusetts, at the middle of the nineteenth century, were strikingly different from the heavily wooded suburbs and mountainsides that dominate New England today. Her poems describe a world of open meadows and large vistas. The birds of her poems are birds of hayfields, gardens, and orchards. A modern poet evoking the New England countryside would more likely describe the songs of the Wood Thrush and other forest birds. Poetry, of course, is not necessarily a dependable source of information about nature; the larks of Dickinson's poems were probably the Skylarks encountered in British poetry, not the inconspicuous Horned Larks that she may have flushed in a winter pasture. Nonetheless, the writing of an earlier century can help describe a lost landscape.

Although Dickinson wrote most frequently about the still-common American Robin, the second most frequent species in her poems, the

PLATE 1. Male Bobolink singing in grassland on a reclaimed strip mine (The Wilds) near Caldwell, Ohio.

Bobolink, has disappeared from much of New England. It has declined for a number of reasons, but principally because of the loss of the open meadows and pastures where it lived. The poem "The Bobolink is gone," which describes the dullness of a meadow following the Bobolink's migration south, has become more poignant as the species has disappeared from most of New England and, indeed, from most of eastern North America.

Nearly all of New England's forests have grown up through the skeleton of abandoned farmland. When walking in a tall, old forest, one constantly climbs stone walls and passes root cellars, stone foundations, and massive piles of rocks (removed one by one to improve fields that no longer exist). The truly giant trees grow along the walls; they have the wide-spreading branches of open-growing pasture trees, not the sleek, columnar shape of trees that have always been surrounded by other trees. If one looks down on the leafless winter forest from an airplane, the gridwork of stone walls maps out the hay meadows, orchards, and vegetable fields of the last century. All of these landmarks point to a past in which a dense and busy human population occupied what is now the interior of an extensive forest.

One can capture the feeling of this lost landscape in the poetry of Emily Dickinson or the early poems of Robert Frost, but the magnitude of the ecological changes in New England is described better by naturalists. The novels and poems of the nineteenth century suggest that the Bobolink was a common bird, but the proportions of its abundance in eastern North America are made more dramatically clear in the detailed descriptions of John James Audubon. In *Ornithological Biography* (1831), Audubon described Bobolinks in New York and Connecticut in mid-May, when "they have become so plentiful, and have so dispersed over the country, that it is impossible to see a meadow or a field of corn, which does not contain several pairs of them."[1] At this time of year they were trapped in "great numbers" for the cage bird trade. At the end of July they would concentrate in huge flocks that "plunder every field, but are shot in immense numbers." According to Audubon, professional hunters shot millions of these birds at their evening roosts during the fall migration along the East Coast and in the rice plantations in the Carolinas, where they spent the winter. Bobolinks and other songbirds were sold for food in city markets.

Bobolinks thrived in the nineteenth century despite the depredations of bird catchers and market hunters. Ironically, after they were protected by the migratory bird treaty in the early twentieth century, their population began a steep decline that continues today. If they had disappeared because of

FIG. I.I. Aerial view of a woodland in southern New England in winter. Note the gridwork in stone walls from an earlier era of cropland and pastures. Photograph by William Niering.

overhunting there undoubtedly would have been an outcry from conservationists. But the decline results from slow changes in the landscape, changes that also have diminished or eliminated the populations of many other species of grassland birds. What Harold Mayfield called the "quiet decline" of grassland birds has attracted surprisingly little notice or concern.[2] This response (or lack of a response) is rooted in the myth that before Europeans cleared the land, unbroken forest stretched from the Atlantic to the Great Plains.

Population Declines in Grassland Birds

Many species of grassland birds were common or even abundant along the East Coast through most of the nineteenth century, but their numbers diminished noticeably between the late 1800s and the middle of the twentieth century. Ludlow Griscom described the Upland Sandpiper, Bobolink, Eastern Meadowlark, and Grasshopper Sparrow as formerly common but declining in Massachusetts.[3] Edward Forbush mourned the virtual disappearance of the Upland Sandpiper from New England: "Our children's children may never see an Upland Plover [Sandpiper] in the sky or hear its rich notes on the summer air. Its cries are among the most pleasing and remarkable sounds of rural life. . . . [Its] long-drawn, rolling, mellow whistle . . . has the sad quality of the November wind."[4]

Although the decline of grassland birds was obvious to any careful observer, only in recent decades have we been able to calculate the precise rate and extent of these population changes. The best evidence for this comes from the Breeding Bird Survey (or BBS), a system of roadside routes scattered throughout the United States and southern Canada where birds are counted each year. In 1994 more than 3,000 of these routes were surveyed. A volunteer proficient at identifying birds and bird songs travels along a 25-mile route, stopping every half mile and recording all the birds seen or heard during three minutes. The results for all of the survey routes east of the Mississippi River indicate that since 1966, when the surveys began, the abundance of 15 of the 19 species of grassland and savanna birds in eastern North America has fallen.[5] Some have shown rapid population changes. Between 1966 and 1994, for example, Grasshopper Sparrows decreased at a rate of 6 percent per year, while the annual rates of decline were 3 percent for Vesper Sparrows, 9 percent for Henslow's Sparrows, and 3 percent for Eastern Meadowlarks. In contrast, only 2 of the 40 species of forest-dwelling migratory birds (a group that has received considerable attention from conservationists) decreased at a rate of more than 2 percent per year during the same period.

Another indication that grassland birds in the northeastern United States are in trouble comes from state lists of endangered and threatened species.[6] Of the 40 species listed as endangered, threatened, or of special concern in three or more northeastern states, 13 are grassland or savanna specialists, and only three are forest specialists. For example, Upland Sandpiper, Northern Harrier, Loggerhead Shrike, Grasshopper Sparrow, Henslow's Sparrow, and Vesper Sparrow are listed in all or most of the New England states.[7] The populations of many of these species have dropped in other parts of the eastern United States, both along the heavily forested East Coast and in the more agricultural Midwest.[8]

Unlike migratory songbirds that live in forests, eastern grassland birds have received relatively little attention from most government wildlife agencies and conservation organizations. One reason for this surprising complacency in the face of well-documented population losses is the general impression that most grassland species are not native to the region but have invaded the eastern states from western savannas and prairies after the

FIG. 1.2. Locations of Breeding Bird Survey routes in North America. From Peterjohn and Sauer, 1994. Reprinted from *Birding* (vol. 26, p. 387), with permission of the American Birding Association and Bruce G. Peterjohn.

clearing of the forest for agriculture. For example, Whitcomb argues that this invasion of the eastern "neosavanna" created by agriculture has been a "failed experiment for many of these species" that are now declining.[9] The implication is that this is a return to ecosystems more similar to those existing before European settlement and therefore should not be a cause for concern. According to Whitcomb, these species could survive only with active management to preserve grassland "in a region where [grassland] is inappropriate as an equilibrium community."[10]

Many historians and botanists have depicted the landscape of the ancient East Coast of North America as carpeted with forest, a forest so continuous that a squirrel could travel from the Atlantic Ocean to the Mississippi River without touching the ground.[11] Since the early 1900s, however,
some botanists have argued that the forest was not always continuous but
was sometimes interrupted by scrubland, barrens, and even, in places, with
prairie-like grasslands.[12] If this is true, then grassland birds may have had
a place in the presettlement landscape.

Were There Grasslands on the East Coast Before European Settlement?

When H. S. Conrad visited Long Island's Hempstead Plains in the 1930s,
much of the area was little bluestem prairie, a yellow-green grassland dotted with the small, bright-green hemispheres of wild indigo plants.[13] In
May, the prairie was blue with the blossoms of birdfoot violets, and in June
it was speckled with the yellow globes of wild indigo in bloom. As Conrad
pointed out, this grassland on the coast of New York was remarkably similar to the tallgrass prairies of Iowa and Nebraska. Moreover, the Hempstead
Plains had a rich community of grassland birds: Upland Sandpipers,
Bobolinks, Vesper Sparrows, and Grasshopper Sparrows were all common
there in the 1920s.[14]

European travelers described the Hempstead Plains as treeless in the
1600s, so this grassland was not a product of European agriculture.[15] The
Hempstead Plains was characterized by thin soil resting on a porous foundation of quartz and granite pebbles, features that, in combination with periodic fires, may have favored the growth of grasses and herbs rather than
trees and shrubs.[16]

The Plains once covered more than 50,000 acres (20,000 ha), and for
many years it was used primarily for sheep grazing and horse races.[17] Large
areas of grassland remained when Conrad visited the area in the 1930s, but
after the Second World War these open areas were subdivided for housing
or plowed for truck farms. Today only a few acres of this prairie survives: a
19-acre parcel belonging to Nassau County Community College and managed by The Nature Conservancy, and a 46-acre parcel managed as a nature
preserve by Nassau County.[18] The smaller preserve has been maintained
with controlled burning.

Although the Hempstead Plains may have been one of the largest and

FIG. 1.3. Hempstead Plains of Long Island in 1909 (from Harper, 1911). From the American Geographical Society Collection, University of Wisconsin-Milwaukee Library.

most distinctive grasslands on the East Coast, it was not the only one. Another grassland, the Montauk Downs, covered part of eastern Long Island, and several large grasslands called glades characterized a plateau in the Allegheny Mountains of western Pennsylvania.[19] Also, in the 1600s a savanna where occasional large oaks broke an expanse of tall, wiry grass (probably bluestem) stretched for 15 miles along the Quinnipiac River north of New Haven, Connecticut.[20] After decades of overgrazing, this area became the almost desertlike North Haven sand plains, and subsequently most of it was developed. Blueberry barrens, which are open expanses covered with low blueberry shrubs and grasses, still cover large areas in eastern Maine, where they are maintained by burning for blueberry production. Some of the largest East Coast populations of Upland Sandpipers, Vesper Sparrows, and other species of grassland birds live in these barrens.[21]

Many of these grasslands may have resulted from the activities of Indians before European settlement. Early explorers and colonists frequently encountered open landscapes created by firewood harvesting, clearing for maize fields, and burning to enhance hunting. For example, Verrazano described the area around Narragansett Bay (Rhode Island) in 1524 as open plains, without forests or trees, for many leagues inland.[22] Champlain and John Smith reported extensive areas of cleared land along the New England coast before Europeans colonized the region.[23] Moreover, an early settler in Salem, Massachusetts, described "open plains, in some places five hundred acres . . . not much troublesome for to cleere for the plough to goe in."[24]

These clearings were not restricted to coastal regions; accounts of early settlers indicate that river valleys had been cleared by Indians for farming and hunting.[25]

Early assessments that Indian agriculture had relatively little effect on the landscape were based on population estimates after European settlement, but population densities were much higher before contact with Europeans triggered epidemics that killed a large proportion of the people in most tribes.[26] As Kulikoff wrote regarding the Chesapeake Bay region, "though English settlers did not find a wilderness, they did create one"[27]; extensive agricultural clearings reverted to forest as Indian populations declined. Pilgrims traveling through the area near Warren, Massachusetts, in 1621 "saw the remains of so many once occupied villages and such extensive formerly cultivated fields that they concluded thousands of people must have lived there before the plague."[28] Maps, drawings, and written accounts of the landscape around Indian settlements in the southeastern United States before European settlement provide evidence of extensive clearings created by farming and of "parklands" maintained by controlled burning.[29] In New York and southern New England, relatively high population densities combined with the slash-and-burn agriculture would have resulted in extensive areas of cleared land in the form of both active and abandoned fields.[30] This would have produced a "mosaic of forests and fields in varying degrees of succession."[31] Another view is that eastern tribes used large permanent agricultural fields from which tree stumps had been removed rather than temporary fields cut out of the forest for slash-and burn agriculture. This permanent farmland would have to be rested occasionally, however, producing "weed-covered" fallow fields of the sort seen by Samuel de Champlain near the site of Boston in 1605.[32] Regardless of whether the Indians used slash-and-burn or permanent field agriculture, their activities would have produced open habitats (abandoned or fallow fields) that could be used by grassland birds.

There also is good evidence that the Indians of the East Coast burned large areas to create open woodlands and grassland for hunting. For example, Roger Williams wrote in the 1640s that Indians in New England "burnt up all the underwoods in the Countrey, once or twice a yeare and therefore as Noble men in England possessed great Parkes . . . onely for their game."[33] In 1818, B. Trumbull reported that the Indians of Connecticut "so often burned the country, to take deer and other small game, that in many of the plain dry parts of it, there was but little small timber. Where the lands

FIG. I.4. Prescribed fire in a bluestem grassland at the Connecticut College Arboretum. Photograph by William Niering.

were burned there grew bent grass . . . two, three and four feet high."[34] In New York, Indians not only burned the woods each autumn to create a more open understory, but they also burned plains and meadows to improve hunting.[35]

Although fires were probably infrequent in most forests that were remote from Indian settlements, fire and other disturbances near settlements provided extensive habitat for early successional species, including potential habitat for grassland birds.[36] An analysis of charcoal deposits in the sediments of 11 lakes in New England demonstrated that before European settlement fires were frequent in densely populated coastal regions, but infrequent in the inland and northern areas.[37] Moreover, Chris Winne's analysis of pollen and charcoal in lake sediments shows that the land around Pineo Pond in eastern Maine has been characterized by frequent, moderate fires and scrubby, fire-adapted vegetation for at least 900 years.[38] Today this area is dominated by blueberry barrens maintained by controlled burning. These support a diversity of breeding grassland birds.[39]

The extent of forest clearing by Indians in the Northeast probably paled in comparison with the extensive agricultural fields created by the Moundbuilders, who lived along the Mississippi River and its tributaries in much of what is now the southeastern and midwestern United States. Mound-building cultures existed in the lower Mississippi River Valley as early as 1500 B.C. The early Moundbuilding cultures probably depended on a mixture of hunting, gathering of nuts, fishing, and small-scale farming based on native plants such as sunflower and marsh-elder.[40] Later, during the Mississippian Period, which lasted from A.D. 700 until the early 1700s, large-scale agriculture supported a dense population living in closely spaced villages. Corn and beans from Mexico replaced indigenous crops, and large areas were cleared for farming.[41]

The largest population center was Cahokia, located on the Mississippi River near its confluence with the Missouri River.[42] This center covered 2,000 acres (800 ha), with 400 acres (160 ha) enclosed in a wooden palisade. The site was dotted with as many as 120 earthen mounds, the largest of which rose to 100 feet (30 m) and covered more than 15 acres (6 ha). The mounds supported wooden buildings, and the area below the mounds was packed with rectangular, thatched-roof houses in which 10,000 to 20,000 people lived.[43] Cahokia and the many villages and towns around it were supported by farming the American Bottom, a 125-square-mile strip of rich alluvial soil in the floodplain along the eastern bank of the Mississippi River. In A.D. 1000 there were 50 villages and eight other large or medium-sized centers within 25 miles of Cahokia. Clearing land for maize fields and cutting trees for fuel and for construction of thousands of houses and large temples and stockade walls must have destroyed most of the bottomland forest and upland oak-hickory forests near Cahokia.[44]

There are no historical accounts of Cahokia because it was abandoned after 1400.[45] Early Spanish explorers visited similar sites that were still occupied in the 1500s, however. A chronicler of Hernando De Soto's expedition (1539–1542) described a Moundbuilder town along the Mississippi River as "in an open field, that for a quarter of a league over was all inhabited; and at a distance of from half a league to a league off were many other large towns, in which was a good quantity of maize, beans, walnuts, and dried plums."[46]

The Moundbuilding centers were abandoned long before Europeans settled the southeastern United States or the Mississippi Valley. This culture may have been destroyed by Old World diseases that swept inland from European outposts on the Florida and Gulf coasts.[47] Before the collapse of this agricultural society, however, there were extensive expanses of open fields in many parts of the Southeast, especially in the lower Mississippi

FIG. 1.5. Above, reconstructions of Central Cahokia, a Mississippian site on the banks of the Mississippi River in A.D. 1150. View between Twin Mounds across Grand Plaza to Monks Mounds, showing a playing field in the plaza, markets around the perimeter of the plaza, and houses of the elite adjacent to plaza. Painting by Lloyd K. Townsend. Below, overview of Central Cahokia, 1000–1150, showing the central ceremonial precinct inside the stockade wall, and agricultural fields in the surrounding region. Painting by William R. Iseminger. Both paintings are reproduced with permission from Cahokia Mounds State Historic Site, Collinsville, Illinois.

River Valley. De Soto's chroniclers did not report bison, but they were present in parts of the Southeast when the first English and French settlers arrived. Alfred Crosby speculated that bison moved from the western prairies into abandoned farmland after the collapse of the Moundbuilder centers.[48]

Grasslands Before Indian Agriculture

The patterns of Indian land use observed in the 1600s began to emerge about 5,000 years ago.[49] Perhaps many species of grassland birds colonized cultivated fields in the East after the initiation of Indian, rather than European, agriculture, and their current decline represents a return to conditions before humans began to modify the vegetation of eastern North America substantially. Undoubtedly, many apparently "natural" open grasslands of eastern North America are the product of human activities. For example, historical accounts and analysis of the pollen record indicate that the extensive heathlands and sand plain grasslands on Nantucket Island, Massachusetts, resulted from the clearing of oak forest and grazing of sheep after Europeans settled that island.[50] Some of the open habitats on the East Coast may predate disturbance by either Indians or Europeans, however.

Smaller shrubby and grassy openings in the eastern forest result from dam-building by beavers. After beaver exhaust the food supply around a pond, they move to another area. The dam on the original pond begins to leak because it is no longer being repaired by beavers, and the pond drains. The bed of the pond often becomes a "beaver meadow," a patch of shrubby vegetation or grassland. This meadow eventually is overgrown with young forest, and after ten to thirty years beavers may recolonize the site and initiate another cycle.[51]

Although beaver meadows are largely restricted to floodplains, their total area was probably extensive before beaver were extirpated in most parts of eastern North America. In Ontario's Algonquin Provincial Park, where beaver are protected, there is a high density of beaver ponds and meadows. After beavers became reestablished at the Quabbin Reservation in Massachusetts in 1952, the population grew rapidly until the density reached 1.3 colonies per mile (0.8 colony per km) of stream. The impact of a dense beaver population can be considerable. In the Adirondack Mountains of New York, beaver dams created patches of disturbance that covered an average area of 17 acres (7 ha), with a maximum area of 30 acres (12 ha).[52] Bela Hubbard, who surveyed land in Michigan before European settlement,

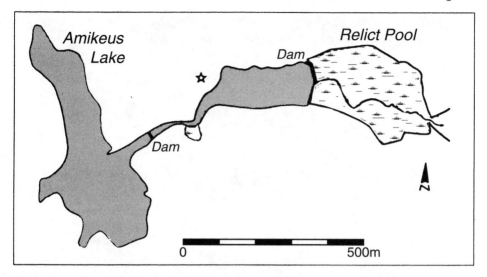

FIG. 1.6. Map of beaver dams and ponds in Ontario, Canada. The "relict pool" is an abandoned pond that has drained, creating a beaver meadow. From Coles and Orme, 1983. Reprinted by permission of J. M. Coles, B. J. Orme, and Antiquity Publications, Ltd.

reported that one-fifth of the area within 12 miles (19 km) of Detroit was covered with "marshy tracts or prairies which had their origin in the work of the beaver."[53] The creation of this type of landscape by a growing beaver population has been documented for the Kabetogama Peninsula in Voyageurs National Park in Minnesota.[54] Beaver density increased from 0.2 per square kilometer in 1927 to 3.0 per square kilometer in 1988. During this period, the amount of land covered by moist and wet meadows increased from 0.6 to 5.5 percent. Coles and Orme argued that ancient forests in England must have been "moth-holed with clearings wherever beaver were present."[55] These "grassy meadows of relict pools" were also an important feature of the presettlement landscape of eastern North America.

Many regions of the East Coast were subject to disturbances that created large openings in the forest.[56] Grassland, savanna, and grassy scrub were probably created by large fires, particularly those that burned following the hurricanes or tornadoes that periodically leveled forests. These catastrophic disturbances were probably most frequent in low-lying, sandy sections of the coastal plain. In many regions farther inland, fires, windstorms, and other disturbances were infrequent, and consequently the forest canopy was

FIG. 1.7. "Beaver meadow" in an abandoned beaver pond. From Hilfiker, 1991. Photograph by Earl Hilfiker, wildlife photographer and lecturer.

almost continuous, with few large openings. For example, the northern hardwood forest of western New York and the White Mountains of New Hampshire probably formed an almost unbroken canopy.[57] Large, grassy openings may have occurred in some of the river valleys in the interior, however. John Winthrop described how one of the first European expeditions to the White Mountains passed through "many thousands of acres of rich meadow" as they paddled their birch bark canoes up the Saco River in what is now Maine.[58] Many of the bird species disappearing from the East today may have lived in these eastern grasslands before the arrival of Europeans and perhaps even before the advent of Indian agriculture.

A Longer View: Pleistocene Steppe and Savanna

When continental glaciers covered much of Canada and the northern United States, the regions immediately south of the glaciers were dominated by spruce parkland, a grassy savanna with scattered spruce trees.[59] Samples of pollen from lake sediments deposited between 18,000 and 12,000 years ago show that this savanna stretched westward from the Atlantic Coast to the Great Plains. Most species of deciduous trees, and presumably the type of

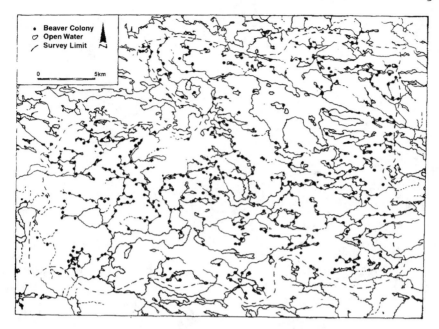

FIG. 1.8. Distribution of beaver families (colonies) in the Lake Elliot region of Ontario. From Coles and Orme, 1983. Reprinted by permission of J. M. Coles, B. J. Orme, and Antiquity Publications, Ltd.

closed forests where these species grow today, were restricted to the extreme southeastern United States.[60] Thus in eastern North America there was a gradient from savanna in the north to dense forest in the south.

Vegetation zones shifted and changed as the glaciers retreated northward beginning about 12,000 years ago. Spruce parkland largely disappeared, and a new gradient, from eastern forest to western prairie, gradually formed.[61] Before this transition occurred, however, the spruce parkland was occupied by a diversity of large open-country mammals: caribou, mastodons, and long-nosed peccaries are frequently found in fossil deposits from the time of the spruce parkland.[62] One of the best samples of spruce parkland animals come from a site called New Paris No. 4 in Pennsylvania.[63] Approximately 11,000 years ago a deep sinkhole acted like a pitfall trap, collecting the skeletons of more than 2,700 animals that fell into the crevice and died. The mixture of small mammals at this site suggests that there was extensive grassland, as does the skeleton of a Sharp-tailed Grouse, a bird now associated with the western prairie. Not many birds fell into this sinkhole, however. A better picture of the birdlife of the postglacial

FIG. 1.9. Reconstruction of spruce parkland in Pennsylvania about 11,000 years ago. The animals depicted here were identified from deposits from this period in a sinkhole called New Paris No. 4. From left: marten, snowshoe hare, long-nosed peccary, sharp-tailed grouse, caribou, and red squirrel. From *Pleistocene Mammals of North America* by B. Kúrten and E. Anderson, © 1980 Columbia University Press. Reprinted with permission of the publisher.

period comes from another site, the caves of Natural Chimneys in Virginia, where skeletons were deposited at about the same time as at New Paris Number 4. The skeletons at Natural Chimneys were deposited in the regurgitated pellets of owls that roosted in the caves, so both small mammals and birds are well represented.[64] Although the remains of these animals may have accumulated over a long period while the vegetation was changing, they provide a glimpse of the bird life of the spruce parkland. The bones of Sharp-tailed Grouse, Northern Bobwhite, Upland Sandpiper, Red-headed Woodpecker, Black-billed Magpie, and Brown-headed Cowbird, along with other grassland vertebrates, such as diamondback rattlesnake and thirteen-lined ground squirrel, point to a landscape with large amounts of savanna or grassland. The remains of woodland species, such as Red-bellied Woodpecker, Eastern Wood-Pewee, and Red-breasted Nuthatch, found in the same deposits suggests either that woodland and savanna occurred in the area at the same time, or that woodland invaded and replaced the savanna while the bones accumulated at Natural Chimneys. In either case, some grassland birds occurred in eastern North America before the spruce parkland receded and disappeared.

A key question is whether grassland birds could have survived the northward spread of closed-canopy forest over eastern North America after the continental glaciers melted away. This is an issue not only for the current warm interglacial period, but also for previous interglacial periods. In all previous interglacial periods and at the beginning of the current postglacial period, large browsers such as mastodons and giant ground sloths may have created and maintained openings in the forest in much the way African elephants maintain savannas in East Africa today.[65] European ecologists have recognized that giant herbivores, particularly the extinct relatives of elephants, probably opened up the forest, creating glades, parklike woods, or even savannas in regions that would otherwise be dominated by dense forest.[66] Such openings would have supported a variety of animal and plant species that depend on grassy habitats. Through most of the past million years, as forests retreated and advanced in response to the shrinking and growing of ice sheets, woodland habitats may have been modified and opened by the giant herbivores. Only in the present interglacial period did the mastodons, ground sloths, and other giants disappear from North America, perhaps as a result of the invasion of the continent by people who had already developed an efficient set of tools and strategies for hunting large animals.[67] Human activities, such as burning and agricultural clearing, may have replaced gigantic herbivores in creating a mosaic of forest and openings, permitting open-country species to persist in the eastern woodlands.[68]

The Origin of Eastern Grassland Birds

The common impression that many species of grassland birds spread eastward from the prairies of the Midwest to the newly cleared farmland of the eastern coast is substantiated by several well-documented examples of range expansion. The prairie subspecies of the Horned Lark (*Eremophila alpestris praticola*), for instance, spread eastward from Illinois and Wisconsin, reaching Michigan and Ontario in the 1870s, New York in the 1880s, New England by 1891, and Pennsylvania and Maryland by 1910.[69] The Dickcissel spread eastward from the tallgrass prairies in the early 1800s, but its range contracted after 1850, and it eventually disappeared as a regular breeding bird along the East Coast.[70] The Western Meadowlark expanded its range into Wisconsin and Michigan after 1900,[71] and the Lark Sparrow spread from the prairies to agricultural areas in the Ohio valley, West Virginia, and western Maryland.[72]

Although these eastward range expansions were well documented, there is no similar evidence for invasion of the East Coast by the species that are most abundant and widespread in eastern grasslands. Upland Sandpipers, Grasshopper Sparrows, Bobolinks, Eastern Meadowlarks, and other common grassland birds were reported by the earliest ornithologists who systematically documented the distribution of birds on the eastern coast of North America. Alexander Wilson's *American Ornithology* (1808–1814) and John James Audubon's *Ornithological Biography* (completed in 1839 to accompany his prints in *The Birds of North America*) were published more than 100 years after most of the eastern seaboard had been cleared, so it is possible that grassland birds colonized the meadows and pastures created by Europeans long before their occurrence was initially documented. Some seventeenth-century European observers, such as John Josselyn[73] and William Wood,[74] described gamebirds and the more conspicuous songbirds, but only a few species are recognizable because the descriptions are sketchy and the names of British birds are frequently used for North American species. Mark Catesby's *Natural History of Carolina, Florida, and the Bahama Islands*, which was completed in 1747, includes descriptions and paintings of many species of eastern birds, including two species of grassland songbirds, the Eastern Meadowlark and the Bobolink.[75] Catesby observed "innumerable flights" of Bobolinks on the Carolina coast in September and, in one of the earliest observations of birds migrating across an ocean, he identified the distinctive calls of Bobolinks coming from large flocks flying high above him at night as he slept on the deck of a sloop off Andros Island in the Bahamas. It is not surprising, however, that there are relatively few descriptions of grassland songbirds from this period. Many grassland birds are small, inconspicuous, and dull-colored, so they could have been overlooked by early observers.

Significantly, the East Coast populations of three species of grassland birds were distinctive enough from western populations to be considered separate subspecies. This suggests that these populations have existed in isolation in the east for many thousands of years, perhaps since unbroken grasslands reached from the Great Plains to the Atlantic during the last glacial period. The Eastern Henslow's Sparrow (*Ammodramus henslowii susurrans*) has a breeding range restricted to central New York and southern New England south to Virginia, eastern West Virginia, and North Carolina.[76] It is darker than the western subspecies of Henslow's Sparrow, with a stouter bill and more buff on the underparts and more yellow in the wing.[77] A subspecies of the Savannah Sparrow called the Ipswich Sparrow (*Passerculus sandwichensis princeps*) is also restricted to the East Coast.[78] It is paler gray than other subspecies of Savannah Sparrow, so it tends to be well camouflaged in the light-colored dunes and beaches where it lives. This population is so distinctive that it was considered a separate species until 1973. During the breeding season, Ipswich Sparrows are virtually restricted to low shrubby vegetation and stands of marram grass on Sable Island, off the coast of Nova Scotia.[79] In winter they are primarily found in a narrow zone of dunes near Atlantic beaches from Nova Scotia to Florida, with the highest densities on relatively undeveloped barrier islands and sandy peninsulas between New Jersey and Virginia.[80]

The eastern subspecies of the Greater-Prairie Chicken was the now-extinct Heath Hen (*Tympanuchus cupido cupido*). During the early years of European settlement Heath Hens were common or even abundant in open grasslands and scrublands on Long Island and around Boston, and they ranged along the coast from southern Maine as far south as Virginia.[81] Because it was an important game species, the Heath Hen was described in many of the accounts of early settlers. In the 1600s, William Wood, Thomas Morton, and other observers wrote that Heath Hens were common in eastern Massachusetts, inhabiting sandy scrub-oak plains, pine barrens, blueberry barrens, and other open habitats.[82] In the nineteenth century, they were common in the open grassland of the Hempstead Plains on Long Island.[83]

The abundance of the Heath Hen at the time of European settlement, and the recognition of the Heath Hen and two other populations of grassland birds as distinct subspecies restricted to the East Coast, suggest that grassland birds inhabited the East Coast long before Europeans arrived or even before Indians started clearing the land for farming. This is consistent with the evidence on grassland plants. Several species of plants are restricted to eastern grasslands (P. Dunwiddie, personal communication), suggesting that

FIG. I.IO. The Heath Hen, an extinct subspecies of the Greater Prairie-Chicken. Painting, by Louis Agassiz Fuertes, from E. H. Forbush, *Paintings of Birds of Massachusetts and Other New England States* (1927) (SC1, 114x), pl. 35. Courtesy of Massachusetts Archives.

they evolved in isolation from the grasslands of the Great Plains. Bushy rockrose is found from Massachusetts to Long Island; sand plain agalinis is found from Massachusetts to Maryland; and sickle-leaved golden aster is found from Massachusetts to New Jersey.[84] In addition, a subspecies of blazing star (*Liatris borealis novae-angliae*) is found only in eastern grasslands.

Because of the paucity of early historical records of small birds, it is likely that only carefully dated skeletal remains could provide definitive evidence of the occurrence of most species of grassland birds before European settlement. However, there is strong evidence that extensive grasslands and savannas occurred in eastern North America when Europeans arrived. We also know that some grassland species were found in the spruce parkland of postglacial times, about 11,000 years ago, and that distinctive eastern subspecies evolved in three grassland species. It therefore is reasonable to conclude that many open-country species are native to the region, not recent invaders from the western prairies. During the eighteenth and nineteenth centuries these species probably became much more abundant than they had been before Europeans cleared the land, but since then they may have

FIG. 1.11. Grassland on the Kennebunk Plains in Maine, looking across a recent burn to an area that has not been burned in more than two years. The endemic East Coast subspecies of blazing star is blooming in the foreground. Photograph by Peter Vickery.

declined far below the level of abundance characteristic of the presettlement landscape. Many of these species are now in danger of regional extinction, and they deserve the same attention from conservationists as birds associated with forests, marshes, and lakes.

Birds of Hay Meadows and Air Fields

Many of the original grasslands, such as beaver meadows and recent burns, were ephemeral. Other areas may have been disturbed frequently enough to create stable grasslands; the Hempstead Plains and some of the barrens of Maine are obvious candidates. Temporary grasslands probably are created much less frequently today because beaver are less abundant and fires are controlled, and most of the more stable grasslands have been developed for agriculture or housing. The blueberry barrens of eastern Maine are an exception; these open habitats have been maintained in a seminatural state by controlled burning to sustain commercial blueberry production.[85]

With the exception of the blueberry barrens and a few other seminatural open areas, the habitats used by grassland birds along the East Coast are highly artificial. Populations of grassland species have diminished primarily because much of the farmland in the Northeast and parts of the Southeast has been abandoned and has then reverted to forest, and because the remaining farmland is now managed more intensively for agricultural production.[86] For example, hayfields have become less suitable as nesting habitat for Eastern Meadowlarks, Bobolinks, and some other grassland species because they are mowed earlier in the summer, before the end of the nesting season, and because they are rotated more frequently.[87] In southern New England, most of the remaining populations of Grasshopper Sparrows and Upland Sandpipers are found in extensive mowed sections at airports and military airfields.[88] The farmland once used by these species has either disappeared or become unsuitable for nesting.

Regional populations of grassland birds can be maintained with proper management of artificial grasslands such as hay meadows, fallow farmland, and the mowed areas near airport runways. Farmland potentially could be managed for grassland birds and early successional species of plants and insects through the Conservation Reserve Program, which pays farmers to take land out of production in order to manage it for conservation of soil and wildlife.[89] Relatively simple changes in airport management (e.g., removal of woody vegetation and changes in mowing schedules to avoid the nesting season) have sustained or improved the habitat for grassland bird species at Westover Air Reserve Base in Massachusetts, Bradley International Airport in Connecticut, and Floyd Bennett Field, a former naval air base, on Long Island.[90] Habitat management at Westover resulted in substantial increases in the abundance of Grasshopper Sparrows and Upland Sandpipers between 1987 and 1994.[91]

Successful management of grassland birds depends on understanding their habitat requirements. Many species of grassland birds are habitat specialists.[92] Horned Larks tend to be in open areas with sparse vegetation, Grasshopper Sparrows are most common in grasslands dominated by bunchgrass with patches of bare ground, and Henslow's Sparrows are found in grasslands dominated by tall, dense grass with little or no bare ground.[93] While recently burned or mowed grassland will provide favorable habitat for Horned Larks and Grasshopper Sparrows, Henslow's Sparrows need grassland that has not been disturbed for several years. Only a mosaic of patches of grassland in different stages of recovery from disturbance will support all of these species.[94]

FIG. 1.12. Mist-netting grassland birds at Westover Air Reserve Base in Massachu-setts. Photograph by Anthony Hill.

Another important feature of habitat quality for many species of grass-land specialists is the area of continuous grassland habitat. In Illinois grass-lands and grassland barrens in Maine, both the density and diversity of grassland birds increased with habitat area.[95] In Illinois, few Grasshopper Sparrows, Savannah Sparrows, Bobolinks, or Henslow's Sparrows were found in patches of grassland smaller than 75 acres (30 ha).[96] Similarly, in Maine, Grasshopper Sparrows were absent on sites with less than 75 acres of grassland.[97] Thus, most of the effort to maintain populations of these species should be directed at extensive areas of grassland. Also, a large area of grassland can be maintained as a habitat mosaic, providing different types of habitat for different species. This is not feasible on small patches of grassland.[98]

Even when grassland birds are absent from an area, it should be possi-ble to create habitat that will attract them. Probably because eastern grass-land birds have always depended on patches of ephemeral habitat, they have a remarkable ability to find and colonize remote sites, even sites far from other bird populations of the same species. When a field in Lincoln, Massa-

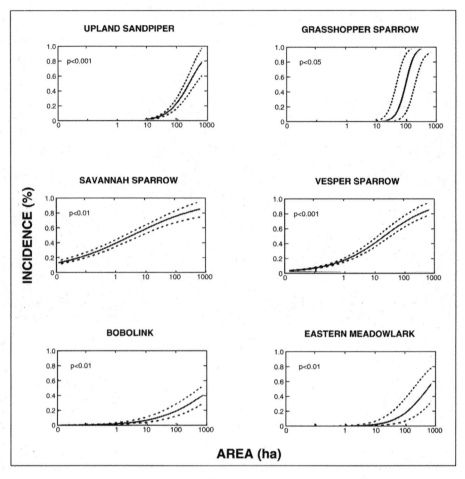

FIG. 1.13. Frequency of occurrence of six species of grassland birds in survey plots in study sites of different sizes in coastal Maine. The 90 study sites ranged from 0.7 to 998 acres (0.3–404 ha) of continuous grassland. From Vickery, 1994. Courtesy of Blackwell Publishers.

chusetts, was managed to maintain tall grass for nesting Bobolinks, it attracted a breeding pair of Henslow's Sparrows, a species that had not been known to breed in Massachusetts during the preceding twenty years.[99] Also, when abandoned strip mines in heavily forested areas of West Virginia were restored and seeded with grass, they were colonized by Horned Larks, Eastern Meadowlarks, Savannah Sparrows, Vesper Sparrows, and Grasshopper Sparrows.[100] These new grasslands were extremely isolated from other areas supporting grassland birds, but they still attracted breeding populations of several species.

Conservation of Eastern Grassland Birds

Populations of grassland specialists in eastern North America are concentrated in open habitats that are sustained by human activities, so we must focus on these highly artificial habitats in order to prevent regional extinctions. In some cases, relatively small changes in management practices, such as shifting mowing schedules to avoid the nesting season, replacing introduced turf grasses with native bunch grasses, or prescribed burning can improve or create good habitat. Expending scarce resources to maintain bird habitat in hay meadows, fallow fields, and airfields may seem unwise to conservationists who are accustomed to protecting forests and wilderness areas. However, artificial habitats are critical for many species of birds, insects, plants, and other organisms because people have destroyed most of the native grassland habitat, including most of the midwestern tallgrass prairies where many of these species were once most abundant.[101] People have not only destroyed natural grasslands directly, but they also have interrupted or dampened many of the natural processes of disturbance, such as fires and beaver activity, that once created the early successional habitats that grassland species need. Eventually these natural disturbances may be reintroduced to extensive areas in eastern North America. In the near term, however, artificial grasslands represent our best hope for maintaining grassland species. These species are an important, and probably an ancient, component of biological diversity along the East Coast of North America.

CHAPTER 2

Another Quiet Decline
Birds of the Eastern Thickets

My first meeting with the Yellow-breasted Chat occurred when as a boy, wandering idly through an old pasture overgrown with bushes near the shore of Lake Quinsigamond in central Massachusetts, I was assailed by a medley of strange sounds which seemed to move from place to place in the bushes about me, while their author kept well concealed. There were turkey calls, whistles, mews and a rapid succession of notes and phrases, musical and unmusical, and all attempts to identify the singer or even to get a fair look at him were unsuccessful. Finally I sat down quietly among some bushes and began to imitate the cry of a bird in distress. At once a Yellow-breasted Chat dashed almost into my face—caution thrown to the winds.

—EDWARD HOWE FORBUSH, Birds of Massachusetts and Other New England States, Part III

ALTHOUGH the Yellow-breasted Chat is classified as a warbler, it acts more like a thrasher or mockingbird, singing a loud, raucous song and emphatically scolding people who intrude on its territory. It usually slips through dense thickets unseen, but on June mornings it suddenly makes itself conspicuous, singing from a high tree or shrub and occasionally launching itself above the singing perch, hovering unsteadily with legs dangling as it sings. Its song is a disjointed series of sounds described by Alexander Wilson as a "great variety of odd and uncouth monosyllables."[1] Unfortunately, this spectacle is increasingly difficult to observe in much of eastern North America: the chat has declined steadily and swiftly in many regions since the early 1900s. It has virtually disappeared from Connecticut, Long Island, and the Hudson River Valley, regions where it was once common, and it has also become less common in many parts of the Southeast.[2] Chats are threatened by the same type of environmental change that threatens many species of grassland birds: the loss of early successional habitat.

Yellow-breasted Chats are attracted to a site by the changes that drive grassland birds away. As an old field or fallow meadow is slowly invaded by

---

PLATE 2. Yellow-breasted Chat singing from a smooth sumac in a thicket maintained along a powerline cut.

woody shrubs and vines, Grasshopper Sparrows and other grassland species decline. When the shrub cover of the field reaches 10–35 percent, the field may be colonized by chats.[3] For a few years, chats may live alongside Grasshopper Sparrows, Eastern Meadowlarks, and Field Sparrows, but these grassland specialists disappear as the shrub cover increases and the entire field becomes a dense, low thicket. Chats thrive in this tangle, but their residence at a particular site typically lasts only a few years. Young trees grow above the shrubs and vines. When the branches of these trees spread outward and interlace, the understory is shaded, killing most of the shrubs and vines, thereby driving away the chats.[4] In the forested landscapes of the East, chats must constantly move from site to site to escape this invasion of trees.[5] They depend on the continual generation of new thickets, and they disappear quickly in regions where shrubby fields are not produced by abandonment of farms or clearing of forest.

Like many grassland bird species, chats require a remarkably specific type of vegetation. They are most common in thickets with shrubs, small trees, or vine tangles less than 6 feet (2 m) high.[6] Occasionally they nest at sites with an overstory, but only in areas with a dense shrub layer (particularly in burned woods).[7] Their breeding territories cover 2–4 acres (0.7–1.5 ha), and they typically do not nest in thickets smaller than 5 acres.[8] A thicket remains suitable for chats for only about nine years,[9] although a dense cover of vines such as Japanese honeysuckle and blackberry can delay this process by preventing trees from becoming established. In most parts of eastern North America, chat habitat is almost as transitory as the habitat of grassland birds, so it is not surprising that chats can quickly disappear from an area. A landscape of mature woodlands, manicured residential areas, and intensively cultivated farmland provides no place for chats.

Several other eastern bird species depend on brushy thickets. This habitat is not usually the subject of Sierra Club calendar photos or poetry, but these birds, and a multitude of other species that depend on low shrubland and vine tangles, are an important part of the biological diversity of North America. Like grassland species, shrubland species are declining and even disappearing from many areas in eastern North America.

Population Declines in Shrubland Birds

Edward Howe Forbush's description of the Brown Thrasher testifies to the abundance of this thicket bird in an earlier landscape of hedgerows and fallow fields: The thrasher "pays little attention, however, to the plowman or the busy farmer, for at planting time he sits near-by on some tree-top and sings—at least so the country people say—'drop it, drop it, cover it, cover it, I'll pull it up, I'll pull it up,' and so some of the country people call the singer the 'Planting Bird.'"[10] Clearly the thrasher was common enough in New England to be well known among farmers, but it is now infrequent and difficult to find in many areas. Like the Yellow-breasted Chat, the thrasher has lost much of its habitat. Since 1966 its population has declined by 1.6 percent per year east of the Mississippi River.[11]

The Breeding Bird Survey shows that most shrubland species, like most grassland specialists, have declined steadily and rapidly in eastern North America since the 1960s. Yellow-billed Cuckoo, Golden-winged Warbler, Prairie Warbler, Painted Bunting, and Field Sparrow all showed statistically significant declines of more than 1 percent per year since 1966. Chestnut-sided Warbler, Yellow-breasted Chat, and Indigo Bunting displayed less severe population decreases.[12] Only 1 of the 16 species of shrubland birds found east of the Mississippi River (Blue Grosbeak) has shown a significant population increase since 1966.

Records of the number of birds captured or recorded at banding stations and bird observatories during migration also show that the abundance of many species of shrubland birds has fallen during the past 20 to 30 years. Fewer Brown Thrashers and Chestnut-sided Warblers were recorded during fall migration at Manomet, Massachusetts, in 1988 than in 1970, and at Long Point, Ontario, in 1988 than in 1961.[13] A longer view is presented by the field notebooks of Ludlow Griscom and Norman P. Hill, who visited the same sites in eastern Massachusetts during spring migration each year between 1937 and 1989.[14] Although this was not a systematic scientific survey, their notes are detailed enough so that the number of individuals recorded per hour in the field could be calculated for a large number of species. Three species that are characteristic of shrubby habitats—Nashville, Chestnut-sided, and Golden-winged warblers—showed a large drop in abundance. Presumably these migration counts represent samples for a large part of the breeding range of these species, although they may also partially reflect loss of brushy second-growth habitat at the migratory stopover sites.[15]

The Eastern Towhee has declined steadily in recent decades in eastern North America. Breeding Bird Surveys results show that the population of towhees has fallen in eighteen states (most of which are east of the Mississippi River) and in one Canadian province.[16] The most severe drop in pop-

FIG. 2.1. Golden-winged Warbler, a shrubland specialist that has declined in most regions of eastern North America. Photograph by Isidor Jeklin. Reproduced with permission from the Cornell Laboratory of Ornithology.

ulation was in New England, where the annual rate of decline between 1966 to 1989 ranged from 5.5–10.2 percent for different states. Banding data for migrating birds at Manomet Bird Observatory also indicate that the abundance of towhees has dropped. An average of 4.5 towhees were captured for every day of banding in the spring of 1970, but by the spring of 1988 the average had dropped to less than 1 individual per day. This was the greatest reduction in abundance for any of the 52 species that are regularly banded at Manomet. Moreover, Christmas Bird Counts show that towhees declined in their wintering area in the southeastern United States. The winter population, which consists both of local residents and migrants from the north, dropped 2–5 percent per year between 1966 and 1990. Although the Eastern Towhee is not restricted to shrubland habitats, it reaches its highest density in low, brushy vegetation, so it may have suffered population losses for the same reason as shrubland specialists. Even if all shrubland habitat disappeared, however, towhees would probably survive in other habitats, such as the understory of young forest and along the forest edge. True shrubland specialists would not have these refuges.

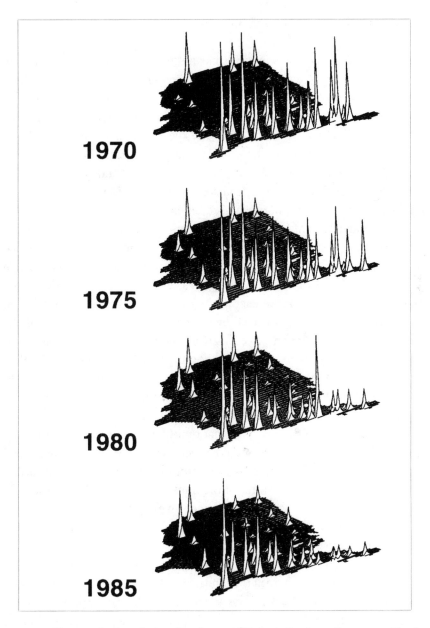

FIG. 2.2. Changes in the relative abundance of Eastern Towhees (in eastern North America) and Spotted Towhees (in western North America) on Breeding Bird Survey routes between 1970 and 1985. The abundance of Eastern Towhees declined substantially along the East Coast of the United States during this period. From Hagan, 1993. Reproduced with permission from *The Auk*.

History of Eastern Shrubland Birds

The decline of shrubland songbirds has not resulted in a flurry of efforts to save their habitat. Except for endangered species like the Kirtland's Warbler, birds that live in early successional woody habitats have received little attention. Few people are aware that the still-common Eastern Towhee may be declining more rapidly than any other bird species in eastern North America.

In some cases the decline of shrubland species is noted but ignored; this change is perceived by some as a return to more "natural" conditions, before the massive destruction and subsequent resurrection of the eastern deciduous forest following European settlement. Shrubland birds, like grassland birds, are frequently seen as interlopers to the East, either colonizing the area or increasing to unprecedented densities after the land was cleared by European settlers. The history of the Chestnut-sided Warbler is often cited in support of this view. This species, which is now widespread and common, was so rare in the East in the early 1800s that only a handful of specimens were seen or collected by the most active ornithologists of the time. Alexander Wilson and Thomas Nuttall observed only a few individuals.[17] John James Audubon describes shooting five Chestnut-sided Warblers during a single May morning in 1808 at Pottsgrove, Pennsylvania. In *Ornithological Biography* (1831) Audubon wrote that he never encountered this species again and had no idea where it might nest.[18] However, Audubon may have obtained additional information before he wrote the text for the octavo edition of *Birds of North America* several years later: here he described the Chestnut-sided Warbler as "rather common" from Texas northward.[19]

Regardless of the exact status of the Chestnut-sided Warbler in the early 1800s, it is clear that its population showed an explosive increase in the second half of the nineteenth century. The proliferation of Chestnut-sided Warblers is frequently attributed to the clearing of the primeval forest after Europeans settled North America, which allowed a species restricted to forest openings to expand across the landscape. Many regions along the East Coast had been settled up to 200 years before the dramatic increase in Chestnut-sided Warbler, however, and it is more likely that the population explosion was triggered not by the clearing of new farms, but by the abandonment of old farms. As Edward Howe Forbush wrote, this species "is not a frequenter of deep woods, nor yet of well-kept gardens, orchards or farmlands, but prefers neglected or cut-over lands, with a profusion of thickets and briers."[20] Throughout the second half of the nineteenth century, farms failed in rocky and hilly regions in the eastern United States, resulting in the steady generation of shrubby, abandoned fields that became the prime habitat for Chestnut-sided Warblers and other shrubland birds.

The rarity of the Chestnut-sided Warbler in the early 1800s remains an enigma, however, because many of the species that share its scrubby habitat were common at that time. Audubon described the habitat of Prairie Warblers in New Jersey as "large openings where the woods had been cut down, and were beginning to spring up," and he recorded Blue-winged Warblers as frequent in the "barrens of Kentucky." He wrote that the Brown Thrasher, which is found in open shrubland, was a "constant resident of the United States. Immense numbers are found in the South, and in all our Eastern States in spring and summer." Audubon described another shrubland species, the Yellow-breasted Chat, as "extremely plentiful in Louisiana, Georgia, and the Carolinas" and "equally abundant in Kentucky, particularly in the barrens of that state; and it ascends the Ohio, spreading over the country, and extending as far as the borders of Lake Erie in Pennsylvania."[21] Moreover, 100 years before Audubon, Mark Catesby described how Yellow-breasted Chats in the Carolinas "frequent the banks of great rivers and their loud chattering noise reverberates from the hollow rocks and deep cane swamps."[22]

Shrublands Before European Clearing

Many natural processes can produce shrublands. Periodic flooding of river banks can create a shrubby fringe along rivers, and grassy beaver meadows and other types of grassland are eventually invaded by shrubs and small trees. Moreover, shrubs, tree saplings, and resprouting trees immediately become established in many of the large clearings created by windstorms and fires, so forest is converted directly to thicket rather than grassland. Thus, the canopy of the presettlement forest was broken not only with grassy openings but also with shrubby tangles. Many of these were small, but others were large enough to provide habitat for shrubland birds.

Where the soil is continually waterlogged, shrublands may be relatively stable for a long period, providing a dependable habitat for shrubland birds. Bogs and other shrubby wetlands embedded in northern forests provide habitat for a diversity of shrubland species. For example, in Michigan, bogs support Willow Flycatchers, Eastern Towhees, and Field Sparrows, and shrub wetlands support Willow Flycatchers (in the southern part of the state) and Chestnut-sided Warblers (in the north).[23]

On the Atlantic coastal plain of the southeastern United States, there are extensive shrubby wetlands called pocosins that often cover thousands of acres. These shrub communities support a diversity of shrubland birds, including White-eyed Vireos and Prairie Warblers.[24] Pocosins are low heath-like bogs dominated by evergreen shrubs such as titi, sweetbay magnolia, redbay, black bayberry, and low gallberry holly.[25] The shrub layer is 1–12 feet (0.5–4 m) high, interrupted by scattered trees or clumps of trees (especially pond pine).[26] Pocosins develop in depressions or sandy flatlands with poor drainage. Their main source of water is rainfall. The waterlogged soils accumulate peat, which blocks drainage and retains even more water, often causing the shrubby bog to expand as the water table rises.[27] Although pocosins remain low and shrubby for long periods in the absence of disturbance, the low vegetation is ultimately sustained by occasional fires. The shrubs and pond pines that grow in pocosins sprout readily after a fire. Also, the cones of the pond pine open more readily when heated by fire, so many pine seedlings become established after a pocosin burns. Without occasional fires the shrubby bog will slowly become a wooded swamp. Before settlement most of the fires that spread across pocosins originated in adjacent pine woodlands and savannas. Pocosins typically undergo large fluctuations in water level, and they are open to fire during dry periods.

Most shrubby habitats in the eastern forest region are much more transient than northern bogs and pocosins. Most are the product of major disturbances of the forest canopy. Severe infestations of defoliating insects can kill the overstory trees, resulting in a low, dense thicket used by shrubland birds. For example, spruce budworm killed nearly all of the fir and spruce in a study area in Baxter State Park in Maine, leading to a decline in forest birds, such as Blackburnian Warbler, and an increase in species associated with shrubby habitats, such as Nashville and Magnolia warblers.[28] Windstorms periodically rip holes in the forest, creating large openings that are soon covered with rapidly growing saplings and resprouting trees. Recently a tornado leveled a 500-acre (200-ha) patch in Allegany State Park in western New York.[29] More than 70 percent of the trees were blown down, and all of the relict trees were heavily damaged. Fallen logs and damaged trees were salvaged during the years following the tornado. Within six years, this tornado site was covered with a low thicket that supported dense populations of Chestnut-sided Warblers and Eastern Towhees, and smaller numbers of Mourning Warblers, Blue-winged Warblers, and other typical shrubland species.

FIG. 2.3. Extensive forest opening created by a tornado in Allegany State Park in New York. The regenerating vegetation supported dense populations of Chestnut-sided Warblers and other shrubland birds. Photograph by Richard White.

Hurricanes can uproot trees over a large area, creating expanses of low, woody vegetation that provides habitat for shrubland birds. The frequency of such events has been revealed by carefully analyzing the age and arrangement of dead trees and logs in old forests. At Harvard Forest in Massachusetts, for example, careful mapping of stumps, fallen trees, dead wood fragments, and the pits created by the uprooting of trees revealed that windstorms had knocked down large numbers of trees four times between 1400 and 1956.[30] Three of these storms were recorded hurricanes. A study in the Pisgah Forest, an old forest in New Hampshire, revealed a similar pattern. The history of this forest was worked out by botanists from Harvard Forest who mapped every live and fallen tree and every fragment of wood on a one-tenth-acre plot (0.04 hectare).[31] The year of death for each fallen tree was determined in a variety of ways. If it injured a living tree when it fell, the date of injury could be determined from the annual growth rings on the living tree. Once the date of the fall of one tree was known, then it could be assumed that trees underneath it had fallen earlier and that trees resting on it had fallen more recently. After the fallen trees had been mapped, the leaf litter was removed and the upper layers of the soil were turned over to reveal

FIG. 2.4. Effect of 1938 hurricane on mature hemlock forest in the Connecticut College Arboretum. The two photographs show approximately the same site before and after the hurricane. Courtesy of the Connecticut College Arboretum.

wood fragments and pieces of charcoal. By piecing together these bits of evidence the researchers could decipher the history of the forest. The forest had begun growing during the years immediately after 1665. Charcoal fragments indicate that an earlier forest was destroyed by fire in that year. The earlier forest consisted of relatively small trees, perhaps because the forest was still recovering from a major hurricane that passed through the area in 1635. After 1665 the forest grew for more than 200 years without any major disruptions, but large numbers of trees blew down in 1898, 1909, 1921, and 1938. These blowdowns coincided with windstorms listed in the weather records: severe thunderstorms in 1898 and 1909, a tornado in 1921, and a hurricane in 1938. The hurricane blew down the last of the old trees that had originally colonized the site in the 1600s, and a new forest began to grow.

Thus, during four brief periods since 1635, large areas in the Pisgah Forest may have been suitable for shrubland birds such as Chestnut-sided Warbler and Brown Thrasher. After a few years, however, a new tree canopy formed and the shrubland birds would have been forced to leave. Shrubland specialists can survive in this situation only if they continually shift to new sites, colonizing areas where most of the trees had been mowed down by tornadoes, thunderstorms, or the occasional hurricane. Pocosins, northern bogs, and regenerating beaver meadows probably provided additional habitat, but most patches of shrubby habitat were more transitory than were these shrubby wetlands. In most presettlement forests, shrubland birds must have been "fugitive species"—species that must constantly shift from one ephemeral patch of habitat to another. Like grassland birds, shrubland birds depended on the continuous generation of treeless patches in the heavily forested eastern landscape.

Although windstorms periodically created large openings in the forest close to the coast, farther inland such disturbances were less common. Major windstorms and fires were particularly infrequent in the hardwood forests of the Appalachian Mountains.[32] These have been called "asbestos forests" because they are resistant to fire. Dead wood and other organic matter decompose quickly in these relatively moist forests, so there is little buildup of fuel. Moreover, except on exposed ridgetops, windstorms tend to have a moderate effect, knocking down single trees or small groups of trees. The result is an almost continuous carpet of forest interrupted by small holes or "gaps" in the canopy, with only occasional large openings created by tornadoes, thunderstorms, or insect infestations. Within the small canopy gaps, numerous tree saplings create a miniature thicket. The saplings quickly grow up to the canopy, spread their branches, and seal the gap. The result is a complex forest with trees of many different ages rather than the large stands of same-aged trees that grow up following large fires and blowdowns. Wherever the frequency of major disturbances exceeds the lifespan of trees, the complex forest generated by canopy gaps will dominate. This is generally true in the core of the eastern forest, far from the hurricane-prone coast and away from the flammable coniferous forests of the southern coastal plain and along the U.S.-Canadian border.[33] These more stable forests are found not only along the Appalachian Mountains but also in the Ohio Valley and parts of Michigan and Wisconsin.

Beginning in 1788 surveyors delineated townships in western New York in anticipation of European settlement of the region.[34] Their notes included precise information on vegetation, including descriptions of blowdowns, burns, and other openings in the forest. These records reveal that most openings were small, suggesting that the forest was relatively stable and that trees reached the canopy primarily by growing up in canopy gaps. There were blowdowns wider than 1,600 feet (500 m) across and burns as large as a mile (1.6 km) across, providing potential habitat for shrubland birds, but most of the disturbances were much smaller. Moreover, only 0.7 percent of the total area showed direct evidence of disturbance, so any shrubland birds living in this heavily forested region must have been highly mobile, moving great distances to colonize new burns and blowdowns.

James Runkle surveyed the vegetation in a number of old-growth woods in inland areas from North Carolina to Ohio and Pennsylvania.[35] He found that the amount of the forest occupied by canopy gaps in different forests ranged from 3 to 24 percent, and that on average gaps created a canopy opening of only about 65 square feet (6 m²), which is much too small to accommodate the territory of a Yellow-breasted Chat or Chestnut-sided Warbler. For example, in Hueston Woods, an old-growth forest in Ohio, 7 percent of the land area was directly under canopy gaps.[36] Most of these gaps were circular and had an area less than 2,100 square feet (200 m²). In Great Smoky Mountain National Park in Tennessee, an old-growth forest that had never been logged had larger canopy gaps than nearby second-growth forests.[37] When one of the gigantic old trees dies in this ancient forest, a large piece of the canopy becomes bare, and dead branches break off and destroy trees in the understory. However, even these holes were smaller than about 3,000 square feet (300 m²).

The largest gaps in an old-growth forest might support such species as Brown Thrashers or White-eyed Vireos, but some of the most specialized shrubland birds probably depended on larger openings. The centers of abundance of many shrubland species originally must have been on the periphery of the eastern forest, not in the protected interior. New thickets must have been constantly generated near the prairie edge, where tornadoes ripped swathes through the forest, and along the coast, where hurricanes periodically flattened entire forests. After European settlement, cities and farmland occupied most of the coastal plain, and the prairie-forest boundary disappeared as both prairie and forest were converted to farmland. Much of the habitat that shrubland birds once depended upon is now permanently secured against the invasion of shrubs and small trees. The remaining large forests are concentrated in the mountainous interior, where large disturbances may be too infrequent to sustain specialized shrubland birds. Moreover, few old-growth forests, with their large canopy gaps, survived the intensive logging of the late nineteenth and early twentieth centuries.[38] A number of other factors, such as fire control and the intensive trapping of beavers, reduced the frequency of disturbances that create early successional habitat. When a shrubland patch grew to forest, there were fewer and fewer patches of new habitat that could be colonized by the birds that once nested there. They became fugitives without a destination.

Creation of Shrublands by People

For at least the past millennium shrubland birds have become increasingly dependent on forest openings created by people. Before Europeans settled the eastern forest, shrublands and thickets resulted from agricultural clearing and forest burning by Indians. Slash-and-burn agriculture, in particular, would have produced a continual supply of habitat for shrubland birds, but even permanent agriculture would generate brushy fields, both fallow and abandoned. Also, periodic burning to improve hunting produced particularly large shrublands. In 1804, Timothy Dwight traveled through an area along the Genesee River in western New York before it was settled by Europeans. He found five openings in the forest, large expanses covered with "grass, weeds and shrubs."[39] Judging by fire scars on the few remaining trees, Dwight concluded that Indians had burned the area to create "pasture grounds" for deer. Before European settlement, both natural and human disturbance produced large areas of early successional, woody habitat.

During the first two centuries of European settlement, intensive grazing and farming probably reduced the amount of habitat not only for birds of the mature forest, but also for the shrubland birds associated with regenerating forest. Beginning in the 1840s, however, many farms were abandoned in hilly and rocky regions as farmers moved into flatter and more fertile regions west of the Appalachians.[40] Since the middle of the nineteenth century, farms have been progressively abandoned in many parts of New England, New York, and the Southeast, creating a continuous supply of brushy

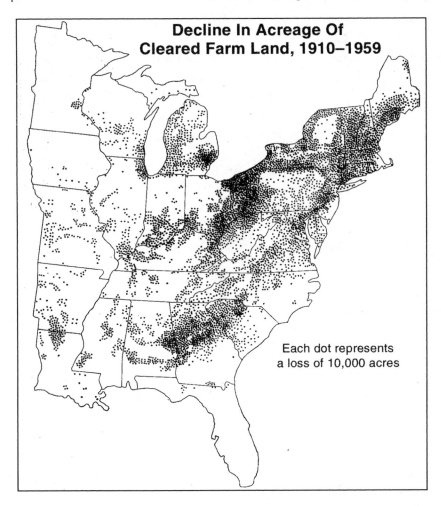

FIG. 2.5. Decline in the area of cleared farmland in eastern North America between 1910 and 1959. From Hart, 1968. Courtesy of Blackwell Publishers.

old fields filled with such birds as Yellow-breasted Chats, White-eyed Vireos, Chestnut-sided Warblers, Blue-winged Warblers, and Bachman's Sparrows. Many of these species undoubtedly experienced population explosions during this period,[41] increasing far beyond population levels characteristic of either the heavily forested presettlement landscape or the heavily agricultural eighteenth-century landscape.

In many parts of the East, the rate of farm abandonment diminished by the middle of the twentieth century. Eventually most of the remaining farms

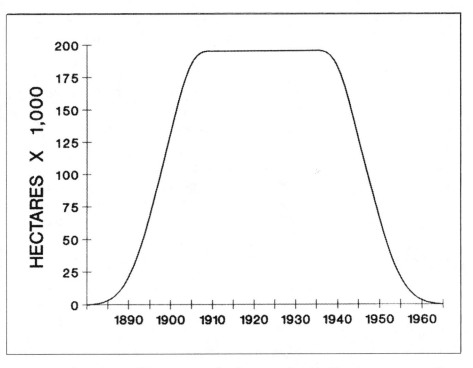

FIG. 2.6. Estimate of the amount of early successional habitat (10–25 years old) resulting from the abandonment of farmland in New Hampshire between 1880 and 1930. An exceptionally large amount of shrubby habitat was available for shrubland birds between about 1900 and 1950. From Litvaitis, 1993. Courtesy of Blackwell Publishers.

were in areas with rich soils and high productivity. In regions with poorer conditions for farming (for example, most of New England and the mountainous Southeast) the number of farms became so small that few were left to be abandoned, and many of these were converted directly to housing developments. Consequently, old fields have become scarce, reducing the amount of habitat for both grassland and shrubland birds. In New Hampshire, for example, the forest cover increased from 47 percent in 1880 to 87 percent in 1980.[42] This dramatic shift occurred largely because farmland was abandoned at a rapid rate during this period. Between 1880 and 1930, farmland in New Hampshire was abandoned at a rate of 32,000 acres (13,000 ha) per year, but the rate of farm abandonment dropped quickly after this period, and by 1960 most of the previously abandoned old fields had become forest. In the five New England states south of Maine, the amount of forest in what foresters call the "seedling and sapling" stage (the stage used by shrubland birds) fell from as much as 29 percent in 1950 to 14 percent in the early 1970s and to 8 percent in the 1980s.[43] Thus, it is not surprising that decreases in shrubland bird populations have been particularly severe in New England, and that two species that require large areas of shrubland, the Yellow-breasted Chat and Golden-winged Warbler, are now listed as endangered or threatened in states where they were previously common.[44] Also, the New England cottontail, a species of rabbit found in areas dominated by shrubs and tree seedlings and saplings, has declined so severely throughout its range that it is listed as a candidate for federal endangered or threatened status.[45] The loss of shrub habitat has been equally rapid and severe in some other eastern states. For example, in New York the amount of commercial forest land in the early successional "seedling/sapling" stage declined from 30 percent in 1980 to 16 percent in 1993.[46]

Kirtland's Warbler:
Generating Habitat for a Species That Depends on Fire

One of the rarest species of birds in North America, the Kirtland's Warbler, depends on early successional woody habitat. In their wintering areas in the Bahamas, Kirtland's Warblers are found in low, broadleaf scrub but are more frequent in open pine woodland with a scrubby understory.[47] During the breeding season they are restricted to dense stands of young jack pine that grow up following an intense forest fire.[48]

Kirtland's Warbler has always been known as a rare bird. The species was first described in 1852 after a specimen was shot during spring migration in Ohio.[49] Subsequently, Kirtland's Warblers were occasionally collected in Ohio and Michigan during migration, and in 1879 the wintering area in the Bahamas was discovered. However, the nesting area in Michigan was not discovered until June 1903, when E. H. Frothingham stumbled upon a group of singing males during a fishing trip on the Au Sable River.[50] Subsequently it became clear that the breeding range is restricted to 13 counties in the northern section of the Lower Peninsula of Michigan, mostly in the Au Sable River drainage. Occasionally singing males are found in jack pine stands in Quebec, Ontario, or Wisconsin, but these invariably turn out to be unmated males.[51]

Like other early successional species, Kirtland's Warblers ultimately depend on habitat disturbances to generate the type of vegetation they need.

FIG. 2.7. Kirtland's Warbler nest in a jack pine. Photograph by Ron Austing.

Kirtland's Warblers are even more specialized than many other early successional species, however; they depend almost completely on the forest fires that create the jack pine plains of northern Michigan. Not only does fire eliminate competing species of trees, but it also causes the tightly closed, long-dormant cones of jack pine to open, spreading seeds on the recently burned forest floor. A new stand of young pines, all of the same age and almost the same height, then grows up. When this stand is about 6 years old and about 6 feet (2 m) tall, it may be colonized by Kirtland's Warblers.[52] They arrive in a stand when the branches of adjacent pines touch, forming a dense, continuous mass of pine needles near the ground. This mass of needles conceals the adults as they approach the nest (a shallow cup on the ground), and later hides the fledglings as they leave the nest.[53] Usually they nest in stands of young pine with occasional scattered openings dominated by sweetfern, sedges, and blueberry. These openings are a source of light that sustains the dense needles on low branches. Nests are typically hidden in clumps of dense grass or low shrubs immediately below these pine branches. Kirtland's Warblers abandon a pine stand after 15 to 20

FIG. 2.8. Jack pine plains, the breeding habitat of the endangered Kirtland's Warbler, in Crawford County, Michigan. Photograph by David N. Ewert.

years, when the lowest branches die, exposing potential nesting sites to predators.

Much about the habitat requirements of Kirtland's Warblers remains enigmatic, however, despite numerous studies. For example, why does this jack pine specialist nest in only a tiny proportion of the range of jack pine? In contrast to the extremely limited range of Kirtland's Warbler, jack pine is widely distributed across the upper Midwest and most of Canada. The Au Sable River drainage has one feature that might make it uniquely suitable for Kirtland's Warbler, however. The extensive stands of jack pine in this region grow on a substrate of Grayling sand, a porous, quick-draining soil.[54] This is critical for protecting the nest (a depression in the sand) from flooding. Grayling sand occurs only near the southern edge of the range of jack pines, so climatic warming might threaten the Kirtland's Warbler population if it resulted in a northward shift in the distribution of jack pine.[55] However, Kirtland's Warblers must have survived similar (though probably less rapid) range adjustments in the past because during the last glacial period large areas of jack pine growing on sandy soil were probably limited to the Atlantic coastal plain in the Southeast.[56]

In more recent times, Kirtland's Warbler populations may have reached

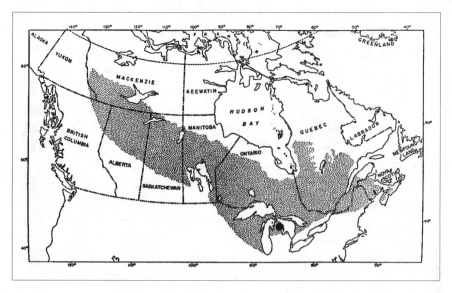

FIG. 2.9. Range of Kirtland's Warbler (dot) and jack pine. From Mayfield, 1960, with permission from the Cranbrook Institute of Science.

a peak after the tall white pine and red pine were logged in the Au Sable River area in the late 1800s. Fires ignited highly flammable logging debris and swept across the region. In 1871 more than 1.2 million acres (500,000 ha) burned in the region where Kirtland's Warblers nest, resulting in great expanses of young jack pine. Although the nesting area of Kirtland's Warbler had not yet been discovered, we have indirect evidence that the population was higher "than ever before or since."[57] During this period a large number of birds were recorded during migration, with an exceptional number occurring outside the usual migratory route. Also, between 1879 and 1897 more than 62 specimens of Kirtland's Warblers were collected on the major islands of the Bahamas.[58] In contrast, recent attempts to find this species in winter have not been very successful; only one individual was detected after an intensive search of 11 islands during two winters.[59] The numbers were also apparently low before the 1871 fires; Henry Bryant collected no Kirtland's Warblers during intensive collecting trips to the Bahamas during the winters of 1859 and 1866.[60]

Harold Mayfield organized the first census of Kirtland's Warblers in 1951. A total of 432 singing males were located, indicating a population of less than 1,000. In 1961, 502 males were counted, but in 1971 only 201 males could be found. This steep decline during a 10-year period precipitated several emergency measures to save the species from extinction.

One recommendation was to increase the amount of breeding habitat. Even before the population decline was detected, the Michigan Department of Natural Resources and the U.S. Forest Service had designated several tracts of more than 3,000 acres (1,200 ha), each to maintain habitat for nesting Kirtland's Warblers. These areas were managed with controlled burning or by carefully simulating the arrangement of pines on a jack pine plain by planting jack or red pines close together and by leaving scattered openings. Both burned and planted areas attracted nesting Kirtland's Warblers.

Research by Harold Mayfield, Lawrence Walkinshaw, and others showed why Kirtland's Warbler populations were declining even though nesting habitat was available. The nest success of Kirtland's Warblers was too low to sustain the population because of parasitism by Brown-headed Cowbirds.[61] Cowbirds lay their eggs in the nests of other species, leaving the foster parents (a pair of Kirtland's Warblers, in this case) to raise their young. Before laying an egg, the female cowbird typically removes eggs that are already in the nest, so the reproductive rate of the Kirtland's Warblers is reduced even if some of its young survive competition with the larger cowbird nestling.

In the nineteenth century cowbirds spread eastward and northward from the prairies as the land was cleared for agriculture. They probably did not arrive on Michigan's jack pine plains until about 1880. Probably because they were not subjected to cowbird parasitism during most of their evolutionary history, Kirtland's Warblers did not evolve the defenses seen in many other cowbird hosts. Unlike those birds, they seldom remove cowbird eggs from their nests, abandon parasitized nests, or chase female cowbirds unless they are already close to the nest. Perhaps as a result, Kirtland's Warblers are the preferred host of cowbirds in the jack pine plains. Although Walkinshaw found several hundred nests of 45 species in the jack pine plains, 93 percent of the cowbird eggs were in Kirtland's Warbler nests. In the 1950s, 55 percent of the Kirtland's Warbler nests were parasitized, reducing the production of fledglings by 40 percent. The incidence of parasitism was 69 percent during the period between 1966 and 1971. Such intense parasitism could easily account for the decline of Kirtland's Warblers between 1961 and 1971, and could potentially drive the species to extinction. Because Kirtland's Warbler is the preferred host of cowbirds, the proportion of parasitized nests may actually increase as warbler nests become more scarce.[62]

To save Kirtland's Warblers from imminent extinction, federal and state agencies and several conservation organizations cooperated in a program to reduce cowbird populations. Traps were baited with sunflower seeds and decoyed with captured cowbirds. Between 1972 and 1981, 40,000 cowbirds were removed from Kirtland's Warbler breeding areas. The number removed per year did not decline between 1972 and 1977, indicating that the program would probably need to be continued indefinitely. The effect of the cowbird control program on the reproductive success of Kirtland's Warblers was immediate. The number of parasitized nests dropped from 69 percent to 6 percent, and nesting success increased from 0.8 fledgling per pair before 1972 to 3.4 fledglings per pair in 1972.[63] Despite the virtual elimination of the cowbird threat, however, the Kirtland's Warbler population did not immediately increase. During the next 17 years, the number of singing males varied between 167 and 242, with no indication of an increasing trend.[64] However, after 1989 the population grew, reaching 347 males in 1991. This resulted from a massive fire that burned about 25,000 acres (10,000 ha) in the jack pine plains in 1980. By 1989 stands of small jack pines had grown on many parts of the burned area, providing good nesting habitat. In 1991 more than half of the males had territories in this burn.

Kirtland's Warblers may have declined after 1961 because of the absence of large burns. Despite efforts to create Kirtland's Warbler habitat in four management areas, more effective fire control since the 1940s had reduced the total amount of young jack pine. Kirtland's Warblers usually nest in stands of jack pine larger than 550 acres (220 ha), and historically much of the population has been concentrated in extensive burns. In 1951, half of the singing males were found in two large burns where fires had occurred in 1933 and 1939; and in 1961, one-quarter of the males were in a single area that had burned in 1946.[65] The decline after 1961 may have been due to the absence of recent large fires as well as the reduction of reproduction caused by cowbirds. After cowbirds were removed, the steep decline in warbler numbers ended, but the warbler population did not increase until a large area of new habitat became available as a result of a massive fire. Like other early successional species, Kirtland's Warblers ultimately depend on the periodic destruction of expanses of mature vegetation.

Kirtland's Warblers also depend on fire to sustain their preferred winter habitat in the Bahamas. They are found most frequently in woodlands of

Caribbean pine that remain open because of frequent summer fires.[66] Kirtland's Warblers declined during two periods when these pines were intensively harvested in the Bahamas (1900–1920, and the 1950s through the
early 1970s), suggesting that shortages of winter habitat may limit the population of this species. It therefore may not be enough to provide good
breeding habitat: we may also need to protect suitable winter habitat. Good
management of the Bahamian pine woodlands will not only help sustain the
Kirtland's Warbler but will also help a number of bird species, such as
Greater Antillean Pewee and Olive-capped Warbler, that are resident in the
Bahamas.

Birds in Clearcuts and Along Powerlines

The management program for the Kirtland's Warbler shows that habitat can
be tailored to meet the needs of an early successional specialist, although
even in this case it was apparently difficult to produce enough habitat without help from an unusually large fire. Creating the right conditions for
species that do not require such large areas of habitat might be easier, but
in most parts of eastern North America, it would still require continual, and
expensive, removal of trees to prevent succession to forest. Except when
threatened or endangered species are at risk, it is difficult to find support for
this type of vegetation management. Consequently, shrubland birds, like
grassland birds, now largely depend on the generation of habitat because of
economic activities. Instead of hay meadows, fallow fields, and air fields,
the critical artificial habitats for shrubland birds are two of the least popu-
lar features of the landscape: clearcuts and powerline corridors. Harvesting
timber by clearing large, continuous areas of forest typically results in a
dense growth of low shrubs, saplings, and stump sprouts that provides
habitat for most of the species of shrubland birds that have been declining
in the East.[67] In Virginia, clearcuts support populations of Yellow-breasted
Chats, Prairie Warblers, Golden-winged Warblers, and Field Sparrows from
the third to the twelfth year after cutting.[68] When the canopy closes, these
shrubland specialists are replaced by forest birds, such as Red-eyed Vireo,
Black-and-white Warbler, and Ovenbird. In northern Maine, recent clearcuts are occupied by a diversity of early successional species, including
Alder Flycatcher, Magnolia Warbler, Palm Warbler, Mourning Warbler,
Chestnut-sided Warbler, and Chipping Sparrow.[69] Thus, rotational clearcutting, in which a small proportion of a forest tract is cut periodically, can

provide good habitat for shrubland birds. This approach will not be appropriate in areas with steep slopes or thin soils because of problems created by soil erosion and runoff,[70] and it may be inappropriate in forests where conservation of woodland salamanders or spring ephemeral wildflowers is a high priority. Clearcutting would never be appropriate for the few remaining patches of old-growth forest in the East; these ancient forests provide some of our best information on how forest ecosystems worked in areas protected from frequent natural disturbance. However, on suitable sites, clearcuts can be used to maintain a wide range of successional stages, especially when the rotation time between harvests is long enough to permit the forest to mature. Although selective cutting may appear less disruptive than clearcutting, it does not create habitats needed by many early successional species.

A study of bird populations in 45 recent clearcuts in Moosehorn National Wildlife Refuge in Maine showed that shrubland birds are equally frequent in small and large clearcuts.[71] These sites ranged in size from 5–277 acres (2–112 ha). Similarly, when Benjamin Zuckerberg, Christopher Schafer, and I surveyed 34 clearcuts that ranged in size from 1.5 to 52 acres (0.6–21 ha) in several state forests in eastern Connecticut, we found that Chestnut-sided, Prairie, and Blue-winged warblers were equally frequent in clearcuts of different sizes.[72] Only Eastern Towhees were significantly more frequent at survey plots in large clearcuts than in similar plots in small clearcuts. In Green Mountain National Forest in Vermont, even tiny clearcuts (called group selection cuts) that cover about one acre (0.4 ha) can support such shrubland birds as Chestnut-sided Warbler and Mourning Warbler.[73] Apparently many shrubland species do not require large, continuous areas of shrubby habitat. In some cases, however, it may be better to cluster or consolidate clearcuts to minimize disruption of the expanses of continuous mature forest needed by many species of forest-interior birds.

Clearcutting has the advantage of paying for itself; there is a financial incentive for harvesting trees. Another artificial habitat in which there is a financial incentive to maintain low, shrubby vegetation is the open corridor underneath powerlines. Power companies traditionally have maintained these corridors by mowing or "bush-hogging" (mechanically chopping up low, woody vegetation), or by broadcast spraying of herbicides. These methods produce grassy corridors with few species of birds or other animals. However, mechanical removal of vegetation is expensive, and blanket-spraying of herbicides has been banned in many communities because

FIG. 2.10. Vegetation surveys in a clearcut in Pachaug State Forest, Connecticut. This habitat supports breeding populations of declining shrubland species such as Chestnut-sided Warbler, Prairie Warbler, and Blue-winged Warbler. Photograph by Leah Novak.

of health and environmental concerns. In the 1970s another approach to powerline management was introduced: the establishment of a thick shrubland that is relatively resistant to the invasion of trees.[74] Initially this is labor-intensive and expensive: workers must cut each young tree and spray each stump or the base of each tree with herbicide. This precisely directed use of herbicide (as opposed to indiscriminately spraying the entire power-

BEFORE TREATMENT

AFTER TREATMENT

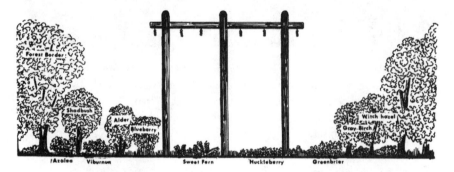

FIG. 2.11. Creation of a stable shrubland under power transmission lines. In the lower drawing, trees have been selectively removed, leaving fast-growing shrubs that spread out and inhibit the establishment of tree seedlings. This can provide good habitat for various species of shrubland birds. From Niering and Goodwin, 1974.

line corridor) can remove trees, allowing shrubs and vines to spread out. These create a thick barrier that shades out most tree sprouts and seedlings. The result is a relatively stable shrub community, often with a diversity of plant species, that requires only occasional maintenance.

The vegetation along powerline corridors typically supports a rich diversity of shrubland birds. For example, one of the first powerline corridors managed for stable shrublands, a demonstration plot established in the Connecticut College Arboretum in 1953, still supported breeding pairs of White-eyed Vireos; Blue-winged, Chestnut-sided, Prairie warblers; and Field Sparrows in 1993.[75] Wide powerline corridors managed by selective

Figure 2.12. Powerline corridor in the Connecticut College Arboretum. Since 1953 this corridor has been managed as a stable shrubland by selective removal of trees. It supports breeding populations of White-eyed Vireos, Blue-winged Warblers, Chestnut-sided Warblers, Prairie Warblers, and other shrubland birds. Photograph by Glenn Dreyer.

tree removal in Pennsylvania and Maryland support an even greater diversity of shrubland birds, including Yellow-breasted Chats and Golden-winged Warblers, two species that require relatively large areas of continuous shrubland habitat.[76] In Maryland, a high proportion of birds nesting along powerline corridors are successful in raising young, indicating that this habitat could potentially support stable populations of shrubland birds.[77] Similarly, birds nesting along a powerline corridor in central Pennsylvania had high rates of nest success because of low rates of cowbird parasitism and nest predation.[78] Although powerline corridors can have a negative impact on forest bird species that require large expanses of uninterrupted forest, they can make a useful contribution by sustaining populations of shrubland birds.

Eventually, as beavers return to more regions in the eastern forest and as the forests age and produce more large canopy gaps, natural processes may provide much of the habitat needed by shrubland birds. Until then, many of these species will continue to depend on the activities of people—on the abandonment of farms and pastures, the harvesting of timber, and the maintenance of powerline corridors and other shrubby openings in the forest.

Conservation of Shrubland Birds

Shrubland species are clearly an ancient part of eastern deciduous forest ecosystems. Most shrubland species can live in relatively small forest openings of the sort created by windstorms, beavers, and outbreaks of leaf-eating insects. Like grassland birds, shrubland birds increased after the land was cleared, especially after many farms were abandoned and became old fields or thickets. Many species have declined steadily for the past few decades, however, and they may eventually disappear from some regions. In order to sustain their populations, relatively stable shrublands, such as bogs and pocosins, should be protected. Also, where feasible, natural disturbances that create shrublands, such as beaver dams, wildfires, floods along wooded streamsides, and localized insect outbreaks, should be permitted. Alternatively, if they are managed appropriately, openings such as those created by timber harvesting or along powerlines can provide suitable habitat for many shrubland bird species. It is important to consider these species in regional conservation plans; the loss of White-eyed Vireos, Yellow-breasted Chats, Chestnut-sided Warblers, and other birds of young second growth would greatly reduce the biological diversity of a forested region.

CHAPTER 3

The Great Plains

Birds of the Shifting Mosaic

Pure colour everywhere. A gust of wind, sweeping across the plain, threw into life waves of yellow and blue and green. Now and then a dead black wave would race across the scene . . . a cloud's gliding shadow . . . now and then . . .

It was late afternoon. A small caravan was pushing its way through the tall grass. The track it left behind was like a wake of a boat—except that instead of widening out astern it closed in again.

—OLE EDVART RÖLVAAG, *Giants in the Earth*

TO European settlers crossing the Great Plains in the nineteenth century, the prairie seemed monotonous, unchanging, and endless. There was a gradual gradient from the wet tallgrass prairies of Illinois and eastern Kansas to the dry shortgrass prairies of Montana and eastern Colorado, but the settlers might travel through the same general type of grassland for weeks. Within each prairie region, however, the grassland consisted of a subtle patchwork of different types of vegetation. This complex pattern of distinct patches of grasses and forbs was created by various disturbances—wildfires, water collection in low spots, intense grazing or vigorous wallowing by bison, or steady cropping of grass and digging by prairie dogs. Each type of disturbance created a subtly different habitat patch, and each patch changed slowly as it recovered from the disturbance. The result was a "shifting mosaic" of habitat patches.

Many prairie bird species are adapted to live in particular types of habitat patches, so they ultimately depend on the disturbances that create their preferred habitats.[1] These ephemeral patches are not as dramatic as the shrubland openings and grassy glades that interrupted the eastern forest, but, like those more distinct patches, they support a distinct group of species.

PLATE 3. Mountain Plover nest in a blacktail prairie dog town.

FIG. 3.1. Mixed-grass prairie on Lostwood National Wildlife Refuge, North Dakota. Sharp-tailed Grouse, Baird's Sparrows, McCown's Longspurs, Chestnut-sided Longspurs, and other prairie species are found in the patchwork of habitats on this refuge. Photograph by Michael Askins.

Fire and Prairie Birds

Before European settlement, fires were frequent on the prairie.[2] Even in the moist tallgrass prairie at the edge of the eastern forest, wildfires probably occurred on a typical stretch of flat or rolling grassland at least every ten years, and perhaps as frequently as every four years.[3] Fires could pose a threat to prairie birds during the summer, when eggs and nestlings might be destroyed, but the absence of fire for a long period could also create problems for them. In some regions unburned prairie is slowly replaced by shrubs or trees, eliminating the habitat needed by grassland birds. In the tallgrass prairies of Illinois and Indiana, all grassland bird species ultimately depend on fire because the forest eventually invades areas that do not burn periodically. In fact, prairie ecologists have theorized that these prairies would have disappeared during the past 5,000 years except for periodic burning by Indians. Farther west, low rainfall and more frequent drought make the grassland more resistant to invasion by trees and shrubs, but even here woody plants may slowly replace grassland in the absence of fire or grazing.[4]

Even when woody plants do not invade the prairie, the grassland will progressively change if it is not burned for several years.[5] On an unburned prairie the soil is shaded by a mat of fallen dead leaves and an overstory of standing dead grass. This dead vegetation intercepts both sunlight and rainfall, creating cool, dry, dark conditions that are unfavorable for the growth of young grass shoots. Moreover, nutrients are tied up in the dead leaves, and microbes encrusting these leaves will absorb the nitrogen in rainwater, reducing the amount of nitrogen reaching the soil. In the years following a fire, there is a gradual buildup of dead vegetation, leading to progressively more shading and blanketing of new growth. The productivity of the grassland declines until the amount of new growth is so low that it balances the amount of dead material that decays each year. At this point, the grassland no longer changes quickly unless it is invaded by trees and shrubs.

When fire burns across a grassland, the blanket of dead grass is suddenly removed, and ash falls to the soil surface. The ash adds some nutrients to the soil, but the main benefit of fire is the removal of dead vegetation. With the dead vegetation gone, the soil now has plenty of sunlight and warmth, and plants have more growing space, so grass shoots grow quickly. Higher soil temperatures and greater light intensity following a burn are particularly important in increasing the growth rate of grass.[6] Denser grassland typically replaces the sparse grass of the unburned prairie. Ironically, this vigorous new growth is easier to walk through than the stunted grass of unburned prairie; it is no longer necessary to plow one's feet through the dense thatch of dead stems that mats the ground.

The young growth in a recently burned grassland is more succulent, more easily digested, and more nutritious than older vegetation in unburned prairie. Consequently, prairie fires typically result in an increase in the density of herbivores, which range from leaf-eating grasshoppers to grazing bison. This, in turn, leads to a better food supply for predators, such as insect-eating birds. From the viewpoint of many species, a prairie fire is a necessity, not a disaster.

Many species of birds decline or disappear when a prairie is never burned. In tallgrass prairies in Illinois, Bobolinks and Grasshopper Sparrows were more common in recently burned prairies than in unburned prairies. In contrast, Henslow's Sparrows were restricted to areas that had not burned for several years, where a dense mat of dead grass had formed below a continuous, thick carpet of standing grass (both living and dead).[7] The pattern was similar, though most of the species were different, for Ker-

FIG. 3.2. Recently burned prairie in J. Clark Salyer National Wildlife Refuge, North Dakota. Photograph by Michael Askins.

nen Prairie in Saskatchewan. Upland Sandpipers were found only in burned prairie, while Baird's Sparrows were more abundant in the dense grass of a grassland that had not burned recently. Savannah and Clay-colored sparrows were also associated with unburned grassland in Kernen Prairie; they thrive in areas that have been invaded by shrubs, such as western snowberry and silverberry, that would be eliminated by fire. The overall abundance of birds can be greater in unburned grassland than in recently burned prairies, but this usually occurs after woody vegetation grows up and such shrubland species as Brown Thrasher, Field Sparrow, and Bell's Vireo become established.[8]

Another species that requires relatively dense grass is the Greater Prairie-Chicken. However, prairie-chickens also need some areas of short grass for their "booming grounds," where males cluster in early spring to defend small territories from other males and to court females. The same display is used for territorial defense and courtship: the male holds his tail erect, partially raises his wings close to his body, and then stamps his feet rapidly on the ground. He then smartly snaps his tail open and shut while producing a tooting or booming sound that is amplified by large vocal sacs on the neck. The booming of groups of displaying males was once one of the signature sounds of the prairie, but Greater Prairie-Chickens have declined severely in almost every part of their range, and they are completely gone in many prairie states and provinces.[9]

After mating, female prairie-chickens need heavy cover to conceal their nests and later to provide hiding places for the mobile brood of young. Consequently, prairie-chickens have declined in regions where this dense cover has been removed by intensive haying or grazing. They also decline in grassland that has been undisturbed for so long that shrubs and trees have invaded. Infrequent burning (every three to five years) is used successfully to maintain dense prairie-chicken populations in North Dakota.[10]

Grazing Lawns of Prairie Dogs and Bison

Although fire was probably the main agent of disturbance in the moist tallgrass prairies of the eastern plains,[11] grazing animals molded the vegetation even more than wildfires in the shortgrass prairies farther west. Prairie dogs were particularly important in the presettlement prairie because they grazed large areas intensively and continuously, creating a distinctive habitat required by many other species. For this reason the five species of prairie dogs are often considered "keystone species."[12] An entire community of organisms depends on a keystone species directly or indirectly, so removal of this single species can result in a cascade of changes. Just as removal of a keystone causes an entire arch to fall inward, removal of a keystone species causes a swift collapse in species diversity.

Prairie dogs are keystone species because their burrowing and grazing transforms the prairie, creating an expanse of close-cropped grass dotted with patches of bare earth surrounding burrow entrances. Because there may be 20–120 burrow entrances per acre (50–300 per ha) in a prairie dog colony, recently excavated earth provides extensive habitat for annual forbs that could not compete where grass is already established.[13] Moreover, the areas between burrow mounds are grazed intensively throughout the year by the numerous prairie dogs. Prairie dogs live in family groups, called coteries, each of which has a small territory with a self-contained burrow system.[14] Individuals defend the group territory against prairie dogs from other coteries, but they greet a member of their own coterie with a "kiss" (probably verifying its identity) and with mutual grooming. Prairie dogs generally stay within the boundaries of their territories, feeding on the vegetation and removing tall plants that might hide approaching predators. In-

FIG. 3.3. Black-tailed prairie dog at Theodore Roosevelt National Park, North Dakota. Photograph by Michael Askins.

tense cropping of grass and forbs within each territory can produce a continuous expanse of lawn-like prairie, with short, compact grass.[15] Because a prairie dog colony or "town" may have hundreds or thousands of closely packed coterie territories, the close-cropped grassland may extend for many miles.

Grazers as different as wildebeest and plains zebra in the African savanna and green turtles in Caribbean seagrass beds produce "grazing lawns" where they have intensively cropped the tops of grass.[16] Ironically, these lawns often provide better grazing conditions than ungrazed grassland. Grass grows from the base of the stem rather than the stem tip, so it can quickly grow back after the top is nipped off by a grazer. The new growth is more nutritious and more easily digested than older vegetation. Moreover, it is typically shorter and denser, resulting in more energy and nutrition per bite, an important advantage for grazers that must process large amounts of vegetation each day to meet their food needs.[17] Thus the act of grazing, if it is not too intense, can improve conditions for grazers.

In Wind Cave National Park in South Dakota, the vegetation on prairie dog colonies is short and grows close to the ground, and the amount of vegetation per acre is two-thirds to one-half as much as in grassland outside of

FIG. 3.4. Close-cropped grass on a prairie dog colony in Theodore Roosevelt National Park, North Dakota. The taller grass in the foreground is outside of the colony. Photograph by Michael Askins.

colonies.[18] Despite the smaller amount of vegetation in prairie dog colonies, however, the grazing lawns attract a disproportionate number of large grazers, such as bison, elk, and mule deer. Compared with prairie outside of prairie dog towns, the cropped grass within towns has a higher nitrogen content, greater digestibility (due to the preponderance of young shoots), and a much lower proportion of dead grass, making it more attractive to grazers. Bison that feed in colonies have higher growth rates than those that feed outside of colonies. Furthermore, periodic, intense grazing by bison also can stimulate the growth of nutritious grass shoots, leading to greater foraging efficiency for prairie dogs. Rather than being competitors, these large and small grazers may benefit mutually from the production of the grazing lawn.

Prairie dog colonies are well known for supporting a set of species that depend directly on the colony residents: burrowing owls that reside in abandoned burrows, and black-footed ferrets and Ferruginous Hawks that are predators of prairie dogs.[19] Other species benefit from the close-cropped, open landscape that prairie dogs create. For example, in Badlands National Park in South Dakota, birds were substantially more abundant on the

FIG. 3.5. Bison grazing on the short grass in a large prairie dog colony in Wind Cave National Park, South Dakota. From Whicker and Detling, 1988, © 1988 American Institute of Biological Sciences.

sparser grass of prairie dog colonies than in the adjacent mixed-grass prairie where there were no prairie dogs.[20] The difference was primarily due to high densities of Horned Larks in the prairie dog towns. Other species, such as Upland Sandpiper and Lark Bunting, were more abundant in the taller vegetation outside of the colony.

The fate of the Mountain Plover may be especially closely tied to grazing. Unlike most other species of plovers, which are found on the shores of oceans and lakes, this plain gray plover is restricted to the high, dry plains of the Rocky Mountain Plateau.[21] It is endemic to shortgrass prairie, and it usually nests in level areas dominated by blue grama, one of the shortest and most drought-resistant of the prairie grasses.[22] Its range largely overlaps the range of the blacktail prairie dog, and it thrives in the open, shortgrass landscape created by prairie dogs. For example, in Charles M. Russell National Wildlife Refuge in Montana, 90 of the 91 Mountain Plovers detected on surveys were in prairie dog towns even though half of the survey routes were on the prairie outside of towns. The ground-dwelling beetles and ants that Mountain Plovers eat are more abundant in prairie dog towns than in surrounding habitats, and close-cropped vegetation is more suitable for their hunting technique, which involves running across open ground with frequent stops to visually scan for prey.[23]

All of the prairie dog colonies in Russell National Wildlife Refuge were grazed by cattle, and Mountain Plovers may benefit from the combined effects of grazing by prairie dogs and large ungulates, regardless of whether the latter are bison or cattle. The behavior of Mountain Plovers near their nests probably reflects a long evolutionary association with ungulates. To protect their ground nests they will rush at a cow or buffalo with outspread wings or even fly at its face to drive it away.[24] Moreover, the main wintering area of Mountain Plovers is in the Central Valley of California, which was originally grazed by dense populations of kangaroo rats and Tule elk.[25]

Mountain Plovers were once abundant enough to support market hunting, but they have become increasingly rare and localized in this century. They have declined on Breeding Bird Survey routes at an average rate of 2.9 percent per year since 1966.[26] Another species that is closely associated with prairie dogs, the black-footed ferret, is on the verge of extinction.[27] These species probably have declined largely because of the eradication or severe reduction of prairie dogs in most parts of the western United States. At the end the nineteenth century, prairie dog colonies covered a huge area—between 100 million and 250 million acres (40 million and 100 million ha).[28] This represented more than 20 percent of the shortgrass and mixed-grass prairie.[29] In the early 1900s, a single colony of approximately 400 million prairie dogs covered almost 25,000 square miles (65,000 km²) in Texas.[30] A government-sponsored poisoning campaign, along with repeated outbreaks of plague, has reduced the population of the five species of prairie dogs by 98 percent, and most of the remaining colonies are small and isolated.[31] Despite the severe decline in prairie dogs and many of the species associated with their colonies, prairie dogs are still frequently poisoned on both private and government-owned land.[32] Prairie dogs are removed from ranching areas because they are thought to compete with livestock for grass and because their burrows can be hazardous to cattle and horses.

The expensive effort to exterminate prairie dogs is based on the premise that they compete for food with cattle because they substantially reduce the amount of vegetation on the prairie. However, recent studies comparing rangeland with and without prairie dogs show that the impact of prairie dogs on cattle is not nearly as severe as one might think after looking at the

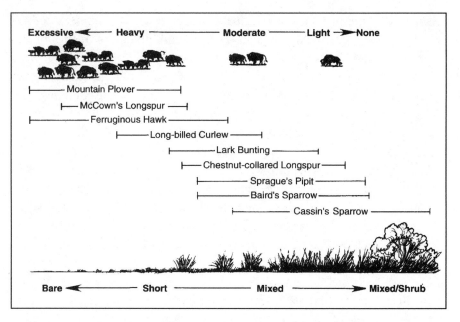

FIG. 3.6. Distribution of different species of prairie birds across grasslands subject to different intensities of grazing. From Knopf, 1996b. Reproduced with permission from F. B. Samson and F. L. Knopf, eds., *Prairie Conservation: Preserving North America's Most Endangered Ecosystem,* © Island Press, Washington, D.C., 1996. Published by Island Press, Washington, D.C., and Covelo, Calif.

sparse vegetation in a prairie dog town.[33] Weight gains are only slightly lower for cattle grazing on prairie dog towns than on the adjacent prairie, probably because cattle, like bison, benefit from the higher nutritional quality and digestibility of the forage in prairie dog towns. Moreover, these studies often involved very heavy grazing by cattle; more moderate grazing pressure might result in the same mutually beneficial relationship that occurs between bison and prairie dogs.[34]

Grazing, like burning, produces a type of grassland needed by many species.[35] Some species require the close-cropped grass found in prairie dog towns or even in fenced pastures that are intensively grazed by cattle, while others depend on the taller grass maintained by infrequent heavy grazing—the type of grazing once provided by migrating herds of bison. For example, the McCown's Longspur favors the "open wind-swept plains and sparse vegetation provided by native shortgrass prairie and overgrazed

FIG. 3.7. McCown's Longspur, one of the prairie species that depends on heavily grazed grassland. Photograph by O. S. Pettingill. Reproduced with permission from the Cornell Laboratory of Ornithology.

pastures," while the closely related Chestnut-collared Longspur is found in taller, denser vegetation.[36] Fire and grazing together produce a mosaic of habitat patches, with different bird species concentrated in different patches. Because grazing is so integral to the ecology of this ecosystem, cattle ranching is not necessarily a threat to the natural processes of the prairie unless it becomes too intensive for too long a time over too large an area.[37] The intricate patchiness of the native prairie may then disappear, causing many species of birds and other animals to decline.

Decline of the Dickcissel

The Dickcissel is one of the most distinctive birds of tallgrass and mixed-grass prairies, and of the hayfields and fallow fields that have replaced them. It looks like a miniature meadowlark, with yellow underparts and black chest patch, and it has a penetrating song that sounds like its name. Although still a conspicuous part of the farmland in some parts of the Midwest, the Dickcissel has declined precipitously. Between 1966 and 1979, its population decreased at an annual rate of 5.4 percent.[38] The decline was so rapid that Stephen Fretwell, who had been studying Dickcissels for many years, predicted that they might become extinct by the year 2000.[39] Between 1979 and 1994 the population increased slightly, but the factors that caused the initial decline are probably still at work.[40] Many of the Dickcissel's problems occur on the llanos, the grasslands of northern Venezuela, where most Dickcissels spend the winter.

Dickcissels have adapted well to the new landscape that has replaced the native prairie. They nest successfully in old fields, hay meadows, and even in oat, wheat, and barley fields. In Kansas they are more abundant in weedy old fields than in native prairie.[41] Like many grassland birds, Dickcissels are polygynous (some males mate with more than one female). Males in old fields usually attract more mates than do males in native prairie, which is another indication of the suitability of old fields for nesting. The dense vegetation in old fields and hay meadows provides good cover to protect nests from summer heat and predators.[42] In the presettlement prairie, Dickcissels may have been concentrated in grassland patches where grass had grown vigorously for two or three years after a major disturbance, such as a fire or intensive grazing by a large herd of migrating bison. They appear to thrive in a variety of agricultural habitats, however, and recent research indicates that their nest success at several sites is high enough to sustain the population.[43]

Dickcissels respond quickly to changes in breeding habitat; entire populations appear to shift from one region to another as conditions change.[44] During the early 1800s the Dickcissel spread eastward, and it became a common nesting bird from the Carolinas to Massachusetts. It began to decline in this region during the late 1800s, however, and by 1900 it had disappeared from the entire East Coast. During the 1920s, Dickcissels were again found nesting in the East, but this much smaller invasion lasted only a few years.[45] Even in the center of their range, in the Great Plains, Dickcissels can be abundant at a site one year and rare the next. For example, in 1968 rains in

FIG. 3.8. Dickcissel, a prairie species that has declined despite its ability to adapt to agricultural habitats. Photograph by W. A. Paff. Reproduced with permission from the Cornell Laboratory of Ornithology.

southern Texas created favorable breeding conditions, and Dickcissels stayed to nest in large numbers.[46] Populations in Ohio, Indiana, and Illinois dropped by as much as 50 percent that year, as would be expected if many individuals stopped in Texas during their northward migration instead of returning to their former nesting areas.

Given the adaptability of Dickcissels during the breeding season, their most vulnerable season may be winter, when they concentrate in the llanos along the Orinoco River in Venezuela. Like their breeding area, the wintering area has been highly modified by ranching and grain production. Fields of sorghum and rice provide winter-resident Dickcissels with a new but potentially deadly source of food. Dickcissels gather in enormous flocks of thousands or tens of thousands of birds. A cloud of birds, banking and turning in synchrony, can drop down on a field and take much of the grain. Consequently, some farmers spray poison on the enormous flocks of Dickcissels (which can number in the millions) that gather to roost in sugar cane fields.[47] Large numbers of wintering birds die because of these control efforts.

Stephen Fretwell hypothesizes that the change in the winter habitat may have had a more subtle effect on Dickcissel populations.[48] In 1986 he reported that there were five males for every female Dickcissel, probably because the larger, heavier billed male can feed on sorghum and rice more effectively than can the smaller female. Females may continue to depend on the smaller seeds in natural grasslands in Venezuela, but because these grasslands are disappearing the number of females has plummeted. The small proportion of females in the population directly depresses reproductive rates, and it also may increase the problem of parasitism by Brown-headed Cowbirds. As the number of female Dickcissels, and hence of Dickcissel nests, declines, a higher proportion of the remaining nests may be parasitized by cowbirds.[49] This reduces the Dickcissel population even further, which may increase the parasitism rate even more. This downward spiral could end in extinction.

Whether the decline of Dickcissels results from direct eradication of the birds in their wintering area or from the subtle interplay of female mortality in winter and the impact of cowbird parasitism in summer, it is clear that we need to understand what is happening to this species both in the prairies where they nest and in the llanos where they spend the winter. This applies to other migratory species of prairie birds as well. Many prairie birds move only a few hundred miles south, from the prairie provinces and adjacent North Dakota and Montana to the southwestern United States and northern Mexico. A study of 20 grasslands in Texas and Oklahoma showed that a patchwork of different grassland types can be important for sustaining grassland birds in winter. Just as in the breeding season, different species were concentrated in different habitats. For example, Lapland Longspurs were mainly in cultivated fields, Smith's Longspurs were in heavily grazed pastures, and LeConte's Sparrows were in lightly grazed pastures.[50]

Restoring the Prairie Mosaic

North American grasslands, like grasslands in other parts of the world, have relatively few species of birds compared to forests and other types of multilayered vegetation.[51] Many of these species are restricted to the Great Plains, however, and preservation of these endemic species depends on the availability of a wide range of grassland types.[52] The tallgrass prairies of the eastern plains have no true endemics, although the Dickcissel and Greater Prairie-chicken are concentrated there. In contrast, the shortgrass and mixed-grass prairies have a large number of species that are found nowhere else, including Mountain Plover, Baird's Sparrow, Lark Bunting, McCown's Longspur, and Chestnut-collared Longspur.[53] Many of these, such as McCown's Longspur and Mountain Plover, are associated with heavily grazed landscapes, suggesting that they evolved alongside grazers (bison and prairie dogs). Some of the more widespread species, such as Clay-colored and Henslow's sparrows, are found in grassland that has not been disturbed recently. Consequently, only a prairie with a range of disturbance patches of different ages will have a full array of grassland species. In Oklahoma, a prairie of this sort might have populations of Horned Larks, Eastern Meadowlarks, Dickcissels, and Henslow's and Grasshopper sparrows. In Montana, a prairie with a diversity of disturbance patches might have Burrowing Owls, Horned Larks, Western Meadowlarks, Chestnut-collared Longspurs, and McCown's Longspurs. The species of birds differ in eastern tallgrass and western shortgrass prairies, and in northern to southern prairies, but the importance of maintaining a patchwork of disturbance patches is the same for all of these bird communities.

In many regions of the Great Plains, the destruction of the native prairie has been almost complete. This is particularly true of the tallgrass prairie. In Illinois, Indiana, and Iowa, only 0.01–0.02 percent of the native prairie remains.[54] These states have only tiny remnants of prairie, in old cemeteries and along railroad tracks. Aldo Leopold's lament applies to much of the Midwest: "No living man will see again the long-grass prairie, where a sea of prairie flowers lapped at the stirrups of the pioneer. We shall do well to find a forty here and there on which the prairie plants can be saved as species."[55] For a time, prairie birds found a place in the prairie hay meadows that were mowed after the land was settled, but most of these meadows were eventually plowed under or converted to more intensively managed hay meadows dominated by introduced species, such as timothy and clover.[56] As the frequency of mowing increased, even the bird species that could nest in these artificial grasslands declined.[57] Farther west, however, much of the native mixed-grass and shortgrass prairie remains in the form of rangeland used to raise cattle and sheep, with expanses of stunted grass similar to those described by early European explorers.[58]

These large areas of native prairie are relics of the great continental grassland that once supported herds of more than 30 million bison.[59] These migratory herds are gone, and massive grass fires no longer sweep across the prairie, so the remaining prairie must be carefully managed to simulate these missing disturbances. Otherwise, the prairie species that depend on disturbances will soon disappear.

The Konza Prairie Research Natural Area in Kansas is a model for the use of grazing and fire to maintain the biological diversity of a prairie. This 8,616-acre (3,487-ha) expanse of grassland, one of the largest areas of tallgrass prairie in North America, was protected by The Nature Conservancy and is managed by Kansas State University.[60] Each of the dozens of watersheds in this preserve is bordered by firebreaks and managed as a separate unit. Some watersheds are left unburned, while others are burned every one, two, four, or ten years. Some watersheds are grazed by bison (which were reintroduced to this prairie in 1987) or cattle, while others are ungrazed. Some watersheds are managed with both grazing and burning. Comparison of watersheds that have been managed differently allows researchers to assess the impact of different burning intervals and different types of grazing on the efficiency of nutrient cycling, the productivity of prairie grasses, and the abundance of different species of plants and animals. At the same time, this experimental manipulation of the prairie produces a large-scale mosaic, providing habitat for most species of grassland birds in the region. Controlled burns prevent the invasion of woody plants and the consequent loss of habitat for all grassland bird species. Watersheds that are burned every four or ten years provide habitat for Henslow's Sparrows, which require dense vegetation, while recently burned grassland accommodates other species.[61]

The densities of some grassland bird species are similar in both recently burned and unburned watersheds on Konza Prairie, indicating that these species are adaptable enough to endure either burning of the vegetation on their territories or some encroachment by woody vegetation.[62] This is promising, because the same species are often found in remnants of native prairie that are too small to maintain a complex patchwork of different types of grassland. Small prairies typically need to be burned periodically to maintain prairie plants, and it appears that this will not be a problem for the Eastern or Western meadowlarks, Grasshopper Sparrows, and Dickcissels that are frequently found in these preserves. Hence, a small prairie that is managed as a single unit can support some grassland bird species, but probably will not be suitable for other species, such as Henslow's Sparrow, Sedge Wren, and Upland Sandpiper, that require a specific type of grassland.[63] These more specialized species depend on large expanses of grassland with mosaics of grassland patches in different states of recovery from burning, mowing, or grazing. Somewhere in this mosaic each species will find its preferred habitat.

Only a few large natural prairies are managed primarily to maintain biological diversity rather than for livestock production, however, and these have relatively little overall impact on the total population of prairie birds. Recent population declines in prairie birds are primarily the result of changes in the use of farm and ranch land, particularly the conversion of grassland (primarily grazing land) into cropland. Compared to grassland, cropland is used only by a few species of prairie birds.[64] This difference has been dramatized by the federal Conservation Reserve Program, under which farmers are paid by the federal government to take marginal cropland out of production. These fields are then planted with perennial grasses to reduce soil erosion. Surveys in North Dakota, South Dakota, and Montana showed that Lark Buntings, Grasshopper Sparrows, Western Meadowlarks, Savannah Sparrows, and many other grassland birds were more abundant in Conservation Reserve Program land than in cropland.[65] Only Horned Lark (a species that has not declined in the Great Plains) was substantially more abundant in cropland.

Impact of Grazing on Prairie Birds

Although prairie preserves and Conservation Reserve Program land can make an important contribution to the preservation of prairie birds, the most important habitat for many species will probably continue to be grassland that is managed for livestock (particularly cattle and, to an increasing extent, bison). As James Brown and William McDonald wrote regarding the arid grasslands of the Southwest, "The best way to preserve the open spaces, arid ecosystems and diverse biota . . . is to keep rural people on the land. Livestock ranching must be both ecologically sustainable and economically viable."[66] Others argue that livestock ranching is incompatible with sustaining natural diversity in arid Western grasslands, and some have even argued that residential development is preferable to ranching because it is more concentrated and thus affects a smaller proportion of the landscape.[67] In a recent paper, Thomas Fleischner presents a long list of studies that illustrate the negative ecological impact of grazing, but his examples are primarily from wooded streamsides, desert, and semidesert grasslands.[68] The species in these habitats did not evolve with intense pressure from prairie dogs or large herds of grazers.[69] In contrast, the species associated with tallgrass, mixed-grass, and shortgrass prairies evolved alongside large grazers, so ranching and preservation of biological diversity are not necessarily incompatible. Until the nineteenth century, prairie birds lived in habitats occupied by millions of bison and prairie dogs, and until about 15,000 years ago they lived in grasslands molded by the activities of a diverse group of grazers, such as horses, rhinos, camels, and mammoths. Given this history, it is not surprising that most native prairie plants are resistant to grazing, and that some species of birds thrive in heavily grazed grassland.[70] This conclusion cannot be extended to all western rangelands, however, because most are not prairies but are forest, chaparral, arid grassland, or desert that did not support large numbers of native grazers.[71]

Grazing, like fire, can be an effective tool to maintain habitat for prairie birds and other prairie organisms. In order to maintain an array of prairie species, however, grazing must be managed to produce a shifting mosaic of recently disturbed and relatively undisturbed grassland.

Conservation of Birds of the Western Prairies

The priorities for conservation differ in the tallgrass prairies of the eastern plains and the shorter grasslands of the western plains. In the upper Midwest, the highest priority is to save relict patches of tallgrass prairie, and to restore prairies by planting native species of grasses and herbs. Many of these relict prairies are important for protecting rare species of plants and insects, but are too small for many species of grassland birds. Grassland birds that require large areas of open habitat depend on agricultural land, so programs that provide suitable habitat in farming areas should be encouraged. The Conservation Reserve Program, which was primarily designed to prevent soil erosion and permit recovery of the soil, could be used more effectively to provide habitat for grassland species.

In the shortgrass and mixed-grass prairies, conversion of prairie to agriculture and other uses is a problem, but large expanses of grassland remain. Most of this grassland is used for livestock production. Grazing, along with fire, creates conditions needed by many species of prairie specialists, so it is not necessarily incompatible with sustaining the diversity of native species. However, grazing must be managed to preserve the complex patchwork of different types of grassland, in different stages of recovery from disturbance, that characterized the presettlement landscape. This will ensure that a full range of prairie birds—from Horned Larks to Baird's Sparrows—will find the specific habitat they need. Another goal should be to restore prairie dogs to a larger proportion of the grasslands in the West by reestablishing them in protected areas and determining how ranges can be managed to sustain both prairie dogs and a cattle industry. Several species of birds and mammals are disappearing primarily because they are associated with the burrow systems or cropped vegetation found in prairie dog towns.

CHAPTER 4

Lost Birds of the Eastern Forest

He stood against a big gum tree beside a small bayou whose black still water crept without motion out of a cane-brake, across a small clearing and into the cane again, where, invisible, a bird, a big woodpecker called Lord-to-God by negroes, clattered at a dead trunk. . . . He only heard the drumming of the woodpecker stop short off, and knew that the bear was looking at him.

—WILLIAM FAULKNER, Go Down, Moses

ALONG with the bear, Old Ben, the Ivory-billed Woodpecker evokes the original wildness of the virgin bottomland forest of the Talla-hatchie River in Faulkner's Go Down, Moses.[1] Faulkner describes the loss of this wildness as Old Ben is killed and the forest is reduced to a tame remnant, a sequence that, for different decades or centuries, describes the history of almost every part of eastern North America. Almost a hundred years before Go Down, Moses was published, Henry Thoreau poignantly de-scribed the loss of wildness in Massachusetts[2]: between the early 1600s and 1855, wolves, black bears, wolverines, mountain lions, lynx, beavers, moose, deer, and turkeys disappeared, leaving "a tamed, and, as it were, emasculated country," a poem from which his ancestors "have torn out many of the first leaves and grandest passages, and mutilated it in many places."[3] By the early twentieth century, nearly all of the great forests of eastern North America had been cleared, eliminating not only large mammals, but also some of the most spectacular birds. The region lost its largest woodpecker, the Ivory-billed, and its only parrot, the Carolina Parakeet. The Bachman's Warbler, never known as common, has probably slipped away, and only vigorous ef-fort will save the Red-cockaded Woodpecker of the southern pine forests. The most sobering loss was the extinction of the Passenger Pigeon, once the most abundant species of bird in North America, if not in the world.

PLATE 4. Carolina Parakeets (two adults below an immature) feeding on cypress cones in a swamp along the Suwanee River in Florida.

These extinctions, particularly the loss of the seemingly limitless flocks of Passenger Pigeons, inspired the early conservation movement in North America. They remain a lesson for other parts of the world, where numerous species are threatened with habitat destruction. Recently, however, skeptics have turned this lesson on its head, claiming that the significance of the experience in eastern North America is that destruction of nearly all of the forest resulted in *so few* extinctions.[4] The implication is that fragmentary remnants of forest, like the small scrap of the Tallahatchie bottomland protected from logging in *Go Down, Moses*, will be sufficient to sustain animal populations.[5]

Destruction of the Eastern Forest

If satellite images of the Ohio Valley had been available in the 1850s, the picture of forest destruction would resemble recent images from Rondonia and other heavily settled regions of the Amazon Basin. Within a few decades, nearly all of the forest outside of extensive swamps had been converted to cropland, and eventually even the swamps (including the huge Black Swamp in the northwestern corner of Ohio) were drained and plowed.[6] The wave of agricultural settlement swept across the midwestern forests and onto the prairie, converting both to rich farmland. Most of the forests not cleared for farming were cleared for lumber.

In contrast to the rapid deforestation of Ohio, the earlier clearing of forest along the East Coast of North America had been a surprisingly slow, measured process. The coastal regions east of the Appalachian Mountains were settled over a period of 200 years. Western expansion was gradual, resulting in an "almost motionless America" until the early 1800s.[7] Even behind the expanding front of settlement, the forest was cleared only slowly. In parts of New England, more than 50 percent of the land was still forested a hundred years after European settlement. Early settlers often received large land grants, more than they needed for subsistence farming.[8] Moreover, due to the tremendous labor and expense required to chop down the forest and remove logs and stumps, it typically required many years to clear a large area.[9] Trees were either laboriously felled with an ax or girdled (killed by removing a strip of bark around the base of the tree to cut off the flow of sap). Felled logs were usually burned on the spot if they were not needed for construction, and the standing dead trees produced by girdling later fell or were cut. Stumps were usually left in the ground until they were rotten

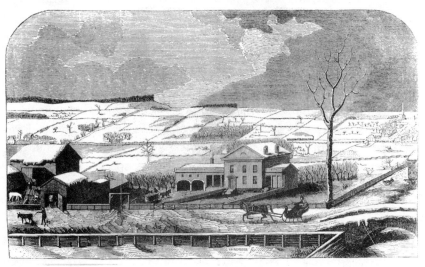

FIG. 4.1. Forest clearing in western New York, from a nineteenth-century history of the region. Above, an early clearing in the forest resulting from construction of the recently completed cabin. Fields will not be cleared until later. Below, the same landscape 45 years later, after "the forest has receded in all directions" and has been replaced with a landscape of orchards, pastures, farmhouses, and villages. From Turner, 1849.

enough to be pulled out by teams of oxen. Using these arduous techniques, a farmer probably could not clear more than 10–15 acres (4–6 ha) per year.[10]

By the time the Ohio Valley and other parts of the Midwest were settled, however, the pace of clearing had accelerated. Better farm implements (reapers, cultivators, and improved plows) and a better transportation system (first canals, then railroads) induced farmers to clear and cultivate larger areas.[11] Midwestern farms were productive enough, and the markets for farm produce were accessible enough, to persuade many farmers to hire laborers to help clear the forest.[12] Moreover, in the late nineteenth century, large-scale harvesting of trees by timber companies contributed to the rapid pace of forest clearing. White pine forests in the Great Lake states and coastal pine forests in the Southeast were quickly destroyed. A logging company would build a sawmill, remove the trees in the surrounding region, and move on to another area when the timber was exhausted.[13] Forests were clearcut with no thought of regeneration, and branches, rotten wood, and other logging debris were left to dry and often to burn. The combination of agricultural clearing and large-scale lumbering left only remnant patches of old forest.

By the end of the nineteenth century, only a small proportion of the original deciduous forest of eastern North America remained uncut. In many regions, old-growth forest was completely eradicated. When a similar pattern of forest destruction occurs in the tropics today, conservationists raise the alarm about the imminent extinction of many species of forest animals and plants. Massive destruction of the temperate-zone forest was not followed by numerous extinctions, however. A few birds became extinct, but these represent a small percentage of the total number of species. To critics of the environmental movement, this demonstrates that concern about high extinction rates in the tropics is exaggerated. If a few remnant patches of the original temperate-zone forest could preserve most of the species that lived in that forest, perhaps a few protected areas will suffice for preserving the diversity of the Amazon Basin and other tropical regions.[14] Despite the superficial similarity of deforestation in nineteenth century Ohio and twentieth century Brazil, however, there are good reasons to expect that destruction of tropical forests will generate many more extinctions.

Rebirth of the Forest

The destruction of the forest in eastern North America was not necessarily permanent. Both coniferous and deciduous woodlands were surprisingly

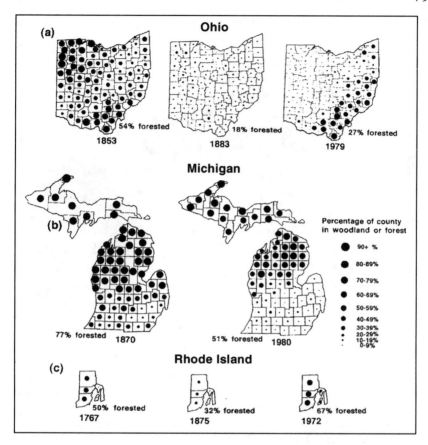

FIG. 4.2. Changes in the forest cover of Ohio, Michigan, and Rhode Island, showing the extent of deforestation as well as the regrowth of forest in many regions. From Whitney, 1994. Reprinted with permission from Cambridge University Press.

resilient. Young forest soon covered areas where burned stumps or logging debris had once stretched for miles. In many regions near the Great Lakes, the tall white pine and eastern hemlock forests did not regenerate because of repeated fires fed by logging debris, but they were replaced by oak, birch, and aspen forests.[15] Even farmland often reverted to forest after two or three generations of farmers had plowed the soil. Many of the settlers who moved to the fertile land of Ohio and western Kentucky left behind farms on the East Coast where the terrain was hilly or rocky, or where the soil had been exhausted. Derelict cropland typically was used as pasture, but frequently these pastures were invaded by thicket or stands of young pine that grew up

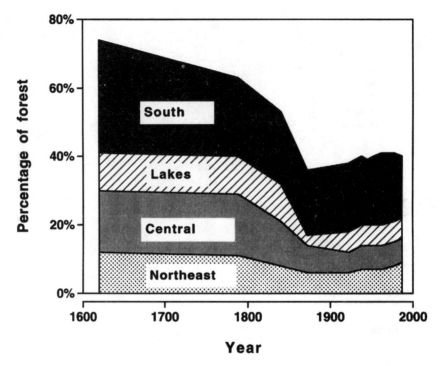

FIG. 4.3. Changes in the percentage of land covered with forest in different regions of eastern North America. South: southeastern states north to Virginia and Kentucky; Lakes: Michigan, Minnesota, and Wisconsin; Central: Ohio to Iowa and Missouri; Northeast: New England south to Maryland. From Pimm and Askins, 1995. © 1995 National Academy of Sciences, U.S.A.

into forest.[16] Thus a wave of forest regeneration followed the westward-moving wave of forest destruction.[17] During the decades when forest was melting away in Ohio, vigorous second-growth woods were spreading across the hills of New England and New York. As the hardwood forest fell in southern Wisconsin and Minnesota, forests were growing back in southeastern Ohio and in West Virginia.[18] After 1910, abandoned cotton and tobacco fields in the southeastern states became forested.[19] Almost from the beginning of European settlement, the loss of forest was offset by the growth of forest. For the first 200 years, the losses exceeded the gains, but after 1870 the total amount of forest slowly increased.

In most regions an old forest was replaced with a much younger, sec-

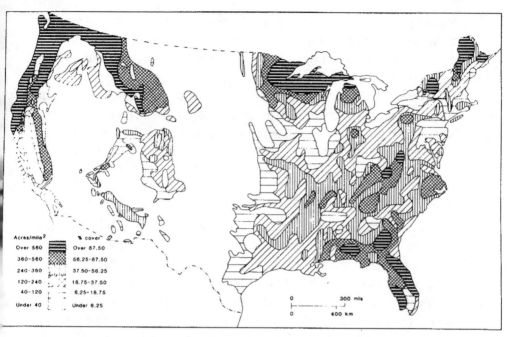

Acres/mile²   % cover
Over 560 — Over 87.50
360-560 — 56.25-87.50
240-360 — 37.50-56.25
120-240 — 18.75-37.50
40-120 — 6.25-18.75
Under 40 — Under 6.25

0   300 mls
0   400 km

FIG. 4.4. Percentage of the land covered with woodland in the United States in 1873. Note that there are several heavily forested regions in the East that would have served as refugia for forest birds. From Williams, 1989. Reprinted with permission from Cambridge University Press.

ond-growth woodland, but this new woodland provided favorable habitat for nearly all species of forest birds. Most species of forest birds establish nesting territories in second-growth woods as the canopy begins to close, and some forest specialists, such as the American Redstart, reach their highest densities in young forest.[20] Only species that require large dead trees for nesting—Pileated and Ivory-billed woodpeckers and Barred Owls—were typically absent or scarce in the new forests that grew on abandoned farmland.

In the early 1870s, the amount of forest dipped to a minimum. The area of the eastern forest had been reduced by about 50 percent, and much of the remaining forest had been degraded because of the practice of grazing cattle in the woods, resulting in the destruction of young trees, shrubs, and ground cover.[21] Even during this period, however, large areas of eastern

North America were heavily wooded. These areas must have served as forest "refugia," sustaining forest bird populations when much of the East was cleared. The largest refugia were in the Appalachian Mountains, where forests on steep slopes remained uncut and many hillside and valley farms had reverted to forest. Also, large sections of the southeastern lowlands had not yet seen the effects of large-scale logging.[22] These refugia may have been critical for some species, sustaining them during the period of most intensive forest destruction.

Species at Risk

Even if all of the eastern deciduous forest and southern coniferous forest had been destroyed, only a small number of species would have been at risk of extinction.[23] Of the 215 species of landbirds in the eastern United States, 138 depend to some extent on mature forest or the early successional stages (shrubland and thicket) that develop into forest. The other species, which live in grassland and marshes, would not be affected directly by deforestation. Moreover, most of the forest-dependent species have extensive geographical ranges outside of the eastern forest (the deciduous forest of the eastern United States and southeastern Canada, and the pine forests of the southeastern United States). Many range across the largely intact boreal coniferous forests of Canada, or into the forests of the western mountains or even the tropical lowlands of Mexico. These species were not in imminent danger of extinction from destruction of the eastern forest.

Surprisingly few species are restricted to the eastern forest. Only eleven species—Chuck-will's-widow, Carolina Parakeet, Red-cockaded Woodpecker, Fish Crow, Carolina Chickadee, Brown-headed Nuthatch, Bachman's Warbler, Yellow-throated Warbler, Worm-eating Warbler, Swainson's Warbler, and Bachman's Sparrow—were found exclusively in the eastern forest. Two of these, the parakeet and the Bachman's Warbler, are extinct. An additional 17 species, including the extinct Ivory-billed Woodpecker and Passenger Pigeon, had more than 75 percent of their geographical range in the eastern forest. Although only 2 percent of the landbirds in eastern North America became extinct, only the species that depended on forests and that were largely restricted to the region were in danger of extinction. Not surprisingly, all of the species that actually became extinct were completely or largely restricted to eastern North America. Two of the eleven endemic

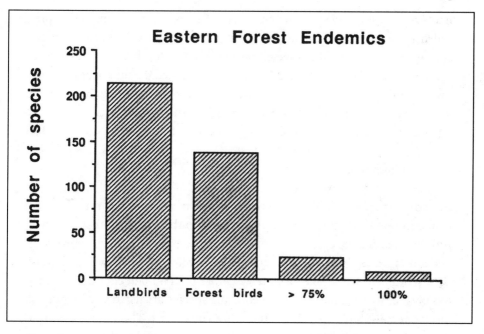

FIG. 4.5. Total number of species of landbirds and forest birds associated with forests in eastern North America, showing the number of species almost entirely (greater than 75 percent of their range) or entirely (100 percent of the range) restricted to the eastern forest region.

species (species found only in the eastern forest) became extinct, giving an extinction rate of 18 percent. Furthermore, of the 28 species with more than three-quarters of their range in the eastern forest, four (14 percent) have disappeared. Thus, a substantial percentage of species that are primarily found in the eastern forest became extinct.

Generally the number of species increases as larger and larger areas of a particular habitat (such as deciduous forest) are surveyed. This relationship between habitat area and number of species can be used to obtain a rough approximation of the number of extinctions that will result from habitat destruction.[24] The assumption is that species will progressively disappear as the habitat shrinks in much the same way as they accumulate as more and more habitat is included in a survey. Just as small islands have fewer species than larger islands in the same island group, a habitat remnant will support fewer species than the larger expanse of habitat that preceded it historically. If its area is further reduced, it will lose even more species. Using an equation based on the relationship between number of species and habitat area, Stuart Pimm and I calculated the number of species that should have become extinct in the eastern forest in the late 1800s, when 50 percent of the forest was gone.[25] The equation "predicted" an extinction rate of 16 percent of the species, which is close to the actual rate both for species restricted to the eastern forest (18 percent) and for species primarily found in the eastern forest (14 percent).

In many regions of the tropics the amount of forest destruction greatly exceeds 50 percent, so an even greater proportion of species are at risk of extinction. Even more important, compared to the eastern forest of North America, particular regions in the tropics have a much higher proportion of endemic species. For example, 199 species of birds are entirely restricted to the Atlantic Forest of coastal Brazil.[26] More than 80 percent of this forest has been destroyed, and 70 of the endemic species are listed as endangered or threatened. Three species are thought to be extinct. The situation is similar for Madagascar and the Philippines.[27] Other parts of the tropics have not suffered such extensive deforestation, but they have numerous endemic species that could be at risk if a large proportion of the forest is cut. Many species, for example, are restricted to particular regions within the Amazon Basin.[28]

Hence, the situation in the forests of eastern North America in the nineteenth century is not comparable to conditions in tropical forests today. The pace of deforestation in the United States was slower—slow enough to be counterbalanced to some extent by the growth of new forest. More important, few species were at risk of extinction. Of the species at risk, a substantial percentage disappeared, a situation that does not bode well for the many tropical species with restricted geographical ranges.

Extinction of the Passenger Pigeon

Four species of forest birds have disappeared from the eastern forest since the 1600s. Although this rate of extinction is predictable given the amount of habitat destroyed, this general type of analysis reveals little about the reason that particular species disappeared. Understanding the history of these extinctions may help prevent extinctions in the future.

The Panglossian view that nature is infinitely resilient and essentially indestructible is not new to critics of twentieth century environmentalism. A similar attitude was prevalent in the late 1800s even as it became clear that populations of many species of birds and mammals in North America were declining at an alarming rate. In 1857, a committee reporting to the Ohio Legislature wrote that the Passenger Pigeon "needs no protection" because "no ordinary destruction can lessen them or be missed from the myriads that are yearly produced." By 1900, the Passenger Pigeon, which once numbered in the billions, was extinct in the wild. The last captive specimen died in 1914.[29]

At the time Europeans reached North America, the Passenger Pigeon may have been the most numerous type of bird on Earth. Alexander Wilson estimated that there were more than 2 billion birds in one migrating flock.[30] Single nesting colonies sometimes covered huge areas: in 1871, a colony in Wisconsin covered 850 square miles and included an estimated 135 million adult pigeons.[31] The destruction of the Passenger Pigeons, along with the reduction of the immense herds of bison to 20 individuals in Yellowstone National Park and a remnant population in the northern wilderness of Alberta, convinced many of the importance of conservation.[32] Even people who had little appreciation of natural beauty and no interest in natural diversity began to realize the folly of eradicating populations that could have been sustained indefinitely with a reasonable level of harvesting.

The disappearance of the Passenger Pigeon has been a mystery. Some have argued that only an infectious disease, perhaps transmitted from introduced domestic pigeons, could account for the eradication of this superabundant species. However, the Passenger Pigeon declined over many decades, beginning in the East in the 1830s and continuing in the central states and Great Lakes region in the 1870s and 1880s.[33] A fatal, highly infectious disease would have swept through the highly nomadic flocks much more quickly.

Harvesting by market hunters surely contributed to the decline of the pigeon. Most pigeons nested in immense colonies. Raccoons, hawks, and other predators were attracted to these colonies, where the closely packed, flimsy stick nests provided easy access to eggs and nestlings, but the predators were satiated before they could attack even a small percentage of the nests. The same applied to hunting by people from the region around the colony. Individual pigeons gained protection by hiding in a crowd, and

the interior of a massive colony was particularly well protected from predation. By the 1870s, however, hundreds of professional pigeoners converged
on every large nesting colony, netting and shooting the adults and taking the
young from nests.[34] Both adult pigeons and their young (squabs) were
packed in barrels and shipped to markets in large cities. In season, wild pigeon was a common source of meat in northeastern and midwestern cities.
For example, an estimated 300,000 pigeons were sold in the markets of
New York City in 1855.[35] The expanding network of telegraph lines and railroad tracks was critical for effectively locating and exploiting Passenger Pigeons because colonies were seldom found in the same state, much less the
same location, in consecutive years.[36]

   Although hunting probably contributed to the decline of Passenger Pigeons, it does not fully explain the extinction of the species. Once the large
colonies were gone, market hunters turned their guns on migrating flocks
of sandpipers and plovers.[37] It is doubtful that all of the remaining small
colonies and solitary pairs of pigeons were sought out by hunters. Moreover, it is doubtful that even the large colonies declined primarily as a result
of hunting. In 1871, an astounding 1.2 million pigeons were taken by 600
professional hunters in a large colony in Wisconsin.[38] As Enrique Bucher
pointed out, however, this represented less than 1 percent of the estimated
135 million birds in this colony. True, the netting, shooting, and felling of
trees in a colony disrupted the nesting of many birds that were not killed,
and some colonies were abandoned early in the nesting season because of
harassment by hunters.[39] However, Eared Doves in Argentina have not declined despite intensive efforts to destroy these birds, which are considered
agricultural pests. Eared Doves nest in colonies with as many as 10 million
adult birds, and trapping and poisoning have had little effect on the size of
the population even though as many as 10 percent of the individuals in a
colony are sometimes killed.[40] Bucher argues that the Passenger Pigeon,
like the Eared Dove, could not have been easily eliminated by harvesting.
The ultimate cause of its extinction was probably habitat destruction rather
than hunting.

   If large forest refugia remained in the eastern forest, as I have discussed
above, then why did Passenger Pigeons disappear? The answer, according to
Bucher, is that the Passenger Pigeon, unlike its close relative the Mourning
Dove, could not adapt to the changed landscape after Europeans had cleared
much of the forest. The enormous flocks of Passenger Pigeon were con-

stantly on the move, and no single region or refugium could sustain them. They could not exist without the great, continental expanse of forest that was disrupted in the nineteenth century.

Although Passenger Pigeons ate some berries and insects, their primary food was nuts, particularly acorns and beechnuts. Nuts, eaten by adults during the spring breeding season, provided the nutrients needed for production of "crop milk," a secretion produced by both male and female pigeons to sustain the young. Both oaks and beeches produce massive amounts of nuts ("mast crops") every few years at unpredictable intervals. Mast cropping aids the trees in seed dispersal. Animals that destroy seeds, such as Wild Turkeys, Red-bellied Woodpeckers, and beetle larvae, take only a small proportion of a mast crop before they are satiated. Nuts are sometimes planted rather than destroyed by other animals, such as Blue Jays and eastern chipmunks. These species store nuts in the ground, where they germinate if they are not retrieved. Burying seeds is particularly common during mast crop years when animals are surrounded with more nuts than they can eat. Thus, a mast crop overwhelms both seed destroyers and seed storers with a superabundance of food, resulting in a higher proportion of seeds that are buried and subsequently survive to germinate.

Mast crops occur synchronously for one or more species of nut-bearing trees across a large region, but not across the continent. This asynchrony in nut production across the eastern forest was the key to the success, and the demise, of the Passenger Pigeon. A flock nesting in Michigan one year because of a mast crop of acorns might nest in Wisconsin the next year because of an abundance of beechnuts. The original Michigan site might not be able to support a large nesting colony again for several years. Thus, the pigeon flocks needed to move across great distances to find good nesting sites.

Flocking was probably critically important for finding the unpredictable concentrations of beechnuts and acorns that resulted from regional mast crops. When moving north from their winter quarters in the southern states, passengers pigeons typically flew in a broad front, about 20 deep and extending east and west from horizon to horizon.[41] If birds in any part of this front detected food, they began to circle, attracting the rest of the flock to the area. Moreover, birds feeding on the ground gave a distinct feeding call (a loud "tweet") that attracted other pigeons.[42] In addition, nocturnal roosts and nesting colonies may have served as "information centers," per-

mitting birds that had not found food the previous day to follow more successful birds to food.

The specialization of Passenger Pigeons on mast crops made them particularly vulnerable to large-scale forest destruction. Beech and oak forests
were cleared throughout eastern North America for agriculture. Beech forest, in particular, was considered good potential farmland. Large areas of
forest remained in some regions, and these might have been suitable for
nesting every few years, but during the intervening years the pigeons may
not have been able to find large concentrations of nuts. Although the pigeons could nest in any oak or beech forest between the Atlantic and the
prairie, nesting was restricted to northern states, where the snow cover was
deep enough to protect nuts over the winter, until the pigeons migrated
north in March or April.[43] Moreover, in the late nineteenth century many
oak forests were dominated by second-growth trees that were too young to
produce large numbers of acorns.[44] Over the long term, Passenger Pigeons
depended on numerous, widely scattered forests that produced large
enough crops of nuts to support nesting. As the number of potential nesting sites declined, the number of years with poor nesting success must have
increased, leading to a steady decline in the population.

As the size of the flocks dropped, they would become less efficient at
finding the remaining concentrations of nuts. Also, the nests of solitary pairs
and small flocks would be subject to high rates of predation because they
would not be "hidden" among millions of other nests.[45] Passenger Pigeons
may have become so highly adapted to living in immense groups that they
were not capable of persisting in small groups. This explanation of the pigeon's disappearance was anticipated, and elegantly described, by Aldo
Leopold: "Yearly the feathered tempest roared up, down, and across the continent, sucking up the laden fruits of forest and prairie, burning them in a
traveling blast of life. Like any other chain reaction, the pigeon could survive
no diminution of his furious intensity. Once the pigeoners had subtracted
from his numbers, and once the settlers had chopped gaps in the continuity
of his fuel, his flame guttered out with hardly a sputter or a wisp of smoke."[46]

Birds of the Southern Bottomland Forests

While the Passenger Pigeon ranged across the eastern forest, the Carolina
Parakeet, Ivory-billed Woodpecker, and Bachman's Warbler were all concentrated in a single, restricted habitat, the bottomland hardwood forests

FIG. 4.6. Flooded bottomland hardwood forest in the southeastern United States. Photograph by William Niering.

and swamps of the southeastern United States. These bottomland forests grow along rivers, in low areas that are prone to seasonal flooding. The largest expanse of southern bottomland forest is in the Mississippi River Delta, particularly in the Atchafalaya Basin and Barataria Bay estuary. This type of forest is also found along rivers of the coastal plain between Maryland and Florida, and along many of the rivers along the Gulf Coast.[47]

The rich soils of the bottomland hardwood forest have always been favored for farming. Most of the Moundbuilder settlements were located in the river floodplains, and chronicles of early Spanish expeditions in the lower Mississippi Valley report that large expanses of the bottomland were cleared of trees.[48] Also, William Bartram describes a large area that had been cleared along the Alabama River in 1775: "We continued over these expansive illuminated grassy plains, or native fields, above twenty miles in length, and in width eight or nine, lying parallel to the river, which was about ten miles distance; they are invested by high forests, extensive points or promontories, which project into the plains on each side, dividing them into vast fields."[49] If any forest songbirds became extinct because of human activities before the arrival of Europeans, they probably were species that depended on the bottomland hardwood forest, but it is more likely that the first such extinctions occurred after European settlement.

After Europeans colonized eastern North America, the bottomland forests and swamps were quickly cleared and drained, first for growing cotton, sugar cane, and rice, and more recently for soybean production. The area of southern bottomland forest in the Mississippi River Valley has been reduced by 77 percent, from an estimated 21 million acres (8.5 million ha) before European settlement to only about 4.9 million acres (2 million ha) today.[50] In contrast, only about 50 percent of the entire eastern forest was cleared at the time of maximum deforestation. Moreover, most of the remaining southern bottomland forest has been severely fragmented. Great stretches of riverside forest have been broken into remnants of less than 250 acres (100 ha). It is not surprising, therefore, that a disproportionate number of the bird species that became extinct were associated with the floodplain forests of southern rivers.

The Carolina Parakeet was observed most frequently along the forested banks of rivers and in heavily wooded swamps. Alexander Wilson wrote that they are "particularly and strongly attached" to "low, rich, alluvial bottoms, along the borders of creeks, covered with a gigantic growth of sycamore trees, or button-wood; deep and almost impenetrable swamps."[51] In his description of cypress trees along the St. Johns River in Florida, William Bartram described how parakeets "are commonly seen hovering and fluttering on their tops: they delight to shell the balls [cones], its seed being their favorite food."[52] Highly mobile flocks of hundreds of parakeets traveled along heavily wooded river banks or through cypress swamps in search of seeds (particularly cocklebur and thistle) and berries.[53] As with the Passenger Pigeon, introduced disease has been suggested as the cause of the disappearance of the Carolina Parakeet,[54] but the parakeet's range contracted slowly over a period of 90 years as it disappeared from one region after another as the forest was cleared.[55] Parakeets were killed in large numbers by farmers protecting their grain and fruit, and were shot by hunters for both food and the feather market.[56] Large numbers were also captured for the cage bird trade. The concentration of parakeets in nesting colonies in hollow trees or in clusters of stick nests in cypress trees, and in winter roosts in hollow trees, made them vulnerable to capture. They disappeared from the lower Mississippi River Valley by 1880, when large tracts of bottomland forest remained uncut, suggesting that direct persecution, not habitat destruction, eliminated them. However, Carolina Parakeets, like tropical fruit-eating parrots, may have been highly nomadic, moving between widely separated feeding areas in response to the seasonal availability of fruit and seeds. If this were the case, then they would have been severely affected by widespread destruction of their bottomland habitat even if some large forests remained. Moreover, their concentration into fewer and fewer areas would make them more vulnerable to hunters and trappers.

The largest North American woodpecker, the Ivory-billed Woodpecker, had almost the same habitat requirements as the Carolina Parakeet.[57] Ivory-billed Woodpeckers probably survived longer in the lower Mississippi Valley than the parakeets because they were more sedentary, with pairs remaining on a large territory throughout the year. Ivory-billed Woodpeckers were about 20 inches (50 cm) long, with striking black and white plumage. Their loud, honking calls and distinctive drumming—a loud double rap of the bill on wood—were once signature sounds of the southern bottomland hardwood forest.

James Tanner's careful study of this species from 1937 to 1939 provides the best description of its behavior and habitat requirements.[58] The greatest concentration of Ivory-billed Woodpeckers originally must have been in

FIG. 4.7. Ivory-billed Woodpecker habitat (sweetgum and oak forest on a river terrace) in the Mississippi Delta, Madison Parish, Louisiana. Photo by James T. Tanner. From Tanner, 1942, © National Audubon Society. Reprinted by permission.

the alluvial bottom of the Mississippi River, a 40–80-mile-wide (65–125 km) floodplain that stretches from the Ohio River to the Gulf of Mexico. They were concentrated in the higher parts of the first river terrace in forests dominated by sweetgum, ash, and various species of oaks, and were less common in permanently flooded swamps dominated by cypress and water tupelo. They were usually found in forests that were flooded for less than a few months per year. In Florida, however, they were also found in forest dominated by cypress, including deepwater swamps away from river floodplains, and they frequently foraged on dead trees (usually trees killed by fire) in pine forests close to swamps. In bottomland hardwood forest, each pair defended a territory of 2.5–3 square miles (6–7 km²). The dependence of these woodpeckers on large areas of bottomland forest or swamp made them vulnerable to extinction.

Although ivorybills fed on berries and southern magnolia seeds, their diet primarily consisted of the larvae of wood-boring beetles (Cerambycidae) and engraver beetles (Scolytidae).[59] In this respect, they differ from the other large woodpecker of the Southeast, the Pileated Woodpecker, which primarily feeds on carpenter ants. The ivorybill foraged on dead trees or branches, sometimes digging deep within the wood to extract large wood-boring larvae, but usually scaling off bark with glancing blows to reveal the beetle larvae underneath.[60] These larvae are especially abundant in the large, dead trees and branches characteristic of old-growth forests. Large snags were also needed for construction of nest cavities.

Another species associated with bottomland habitats was the Bachman's Warbler. Although this species is probably extinct, there are records from the wintering areas in Cuba from as recently as 1984. Intensive searches for this species have failed to detect breeding populations, however, so the prognosis for its survival is not good. Originally Bachman's Warbler was found in many of the same forested bottomlands as Carolina Parakeets and Ivory-billed Woodpeckers, in widely scattered localities along the southeastern coast south of Maryland, and along the Gulf Coast to East Texas.[61] Its range extended northward along the Mississippi River to at least Missouri. The species has been rare since its discovery in South Carolina by John Bartram in 1833. Subsequently it was only recorded in its wintering habitat in Cuba until a specimen was taken by a feather hunter in Louisiana in 1868. Only identification of this specimen by a local ornithologist prevented it from becoming an ornament on a woman's hat: its feet had been cut off and its wings stiffly extended in preparation for this fate.[62] Between 1890 and 1920 Bachman's Warbler was observed and collected frequently, but after that it grew scarce.

Bachman's Warbler apparently increased at a time when floodplain forests were being harvested at an accelerating rate and at a time when Carolina Parakeets had disappeared and Ivory-billed Woodpeckers were steadily declining. Ironically, logging may have initially created habitat for Bachman's Warbler, which used disturbed areas with an open canopy for nesting.[63] Most nests were found in openings in the bottomland forest where there were canebrakes (thick stands of low bamboo or cane) or dense tangles of blackberry.[64] Originally these openings must have been created by fires, hurricanes, tornadoes, insect damage, and the collapse of large trees. Also, continual undercutting of river banks by flood waters provided a steady supply of recently disturbed habitat. The initial phase of logging, which often involved selective removal of the most valuable trees, may have generated habitat for Bachman's Warbler.[65] Some of the first nests discovered were in dense brambles in forests that had recently been logged.[66] Moreover, an exceptionally large number of Bachman's Warblers were recorded during both migration and the breeding season between 1890 and 1920, when logging in the southern bottomlands was at a peak. Subsequently, most of the forest was completely cleared and the land was drained for agriculture, eliminating both the mature floodplain forest and the shrubby openings embedded within it.[67] The forest openings and canopy gaps needed by Bachman's Warbler disappeared in these areas as thoroughly as the ancient trees needed by Ivory-billed Woodpeckers. The remaining patches of mature floodplain forest may have been too small and fragmented to contain enough openings and canebrakes to support Bachman's Warblers.

Canebrakes, which originally covered large areas of the Mississippi River bottomlands and river bottoms along the coast of the Carolinas, may have been an important habitat for the Bachman's Warbler. J. V. Remsen hypothesized that the Bachman's Warbler, like some tropical songbirds, was a bamboo specialist, and that the decline of the warbler was the result of the virtual disappearance of large stands of the bamboo (cane) on which it depended.[68] It is noteworthy that most descriptions of Bachman's Warbler nests mention the presence of cane in the understory. Large canebrakes were virtually impenetrable for people unless they hacked their way through the dense stems, which might explain why so few nests of Bachman's Warbler were discovered. William Bartram described a canebrake in northern Florida so large that "there appears no bound but the skies," where the canes were "ten or twelve feet in height, and as thick as an ordinary walking staff; they grow so close together, there is no penetrating them without previously cutting a road."[69]

The decline of the warbler may have resulted from the direct and indirect destruction of nearly all large canebrakes.[70] Cane grew on rich, alluvial soil that was often cleared for farming. Also, cane ultimately depends on seasonal flooding and occasional fires. Thus, flood and fire control may have led to a quick disappearance of canebrakes. An additional complication for a bamboo specialist is the synchronous flowering of bamboo across a particular region, followed by a complete die-off of the plants in that region.[71] When this happens, bamboo specialists need to move to another region where a bamboo die-off has not occurred. Like the Passenger Pigeon and perhaps the Carolina Parakeet, the Bachman's Warbler may have needed blocks of appropriate habitat spread across an extensive geographical area due to the erratic nature of appropriate food or nest sites at any one location. This would have made it vulnerable to widespread habitat destruction.

Destruction of natural habitat in Cuba, the only known wintering area of Bachman's Warblers, also may have contributed to the disappearance of this species.[72] If they depended on lowland forest in winter, then extensive clearing for agriculture could have caused the decline. However, the few descriptions of their distribution in winter indicate that they used a wide range of habitats, including planted forests, gardens, and both the lowlands and mountains. The decline of the species after 1920 also has often been attributed to several intense hurricanes that hit Cuba in the 1930s.[73] Recent studies have shown that intense hurricanes have a relatively mild impact on populations of many species of winter-resident warblers in the Caribbean Basin, however.[74]

So little is known about the history and ecology of these species (especially the parakeet and the Bachman's Warbler) that the debate about the reasons for their disappearance will probably be endless. It is significant, however, that most of the species primarily associated with the bottomland hardwood forests and alluvial swamps of the southeastern United States are now extinct. Of the songbirds that were bottomland and swamp specialists, only two species survived: the Prothonotary Warbler (a species found in "damp and swampy river bottoms and low-lying woods, which are flooded at times"[75]) and the Swainson's Warbler (which is commonly found in or near canebrakes under a dense canopy in river bottoms,[76] but is also found in rhododendron thickets in the Appalachian Mountains). An extinction rate of three out of five species (60 percent) is even greater than the rate of extinction predicted by the species-area relation. Because about 77 percent of the hardwood bottomland forest has been cleared, the predicted rate of extinction is only 30 percent.

Conservation of Birds of the Southeastern Bottomland Forests

If there had been a public policy to protect endangered species in the 1870s and 1880s, the three extinct species of the bottomland forests probably could have been saved. A few large, representative tracts of mature bottomland forest could have been preserved with negligible impact on the industrial growth of the United States and with few long-term effects on nearby communities (which were frequently abandoned by the logging companies after the trees were stripped from the surrounding region in any event). Many of these communities ultimately could have derived more income from natural history tourism and sustained forestry than from clearing and farming the floodplain.

Today, the remaining areas of mature bottomland forest should be carefully protected. Old-growth stands are now so rare that they warrant complete protection to preserve their beauty and their value as a source of scientific information. Second-growth bottomland forests are often easier to protect if they have a direct economic benefit, such as sustainable timber-harvesting, but they will be more effective in sustaining biological diversity if some old trees and dead wood are retained. In floodplains that have been used extensively for farming, the forest can be replanted. This has already been recommended for areas where soybean farming is not economical because of frequent flooding; growing oaks and other high-value hardwoods on this land could provide an alternative source of income for landowners.[77]

Forests with a natural flooding regime are particularly valuable because they have the disturbed patches, such as canebrakes, that support a set of species not found in the interior of an old-growth forest. Also, wide strips of bottomland forest sustain species that are infrequent along narrow strips. In South Carolina, Swainson's Warbler, Northern Parulas, and Prothonotary Warblers were more frequent along rivers with wide bands of floodplain forest than in narrow bands of forest, and Mississippi Kite and Swallow-tailed Kite were detected only in strips wider than 3,000 feet (1000 m).[78]

Although three of the specialized species of the floodplain forest have been irretrievably lost, other species that are concentrated in this habitat, such as the Prothonotary Warbler and Swainson's Warbler, can still be retained as part of the rich natural heritage of the South if the remaining floodplain forests and protected and restored.

CHAPTER 5

Deep-forest Birds and Hostile Edges

For a moment Sayward reckoned that her father had fetched them unbeknownst to the Western ocean and what lay beneath was the late sun glittering on green-black water. Then she saw what they looked down on was a dark, illimitable expanse of wilderness. It was a sea of solid treetops broken only by some gash where deep beneath the foliage an unknown stream made its way. As far as the eye could reach, this lonely forest sea rolled on and on till its faint blue billows broke against an incredibly distant horizon.

—CONRAD RICHTER, *The Trees*

ALTHOUGH many ecologists and historians have underestimated the extent of open grassland and scrubland in the presettlement landscape of eastern North America, there is no question that dense, uninterrupted forest covered huge areas. Along the Appalachians, in the Ohio Valley, and even in parts of the coastal plain, a "sea of forest" stretched unbroken for hundreds of miles. These immense deciduous forests were home to a great diversity of specialized woodland birds. Many species of warblers, vireos, thrushes, and tanagers thrive only deep inside the forest, far from open habitats. Their populations tend to melt away in the suburban woods, farm woodlots, and small nature preserves left behind by forest clearing. In many regions, a growing network of roads, power-lines, and housing developments are isolating and degrading their habitat, leading to reduced populations and even local extinction.

Population Crashes in Forest Islands

Forest birds probably declined in the remnant patches of forest left after Europeans initially transformed the landscape of eastern North America in the

PLATE 5. Hooded Warbler feeding two of its own nestlings and a larger Brown-headed Cowbird nestling in the understory of an eastern deciduous forest. A jack-in-the-pulpit is in bloom behind the nest.

eighteenth and nineteenth centuries, but these changes were never documented. Local population declines could be substantiated only in the twentieth century, after the initiation of standardized censuses of breeding birds. The standard method of estimating populations of breeding birds, which is called spot-mapping, was developed by A. B. Williams in 1927 to monitor bird populations in a beech-maple forest in Cleveland, Ohio.[1] In 1937, the National Audubon Society initiated the Breeding Bird Censuses by encouraging volunteers to select a favorite study site and monitor bird populations using spot-mapping.[2] Most of these censuses continued for only a few years, providing little or no information about population changes, but a few were sustained for decades. These long-term censuses provided the first warning that forest birds—particularly migratory forest birds—might be in trouble.

Spot-mapping is used to estimate the number of territorial birds found in a relatively small site. At the beginning of the breeding season, males begin to defend the boundaries of their territories. They sing persistently, announcing their possession of a territory both to other males, which are potential intruders and competitors, and to females, which are potential mates.[3] Singing also makes birds conspicuous to census-takers, so bird populations are relatively easy to count. In contrast, most mammal and reptile populations can be estimated accurately only by repeatedly catching and marking individuals. The most accurate methods of estimating bird populations also involve catching and marking individuals (in the case of birds, with numbered metal bands and individually distinctive combinations of colored leg bands), but a reasonably accurate estimate of their populations can be obtained more easily by mapping the positions of singing males. Particularly during the early breeding season (May and June in eastern North America), males make this process easy because they sing frequently during the morning hours. The same Wood Thrush or Red-eyed Vireo will sing from the same general area morning after morning, so after repeated visits to the site, the census maps will show a tight cluster of observations of a particular species at a particular location. A cluster of Wood Thrush records can be considered a separate territory if the individual at this site was heard singing at the same time as Wood Thrushes in neighboring clusters. This demonstrates that the adjacent clusters represent separate territories rather than two centers of activity for a single individual bird.

Spot-mapping has drawbacks. For most species it is useful only for estimating the number of males. Females are generally inconspicuous and do not sing, so they are not counted accurately. Consequently, it is usually un-

FIG. 5.1. Interior of upland hardwood forest in Maryland. Photograph by Robert Askins.

clear whether particular males are mated or unmated. In a few species, such as Downy Woodpeckers and Great Crested Flycatchers, males and females use similar calls to defend their joint territory, making it difficult to interpret clusters of observations and records of two individuals vocalizing at the same time. Two Downy Woodpeckers calling simultaneously could either be a mated pair or rivals of the same sex. The method is even less effective for species such as Blue Jays that do not defend territories away from the immediate vicinity of the nest. In some territorial species, the position of the territory may shift during the course of the breeding season, and males that are unsuccessfully searching for a territory ("floaters") may appear in the study area briefly. A more serious pitfall is that different observers vary in

how well they detect singing birds and how they interpret the clusters of observations on maps.[4]

Despite these problems, spot-mapping is still one of the most accurate
methods for estimating the number of territorial birds in a prescribed study
area, and it requires much less time than capturing and color-marking individuals.[5] Even when different observers have censused a site during different years, the spot mapping method is accurate enough to reveal major population changes such as range expansions, severe population declines, or
local extinctions.

Changes of this magnitude confronted Greg Butcher when he mapped the
territories of birds in a Breeding Bird Census site in the Connecticut College
Arboretum in 1976.[6] The site, an oak-hemlock forest in New London, Connecticut, had been surveyed intermittently since 1952. During the early years of
the census, bird populations had fluctuated, but most species had not shown
an overall increase or decrease. In the 1970s, however, the populations of
many species plummeted. Some species, such as Black-throated Green Warbler, American Redstart, and Canada Warbler, disappeared entirely from the
site. Species that had once had dense populations, such as Hooded Warbler
and Red-eyed Vireo, were represented by only a few scattered individuals. For
example, Red-eyed Vireo declined from more then 20 males in the 1950s to
only four in 1976. The study site was protected by a fence and carefully shielded
from human disturbance, yet some of the most distinctive forest birds had disappeared or almost disappeared. Moreover, periodic vegetation surveys
showed that the structure of the forest had changed relatively little since 1952.
The canopy had become more continuous and the shrub layer had thinned, but
this hardly seemed to account for the major changes in the bird community.

If the loss of forest birds had been recorded only at the Connecticut College Arboretum, the pattern could have been dismissed as a purely local
phenomenon. Unfortunately, however, observers at other sites were reporting similar declines in forest migrants. Populations of numerous species
crashed at most of the Breeding Bird Census sites in the eastern deciduous
forest where censuses had been taken for more than 20 years.[7] For example,
at Cabin John Island in Maryland, six species (Wood Thrush, Yellow-
throated Vireo, American Redstart, Kentucky Warbler, Hooded Warbler,
and Scarlet Tanager) disappeared and two species (Red-eyed Vireo and
Northern Parula) declined by more than 50 percent between 1947 and 1988.[8]
Six species of forest migrants were also lost at Greenbrook Sanctuary in
northern New Jersey during approximately the same period.[9] In 1947, 20

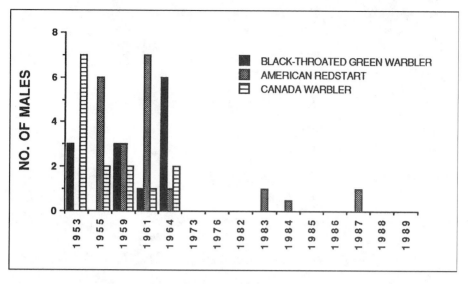

FIG. 5.2. Decline in the number of territorial males of three species of warblers in the Connecticut College Arboretum. From Askins, 1990. Reproduced with permission from the Connecticut College Arboretum.

Ovenbirds and 18 Hooded Warblers were banded in the 160-acre (67-ha) study site, but by 1987 both of these species were absent. The pattern was remarkably similar for many other regions in eastern North America: three additional sites in Maryland, another site in northern New Jersey, and a site in upstate New York.[10] The same species—Yellow-throated Vireo, Red-eyed Vireo, Northern Parula, Black-throated Green Warbler, Black-and-white Warbler, Ovenbird, Kentucky Warbler, and Scarlet Tanager—declined or disappeared at all of the sites where they were originally present.[11] Some of these censuses had been repeated every year by the same observer or group of observers, so it is unlikely that these species were overlooked in later years. The pattern was alarming; an entire group of forest species had virtually disappeared before anyone suspected they were in trouble.

Cause of the Declines

The search for causes began immediately. Researchers focused on two features of the population declines: they occurred primarily in islandlike patches of forest, and most of the declining species were neotropical mi-

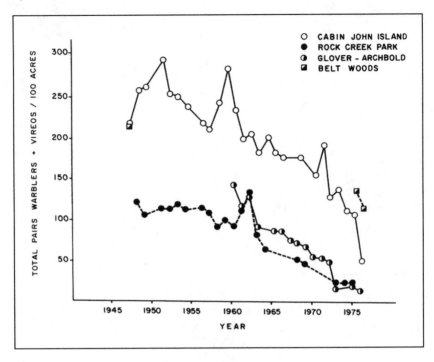

FIG. 5.3. Decline in the abundance of warblers (Parulinae) and vireos (Vireonidae), two of the major groups of neotropical migrants, at three sites in the Washington, D.C., metropolitan area. From Lynch, 1987. Reproduced with permission from Surrey Beatty & Sons Pty., Ltd.

grants, species that nest in the north temperate zone and spend the winter in Mexico, Central America, South America, and the West Indies. The original explanations for the population declines emphasized these common elements: numerous species had declined either because they could not sustain populations in small remnant patches of nesting habitat or because of the destruction of rain forests and other tropical habitats that they need during winter. Both suburban sprawl in the temperate zone and forest destruction in the tropics had accelerated since the early 1950s, so both explanations are logical. An intense debate about these hypotheses spurred research on forest migrants in many parts of the Western Hemisphere: in large national forests and parks in the Appalachian and Ozark mountains, remnant woodlots in Illinois and Pennsylvania, migratory way stations on the barrier islands off the Mississippi coast, mountain rain forests of Jamaica, and tropical dry forests of the Yucatan Peninsula. After 20 years of research, we now know much more about the habitat needs of forest migrants at both ends of their migratory journey.

Forest Fragmentation in the Breeding Areas

The numerous sites where populations of vireos, warblers, thrushes, and other neotropical migrants went into a nosedive in the 1960s and 1970s shared one common denominator: all were small patches of forest surrounded by suburban and urban areas or farmland. Most of these study sites are located in forests smaller than 100 acres (40 ha).[12] Forest migrants also declined in isolated forests in Wisconsin between 1954 and 1979, and, again, most of the forests were smaller than 100 acres.[13]

If migratory birds were showing a general, continentwide decline, then it would not be surprising to detect it first in small, mostly suburban nature reserves, parks, and wildlife sanctuaries. In the 1960s the best early warning system for a decline of forest migrants was the system of Breeding Bird Census plots. Only a few of these had been run long enough to be useful for assessing population trends, and most of the long-term study sites were located in places convenient to the metropolitan areas and college and university campuses where most birders and professional ornithologists were based. This situation immediately caused people to question the meaning of the results, however. Were these small islands of forest really representative of the landscape of eastern North America? What was happening in the great expanses of forest in the southern Appalachians, the upper Midwest, northern New England, and southeastern Canada?

Population Trends in Large Forests

During the summers of 1931 and 1932, Aretas Saunders walked through Quaker Run Valley in Allegany State Park in western New York, counting birds that he saw or heard within 250 feet of his transect lines.[14] This is the largest state park in New York state, and Saunders' study area covered nearly 17,000 acres (6,900 ha) of virtually continuous forest. More than 50 years later, in the early 1980s, Timothy Baird carefully repeated Saunders' survey at the same study site. The large expanses of young, heavily cutover forest that Saunders traversed had grown into mature deciduous forest, but the study area was large enough and diverse enough so that Baird was still able to compare the abundance of birds in roughly comparable habitats for the

FIG. 5.4. Aerial view of the Connecticut College Arboretum in New London, one of the small forest preserves that suffered severe declines in migratory birds. The highway interchange and apartment complex were built during the period of the population declines. Photograph by Virginia Welch, 1982. Reproduced with permission from the Connecticut College Arboretum.

two periods. In contrast to the almost uniformly catastrophic declines of neotropical migrants in small forests, the patterns of population change in Allegany State Park were complex, and many of the changes probably reflected obvious habitat changes, such as the thinning of the understory due to heavy browsing by white-tailed deer. For example, in mature oak-hickory forest, the density of many species of neotropical migrants—such as American Redstart, Blackburnian Warbler, Hooded Warbler, Veery, and Cerulean Warbler—had increased appreciably, while other species—Ovenbird, Wood Thrush, Magnolia Warbler, and Black-throated Green Warbler—had declined. In contrast, mature forests dominated by maple, beech, birch, and eastern hemlock showed a more consistent reduction in the abundance of neotropical migrants: Red-eyed Vireos, Black-throated Green Warbler, Magnolia Warbler, Blackburnian Warbler, and Ovenbird all showed substantial declines. Species that nest in the understory were particularly hard hit, perhaps because of the heavy browsing by deer.

The story at another large forest, Hubbard Brook Experimental Forest in New Hampshire, is also complex. Richard Holmes and his colleagues monitored bird populations in a plot in this forest between 1969 and 1986.[15] They found that different species of neotropical migrants showed different population trends: five declined, one increased, and eight showed little change. Some of the declines probably resulted from unusually high bird densities during the first three years of the study due to an outbreak of saddled prominent caterpillars. This provided a rich source of food for insect-eating birds, and it is not surprising that the density of these birds declined when caterpillars were no longer superabundant.

More species of neotropical migrants declined than increased in a large area of forest in the Cheat Mountains of West Virginia,[16] but again local habitat change may be involved. The study site is an old-growth, spruce–northern hardwood forest, and red spruce has been dying at this site, just as it has at many other high-elevation sites in the Appalachian Mountains.[17]

In other large forests, migrant populations have shown little change or have increased. For example, in the early 1980s David Wilcove used the same methods employed by Ben Fawver in 1947 and 1948 to survey birds at ten sites in Great Smoky Mountains National Park, on the border of North Carolina and Tennessee.[18] None of the species of neotropical migrants showed consistent declines between the two surveys at the different sites, and some species increased. A more perplexing pattern was documented in Breeding Bird Censuses at two sites, an old-growth eastern hemlock–white pine forest and a young hardwood forest, both in the extensively forested White Memorial Foundation property in northwestern Connecticut.[19] The three most abundant species of neotropical migrants—Veery, Red-eyed Vireo, and Ovenbird—all increased substantially between 1965 and 1988, while the populations of most other species showed little change. The reason for the increase of neotropical migrants at these two sites remains an enigma, but the work at Hubbard Brook indicates that populations of migratory songbirds can rise when caterpillars are particularly abundant. Unfortunately, however, we have no information on insect abundance from White Memorial Foundation or most other census sites.

This confusing variety of population trends for neotropical migrants in large forests contrasts with the distinct and consistent pattern of declining populations documented in small forests. Migratory songbirds might be declining in some large forests, but they appear to be in the greatest trouble in small, isolated forests. If this is the case, then neotropical migrants should be more common in large forests than in small patches of forest in the same region. This prediction had been tested in many regions in the eastern United States and Canada.

Songbird Populations in Forests of Different Sizes

Between 1979 and 1983, Chandler Robbins and his colleagues from the Patuxent Wildlife Research Center visited 271 forests in Maryland and adjacent states.[20] In each forest they established survey points which they visited three times at intervals during the breeding season. During each visit they counted all birds they saw and heard from a survey point during a 20-minute period. The forests were carefully chosen so that small, medium, and large forests were represented. The smallest forest patches were less than 5 acres (2 ha), and the largest covered more than 3,600 acres (1,450 ha). With this impressive sample of forests from a large region, Robbins and his group could determine whether the abundance of neotropical migrants is lower in small forests than in large forests.

The pattern that emerged from this study is dramatic: both the density of individuals and the number of species dropped off precipitously in forests smaller than about 240 acres (100 ha). Of 38 species of neotropical migrants, 22 were less abundant in smaller forests. For example, in forests in the 25- to 75-acre range, there was an 80 percent chance of detecting a Red-eyed Vireo and a 50 percent chance of detecting a Scarlet Tanager at any randomly located survey point. For forests larger than 250 acres, the chances increase to close to 100 percent for the Red-eyed Vireo and about 70 percent for the Scarlet Tanager. Moreover, most species of forest warblers were more frequent in larger forests; the probability of finding them increased steadily with forest size.

A few species were completely absent from small and even medium-sized forests. For example, Black-throated Blue Warblers were found only in forests larger than 3,000 acres (1,200 ha), and Cerulean Warblers and Canada Warblers were only found in forests larger than 300 acres (125 ha).

Several other regional surveys have shown that neotropical migrants are most abundant in large forests. Both the density and the number of species were greater in large forests than in small forests in New Jersey, Maryland, Wisconsin, southern Ontario, Missouri, and Connecticut.[21]

Recently the Cornell Laboratory of Ornithology has sponsored a nationwide study of the distribution of tanagers in forests of different sizes

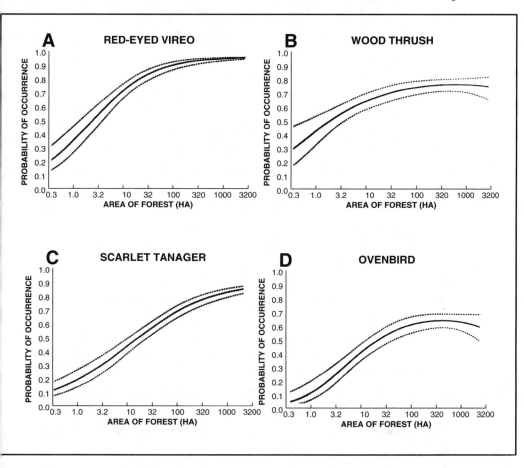

FIG. 5.5. Probability of detecting four species of migratory birds at survey points of different sizes in Maryland and adjacent states. From Robbins et al., 1989a, with permission from *Wildlife Monographs*.

throughout the U.S. and Canada.[22] In 1994, hundreds of volunteers completed surveys of about 2,200 sites. The results confirmed the importance of large forests for the Scarlet Tanager. The probability of finding breeding Scarlet Tanagers in survey plots increased from 30 percent for 1-acre forests to 68 percent for 100-acre forests and 83 percent for 1,000-acre (417-ha) forests. Their occurrence also depended on the amount of forest in the landscape around the study site; they occurred more frequently at sites in heavily forested regions. This detailed analysis of a single species confirms the importance of forest fragmentation for bird conservation: if a forest is split into small pieces, Scarlet Tanagers will become substantially less abundant.

Most of the studies of birds in forests of different sizes have been completed in agricultural or suburban landscapes in which forests constitute discrete islands surrounded by open, artificial habitats. Forest area is important even in more heavily forested landscapes, however. Margarett Philbrick, Dave Sugeno, and I surveyed bird populations in southeastern Connecticut, a region that is almost 70 percent forested.[23] Even in this situation we found that many migratory forest birds were less abundant in small forests and that some species were restricted to large forests. For example, we found Cerulean Warblers and Yellow-throated Vireos only in forests larger than 800 acres (325 ha). We were surprised to find that forest area had such a strong impact because small forests often were separated from other forests only by openings along roads or powerlines.

Small forests apparently do not provide good habitat for many species of neotropical migrants: populations of migrants have generally declined in small forests over the past few decades, and both the density and number of species are substantially higher in larger forests, particularly in forests larger than about 250 acres (100 ha).[24] Small forests often have vegetation similar to the interior of more extensive forests.[25] Why are they such an unfavorable environment for forest migrants? The answer, it seems, is that the landscape surrounding small forests creates problems. The close proximity of cropland, pasture, roadsides, or residential areas frequently leads to low nesting success for forest birds.

Hostile Edges

Soon after it was discovered that many forest migrants are less frequent in small forests than in large forests, biologists initiated detailed studies to find the cause. The agricultural or suburban areas around a small nature reserve or park often have high densities of predators such as raccoons, dogs, cats, opossums, and crows. Perhaps these predators penetrate the forest. Dogs and cats are introduced into the environment by people, while small and medium-sized wild predators may be favored by the more open habitats, new food sources (such as garbage), and absence of large predators in settled environments. These predators usually present little threat to adult songbirds, but eggs and nestlings are vulnerable. Many neotropical migrants build cuplike nests on the forest floor or in the shrub layer, where they are particularly likely to be found by such predators as raccoons and eastern chipmunks.

Another potential threat is the Brown-headed Cowbird, which lays its eggs in the nests of other birds. Cowbirds feed in pastures, stockyards, grain fields, and suburban neighborhoods, but they may penetrate nearby forests hunting for nests.[26] As a result, forest birds may raise cowbird young instead of their own young. If predators and cowbirds are penetrating the forest from surrounding habitats, then birds nesting near the edge of the forest may have particularly low rates of nest success, and in a small wood-lot, virtually all nests would be vulnerable.

Numerous studies have shown that nest predation is lower in the forest interior than in the strip of forest adjacent to agricultural or suburban areas. For example, when 276 nests of 13 species of birds were monitored in Wisconsin, it was discovered that those close to the forest edge produced few young.[27] The percentage of nests that fledged young increased from 18 percent within 330 feet (100 m) of the forest border to 58 percent in a zone between 100 and 200 meters of the border, to 70 percent beyond 200 meters. In a similar study at Rose Lake Wildlife Research Area in Michigan, researchers searched for nests near the boundary of abandoned fields and hardwood forests and in the interior of both of these habitats.[28] Both nest predation and cowbird parasitism were more frequent along the boundary than in the interior of the two adjacent habitats, resulting in lower reproductive rates along the forest/field edge. Ironically, the density of nests was considerably higher along the forest boundary because of the abundance of edge-loving species such as Indigo Bunting, Field Sparrow, and Song Sparrow. This concentration of nests may attract predators and cowbirds, contributing to low nest success near the edge.

In some cases, even narrow, open corridors through the forest can depress the reproductive success of birds in the adjacent forest. This was shown by Gregory Chasko and Edward Gates in a study of nests at various distances from two powerline corridors in Maryland.[29] There was no increase in nest predation near the forest edge along one of the corridors (a grassy strip maintained by mowing). However, along the other powerline right-of-way (which was covered with a low thicket), nest predation was lower for nests deep in the forest, far from the corridor, than for those close to the corridor. Several studies of both natural nests and artificial nests show that the frequency of predation on eggs and nestlings is higher for birds that nest in the section of a forest closer than 165 feet (50 m) to the

FIG. 5.6. Forest edge in West Virginia. Photograph by Robert Askins.

boundary with more open habitat, such as farmland, a powerline corridor, or a housing development.[30]

A large proportion of a small, isolated patch of forest will be in this zone, so the overall nest success for small forests will be lower than for large forests. David Wilcove was one of the first researchers to test this prediction.[31] In 1983 he placed artificial nests (wicker baskets with eggs from domesticated Japanese Quail) in forests of different sizes. In the largest forest (Great Smoky Mountains National Park in Tennessee), only 2 percent of the artificial nests were found by predators during a seven-day period. A large forest in Maryland also had a relatively low predation rate: 18 percent of the nests were found by predators. Many of the smaller woodlots (less than 40 acres) had much higher predation rates. This was particularly true of small woodlots in suburban areas, where 71 percent of the nests suffered from nest predation. Wilcove could identify many of the predators by the tracks they left on cardboard squares coated with black powder that were placed next to the artificial nests. Tracks of dogs, cats, raccoons, and striped skunks were found at nests that had lost their eggs.

Although artificial nests permit researchers to complete highly standardized experiments, they may not be comparable to real nests. Compared to wicker baskets and other types of artificial nests, natural nests may be better hidden and may provide different cues (such as different scents or the feeding activities of parents) to predators.[32] Also, the use of quail eggs (which is standard in artificial nest experiments) may be a problem because these eggs are too large to be consumed by eastern chipmunks and mice, both of which are known to raid the nests of small songbirds.[33]

The best solution to this problem is to monitor a large number of real nests. Finding the nests of forest birds is extremely difficult, however, and most of the successful studies have relied on large teams of undergraduate assistants to comb the woods. One of the most thorough studies of this type showed that few of the forest birds in six small woodlots at Lake Shelbyville, Illinois, were reproducing successfully. For 25 species that build open, cup-shaped nests, 80 percent of the nests were raided by predators. Moreover, 76 percent of the nests of neotropical migrants contained at least one cowbird egg. With these levels of predation and parasitism, populations of forest birds cannot compensate for annual mortality. Without immigration from other populations, they would quickly disappear from these woodlots.[34]

Research near Hawk Mountain, Pennsylvania, suggests that birds living in large forests have much higher reproductive rates than do those living in small woodlots. In separate studies of Ovenbirds and Wood Thrushes, re-productive success was compared in plots within a large, continuous forest and in relatively small forest woodlots in nearby farmland. For Ovenbirds, the contrast in reproductive rates between the large forest and the small woodlots was dramatic: 59 percent of the males in the large forest success-fully raised young compared with only 6 percent of the males in small patches of forest.[35] Wood Thrushes showed a similar pattern. The percent-age of nests that successfully produced young was 43 percent in small forests (less than 80 ha), 72 percent in medium-sized forests (80–127 ha), and 86 percent in the large continuous forest.[36] Predation accounted for 95 percent of the nest failures in Wood Thrushes.

Numerous studies show that nest predation is considerably more fre-quent close to the forest edge and in small forest patches than in the interior of large forests.[37] The pattern for cowbird parasitism is not as consistent, however. In one of the first studies to address the relationship between for-est edges and cowbird parasitism, Margaret Brittingham and Stanley Temple surveyed cowbirds and monitored nests in a large forest in the Baraboo Hills of Wisconsin.[38] They found that the density of cowbirds decreased progres-sively with distance from forest openings. The percentage of nests para-

FIG. 5.7. Ovenbird at nest. Ovenbirds, like many forest-interior migrants, nest on the forest floor, where they are vulnerable to predators. Photograph by Mike Hopiak. Reproduced with permission from the Cornell Laboratory of Ornithology.

sitized also declined with distance from the edge; it was 65 percent within 100 meters (330 feet) of an opening, and only 18 percent in the interior of the forest, more than 300 meters away from an opening. A study in southern New Jersey revealed a similar pattern for the edges created by roads traversing large forests.[39] Cowbirds were recorded less frequently in the forest interior than along roads, especially roads with shoulders of mowed grass.

If cowbirds are most abundant along forest edges, then birds living in small patches of forest would be particularly vulnerable to cowbird parasitism. Scott Robinson found high rates of cowbird parasitism in forest patches in Illinois that were so small and narrow that no site within the forest was far from the edge.[40] Later studies showed, however, that cowbird parasitism is not a severe problem for forest birds in some small forests, while it is a serious problem in the interior of some large forests. The tendency of cowbirds to occur in forests of different sizes apparently varies from region to region. Although Robinson found that 76 percent of the nests in small patches of forest in a heavily farmed landscape in Illinois were parasitized, the cowbird parasitism rates in many small patches of forests on the East Coast are considerably lower. In small woodlots near Hawk

Mountain, Pennsylvania, virtually none of the Ovenbirds with fledglings were raising cowbird young, and only 13 percent of the Wood Thrush nests contained cowbird eggs.[41] In a small woods at the University of Delaware, the percentage of parasitized nests varied greatly from year to year. Although it reached 65 percent in some years, in most years the incidence of parasitism was so low that it was not a substantial factor in determining nest success.[42] The impact of cowbirds is equally variable among large forests. In many years of locating and monitoring the nests of several species of neotropical migrants at Hubbard Brook Experimental Forest in New Hampshire, researchers never found a parasitized nest.[43] This study site is part of the heavily forested White Mountain National Forest. In contrast, at Devil's Den Preserve, a Nature Conservancy Preserve that is part of an approximately 10-square-mile (25 km²) expanse of forest in Connecticut, parasitism of Worm-eating Warbler nests by cowbirds varied between 5 and 29 percent in different years. Moreover, a large forest in Dutchess County, New York, had surprisingly high rates of cowbird parasitism (33 percent for neotropical migrants).[45] Parasitism rates were even higher in the interior of large forest tracts in Shawnee National Forest in southern Illinois; 50–60 percent of the nests of neotropical migrants had cowbird eggs or young,[46] and more than 75 percent of Wood Thrush nests were parasitized.[47]

Although there is a general tendency for cowbirds to be more active near the forest edge and in small woodlots, the abundance of cowbirds in these situations varies greatly in different parts of the continent.[48] In some regions, even small forests have low numbers of cowbirds, while in other regions cowbirds penetrate deep into extensive forests. These differences result from the nature of the surrounding landscape: heavily forested landscapes support few cowbirds, while landscapes dominated by grain fields and feedlots support large flocks of cowbirds. Cowbird females equipped with miniature radio transmitters commute as far as 4 miles (7 km) from agricultural feeding areas to woodland territories where they hunt for host nests.[49] Consequently, the intensity of cowbird activities is related not only to the immediate surroundings of a forest but also to the patterns of land use in a much larger region.

Another negative effect of forest edges—a reduction in the abundance of insect prey—has been documented only recently. Dawn Burke and Erica Nol studied both Ovenbirds and the forest-floor insects they feed upon in 31 forests of different sizes in southern Ontario.[50] They found that the biomass (or weight) of leaf litter insects in samples from Ovenbird territories was 10 to 36 times higher in large forests than in small forests. The proximity of the edge in small forests may result in increased exposure to sun and heat, leading to desiccation and a reduction in the abundance of leaf litter insects. Given this contrast in prey abundance, it is not surprising that the density of Ovenbirds was lower in small forests than in large forests, and that a higher proportion of males in small forests were unmated. Of course higher rates of nest predation or cowbird parasitism in small forests could contribute to these differences, but evidence that smaller forests may have lower food abundance strengthens the case for preserving large areas of unbroken forest.

Source and Sink Populations

Regional differences in the health of forest bird populations was made dramatically clear by a set of coordinated studies in several Midwestern states.[51] Five teams of researchers monitored nests in nine widely separated sites. Some sites were in parts of southern Missouri and northern Wisconsin, where forest covers more than 90 percent of the land, while others were in predominately agricultural areas of Illinois, central Wisconsin, and northern Missouri, where more than 60 percent of the land is agricultural. During the four years of the study, more than 5,000 nests were located at the different sites. In regions dominated by farmland, both cowbird parasitism and nest predation were extremely high for many neotropical migrants.[52] At many of these sites, Ovenbirds, Hooded Warblers, and Wood Thrushes produced so few young that it is unlikely that adults are replacing themselves during their lifetimes. They did not raise enough young each year to offset yearly mortality among adults. Given the short lifespan of small songbirds, this should result in a rapid decline of populations in these regions. These populations may be sustained by immigration from heavily forested regions, however. Nest success in the heavily forested regions was high enough to provide a surplus of young birds. Although the evidence is still circumstantial, it appears that the large national and state forests of northern Wisconsin and southern Missouri may be "population sources," providing a steady flow of immigrants to forests in regions with high levels of nest predation and cowbird activity. The latter would be "population sinks," where migratory birds establish breeding territories but do not produce many young. The implications for forest management are profound. If nest success in the source areas drops because of forest clearing, the results might be seen in falling bird populations in woodlots hundreds of miles away. Thus, for example, the abundance and diversity of migratory forest birds in central Illinois may depend primarily on how forests in northern Wisconsin are managed. Similarly, the heavily forested Appalachian Mountains may support source populations of neotropical migrants for the East Coast. For example, in the White Mountains of New Hampshire, female Black-throated Blue Warblers produced an average of 4.3 fledglings per year, which is well above the level needed to sustain the local population.[53]

Forest Management and Migratory Birds

Most of the work on forest migrants has been done in agricultural or suburban landscapes, where forest is restricted to islandlike patches of different sizes. These studies show that fragmenting a large forest results in a decline in migratory birds not only in the cleared areas but also in the remaining forests. Forest migrants also tend to decline as the density of housing on the border of small forests increases, as a survey of birds in 72 woodlots in southwestern Ontario recently showed.[54] Wood Thrushes, for example, were absent from even fairly large woodlots that were surrounded by numerous houses, but they were frequently found in smaller woodlots with few or no houses on their borders. Houses not only harbor dogs and cats but also may provide habitat and food for cowbirds, raccoons, Blue Jays, and other species that depress nest success in forest birds. Forest birds are also known to suffer higher rates of nest predation and cowbird parasitism along roads and powerline corridors. Thus, second-home development, highway construction, and resort development will degrade the quality of nesting habitat in large forests.

Many of the forests that are protected from this sort of development are managed for wood production. Harvesting of trees in these forests, unless it is done through selective cutting of individual trees, creates openings and forest edges. A critical question is whether these temporary openings in the forest depress the nesting success of neotropical migrants in the same way as permanent openings created by houses, roads, and powerlines.

Openings created by timber harvesting are different from permanent openings in two important ways: they are ephemeral, and they do not necessarily improve access to the forest interior for open-country predators. Regeneration typically occurs quickly in eastern deciduous forests. Ten years after harvesting, forest-interior birds may recolonize the area.[55] Also, clearcutting usually creates edges inside the forest rather than on the boundary of the forest. These internal edges do not necessarily connect to agricultural or suburban habitats, where the density of nest predators and cowbirds is high.

A systematic evaluation of the effects of timber-harvesting was completed in mixed oak, hickory, and pine forest in Mark Twain National Forest, Missouri.[56] Frank Thompson and his colleagues compared bird populations in nine harvested sites that contained recent clearcuts (patches where all trees have been cut) and nine unharvested sites that were a least one kilometer from the nearest clearcut. The clearcuts, which had grown up to thicket or saplings, covered no more than 20 percent of the harvested sites. The bird populations were surprisingly similar on transects that traversed the two types of sites. Scarlet Tanager, Red-eyed Vireo, and Pine Warbler were less common in the harvested forests because they tended to be absent from regenerating clearcuts. Not surprisingly, early successional species such as Prairie Warbler, Yellow-breasted Chat, and Indigo Bunting were more abundant in the harvested sites. The same was true, however, of some forest-interior species, such as Black-and-white Warbler, Worm-eating Warbler, and Kentucky Warbler. These species are probably attracted to the dense understory found in young, regenerating forests. Overall, the differences in relative abundance for forest-interior species in harvested and unharvested sites were minor. Moreover, the abundance of Brown-headed Cowbirds and two nest predators (Blue Jay and American Crow) were similar in harvested and unharvested sites, suggesting that nest success might not be depressed in harvested areas.

A study in a northern hardwood forest in White Mountain National Forest in New Hampshire yielded similar results.[57] In this study, six harvested areas were compared with six unharvested areas. The harvested areas contained recent clearcuts smaller than 50 acres (20 ha), and low vegetation (tree seedlings and saplings) covered 18 percent of these sites. Most of the species characteristic of older forest (including Scarlet Tanager and Red-eyed Vireo) were about equally common in the harvested and unharvested forests. Only the Ovenbird was significantly more abundant in the unharvested sites. The harvested sites had a greater diversity of birds because of the presence of many early successional species, such as Chestnut-sided Warblers and White-throated Sparrow, in the clearcuts. As in Missouri, there were also some forest-interior birds that were more abundant in the harvested areas than in the unharvested areas. Rose-breasted Grosbeak,

American Redstart, and Veery were considerably more common in the sites with clearcuts. These species thrive in younger, more open forests. Brown-headed Cowbirds were not a factor in this forest; only one individual was seen during two years of field work.

Both of these studies indicate that small clearcuts do not have a major impact on the abundance of migratory forest birds, and that some species actually increase in abundance in forests where timber has been harvested. These conclusions do not necessarily apply to other types of timber harvesting, however. The "harvested" areas in both studies were still mostly covered with "sawtimber" (mature) forest after harvesting had been completed. More frequent harvesting or larger clearcuts would probably have a much greater impact on forest birds. Also, although the relative abundance of most forest species did not differ greatly between harvested and unharvested sites, it is possible that the harvested sites are population sinks with low rates of nest success. Although cowbirds are just as frequent (or infrequent) in harvested and unharvested forests, it is possible that nest predation rates are higher near clearcuts.

The possibility that nest predation rates are elevated in forests with scattered clearcuts has been investigated with artificial nests. Richard DeGraaf set up artificial nests with small chicken eggs on the forest floor and in shrubs or saplings in White Mountain National Forest.[58] Microswitches wired to a camera were placed under the eggs so that any predator that raided a nest would be photographed. Artificial nests were set up in six study sites, three that had not been harvested and three with clearcuts smaller than 39 acres (16 ha). Although several cameras were chewed and destroyed by black bears, a large number of predators were photographed in the act of taking eggs. Black bears and fishers accounted for most of the predation. The overall rates of predation were similar in harvested and unharvested blocks of forest, indicating that timber-harvesting does not lead to an increase in the density or activity of nest predators.

In a similar study in Maine, artificial nests containing quail eggs were placed along transects in clearcuts and in the forest at various distances from the clearcut-forest boundary.[59] Predation rates were higher for ground nests in the forest than for those in the clearcuts. Also, there was no tendency for the frequency of predation on nests to increase near the edge.

These experiments with artificial nests indicate that small clearcuts do not increase the frequency of nest predation in areas that are primarily forested. Experiments with artificial nests must be interpreted cautiously, however. The relatively large chicken and quail eggs used in these experiments are too large to be taken by mice, which are potentially important nest predators that might increase in density when openings are created in a forest.[60] More studies with real nests, rather than artificial nests, are needed.

Recently, Ilpo Hanski and his colleagues were able to determine the success rate of a large number of natural nests in and around clearcuts in northern Minnesota.[61] They had two study plots in deciduous (mostly quaking aspen and paper birch) forest and mixed deciduous-coniferous forest. Two to four people searched for nests in the study plots each day during the early summer. They found almost 350 nests, which they monitored every three or four days to determine the fate of eggs and young. Inside the forest, eggs and young were often taken by predators, but the rate of predation was no greater close to clearcuts than deep within the forest, far from clearcuts. Nests inside clearcuts actually had lower rates of predation than those in the forest, indicating that the clearcuts were not attracting large numbers of predators. In contrast, a study in White Mountain National Forest showed that Ovenbird nests within 650 feet (200 m) of a clearcut edge had higher predation rates than did those 200 to 400 m inside the forest.[62]

Ultimately, a conclusive assessment of the effects of clearcuts on nest predation in eastern deciduous forests must await additional research on predation on natural nests in harvested and unharvested blocks of forest. The Missouri Ozark Ecosystem Project, which was organized by the Missouri Department of Conservation, should provide this type of information.[63] A small army of researchers (mostly undergraduate interns) have been monitoring nests in nine blocks of oak-hickory forest. After nest success was monitored in all of these blocks for four years, three blocks were subjected to harvesting with small clearcuts, three others were harvested with selective cutting that creates only small openings in the forest canopy, and the final three were not harvested. The impact of both clear-cutting and selective cutting on the nest success of neotropical migrants should become clear as this study continues.

Loss of Winter Habitat as a Cause for Population Declines

If you have watched migratory birds in the breeding areas in northern deciduous forests, it is a revelation to visit the tropics during the winter and see the same species of birds in a dramatically different context. For several years David Ewert and I have studied winter-resident birds in Virgin Islands

National Park on St. John in the U.S. Virgin Islands. Here we encounter a diversity of familiar birds that have traveled hundreds of miles south from the deciduous and conifer forests of temperate North America. Small flocks of Black-and-white Warblers, American Redstarts, and Northern Parulas move through the moist forest, flitting among the sunlit bromeliads and orchids of the treetops.[64] Northern Waterthrushes hop between the looping prop roots on the muddy floor of red mangrove swamps, and Prairie Warblers hide in the thorny shrubs of cactus scrub. During the winter a remarkably large proportion of the bird species on the island are migrants from the United States and Canada. This is particularly true in the lush moist forests that grow along ridgetops and at the bottom of steep valleys, where you can watch a variety of northern warblers while listening to the morning chorus of resident Bridled Quail Doves and Scaly-naped Pigeons.

Many species of neotropical migrants stay in their nesting areas for only three or four months. They spend some time in transit between breeding and wintering areas, but most of the year is spent in wintering habitats. These tropical habitats are clearly crucial for the survival of these birds. Consequently, when populations of migrants began to decline and disappear from protected nature reserves in the north, it was logical for researchers and conservationists to look to habitat destruction in the tropics as a possible cause. If birds that nest in forests also depend on forests during the winter, then they might be under special stress because large amounts of tropical forest have been destroyed during the past 50 years. In Central America (where migrants are highly concentrated in winter[65]), the percentage of land covered by forest and woodland declined from 60 percent in 1950 to 41 percent in 1980.[66] Over 80 percent of the lowland rain forest has been lost in some regions of Mexico and Central America[67], such as in the Sierra de los Tuxtlas of Veracruz[68] and the Caribbean Slope of Costa Rica.[69] It is difficult to see how these changes could not have an effect on the migratory birds that live in tropical forests for more than six months each year.

Species that nest in forests are not necessarily restricted to forests during the winter, however. Northern birders visiting the West Indies and Latin America are often surprised to find migratory birds that they associate with the deep forest foraging in city parks, coffee plantations, gardens, and along the ragged edges of highly disturbed woodland. Some observers have suggested that forest-nesting migrants might thrive in such habitats in winter and therefore might benefit when continuous tropical forest is replaced with more open habitats.

FIG. 5.8. Tropical forest at Tikal National Park, Guatemala, with 200-foot (65-meter) high Mayan pyramids rising above the canopy. Black-and-white Warblers, Kentucky Warblers, Hooded Warblers, American Redstarts, and other neotropical migrants are common in this forest in the winter. Photograph by Robert Askins.

Although some migratory birds are found in disturbed habitats, others spend the winter deep inside moist tropical forests, where they appear to fit into the diverse community of resident tropical birds. Migrants move through the canopy of the rain forest with mixed flocks of woodcreepers, greenlets, small flycatchers, antwrens, and other resident birds.[70] They also join antbirds and woodcreepers in the shrubs and small trees above swarms of army ants, where they snap up insects flushed by the ants from the leaf litter.[71] Moreover, many species of migrants set up territories in tropical forests, defending them against both males and females of the same species.[72] These migrants appear to depend on the tropical forest just as much as resident rain-forest birds.

In winter, many species of migrants are found both in undisturbed forest and in a range of disturbed and open habitats. The best way to assess the relative importance of these habitats to migrants is to measure the density and winter survival rates of these birds in each habitat. Midwinter surveys of migrants have been completed in a variety of habitats in northeastern Mexico, western Mexico, the Yucatan Peninsula of Mexico, Belize, Costa

FIG. 5.9. Virgin Islands National Park, U.S. Virgin Islands. Winter-resident warblers are infrequent in the dry scrub in the foreground but are common in the tropical moist forest in the ravines and ridgetops of the mountains in the background. Mixed flocks of Northern Parulas, Black-and-white Warblers, American Redstarts, and other warbler species are often encountered in these forests. Photograph by Robert Askins.

Rica, and the West Indies.[73] The same standard methods were used in these studies: either mist nets (fine, semi-transparent nets that are spread between poles) were set up for a prescribed number of days so that birds could be caught and banded, or birds were counted at a series of survey points. Typically, all birds seen or heard within 25 meters of each survey point were recorded during a ten-minute observation period.[74] In some studies, both methods were used because mist nets are more effective at sampling secretive birds of the understory; point counts are superior for sampling species that live in the forest canopy and seldom descend to the level of mist nets.[75]

All of these studies show that there is no single pattern of habitat use for migratory birds; different species are found in different habitats in winter. Moreover, some are habitat specialists while others use a broad range of habitats. A number of species that nest in forests are found in distinctly different habitats in winter, such as citrus groves or plantations where cacao and coffee grow under a canopy of shade trees.[76] While some species depend on tropical forests or even particular types of forest, such as mangrove swamps, others benefit when the forest is replaced with second-growth scrub or other disturbed habitats.[77] Many of the habitat specialists use the same habitat in winter and summer. Yellow-breasted Chats, for example, are found in low scrub or thicket at both ends of the migratory journey, and Kentucky Warblers are found primarily in mature forest in both summer and winter.[78] However, the surveys of migrants revealed some surprises; a few species use distinctly different habitats in the temperate zone and tropics. For example, Eastern Wood-Pewee and Blue-gray Gnatcatcher nest in forests but may spend the winter in scrub, while Blue-winged and Chestnut-sided warblers nest in scrub and typically overwinter in forest.[79]

On the basis of information about use of different habitats by migrants in winter, Diamond estimated that over half of the 66 species of neotropical migratory songbirds that nest in Canada are likely to lose more than 25 percent of their winter habitat between 1985 and 2000, and that 11 of these species may lose as much as 50 percent of their habitat.[80] Species that are concentrated in forests during the winter are especially threatened by habitat destruction. A recent analysis of the literature on habitat use by migrants in winter showed that at least 45 species are likely to fall into this category.[81] This group includes the Cerulean Warbler, which winters in humid forests in a narrow altitudinal zone in the foothills of the Andes.[82] This habitat has fertile soils and high rainfall, making it an ideal environment for growing coffee, cacao, coca, and vegetables. As a result, most of the forests of this zone have been converted to farmland. Not surprisingly, the Cerulean Warbler has declined steadily and rapidly on Breeding Bird Survey routes since surveys were initiated in 1966, a pattern that is not shown by most other species of forest warblers.[83] The Cerulean Warbler may also be threatened by its dependence on large areas of mature deciduous forest for nesting, so its future will depend on how forests are preserved and managed at both ends of the migratory journey.

Even some of the species that are not restricted to forest in winter could be threatened by the destruction of tropical forest because their survivorship through the winter may be higher in moist forests than in other habitats.[84] For example, in Veracruz, Mexico, Wood Thrushes frequent both forest and scrub, but territorial birds, which have a higher survival rate than nonterritorial birds, are concentrated in forest.[85] In Jamaica, American Redstarts showed a decline in body weight during the winter in dry woodland, but not in much wetter coastal mangrove forest.[86] Also, more American

Redstarts persisted through the winter in mangroves than in dry forest. Thus, some of the habitats occupied by migrants in winter appear to be more valuable than others.

As a group, however, neotropical migrants are much less vulnerable to destruction of tropical habitats than are many of the birds that nest in the tropics. Most species of migrants are found in a range of habitats and in many regions of the neotropics in winter.[87] In contrast, many resident tropical species are highly vulnerable to extinction because they are restricted to one or two habitats in a single region, such as the Greater Antilles, northern Amazonia, or the Atlantic forest of eastern Brazil. The handful of migratory species restricted to a single habitat in a single region during the winter are also particularly vulnerable to extinction, and they deserve special attention from conservationists. These include the Cerulean Warbler, which is found primarily in mountain forests in the Andes, and the Kirtland's Warbler, which is concentrated in pine woods in the Bahama Islands.[88]

Control of Migrant Populations: Winter or Summer?

Researchers have disagreed about whether population declines in forest migrants result from degradation of breeding habitat due to forest fragmentation or to destruction of winter habitat in the tropics.[89] These population declines have occurred primarily in temperate-zone sites where the forest is fragmented by suburban or agricultural openings, and they mainly involve species that are concentrated in tropical forests during winter, so disentangling the evidence for the two competing explanations is difficult. In the White Mountains, population changes in both Black-throated Blue Warblers and American Redstarts can be predicted from nest success during the preceding year, indicating that breeding season conditions are paramount in driving population changes.[90] On the other hand, a disproportionate number of species that have declined since 1978 on the Breeding Bird Survey routes across North America are species that depend on forests during the winter, suggesting that tropical deforestation is driving population changes.[91]

Thomas Sherry and Richard Holmes developed a graphical model that shows how breeding and wintering habitat could simultaneously cause changes in populations of migrants.[92] This model depends on the assumption that habitat varies in quality in both the breeding and wintering areas. Several recent studies of migrants in breeding and wintering areas have shown that migrants prefer some habitats over others and that they tend to do better (in terms of survival, weight gain, or reproductive rate) in the preferred habitats.[93] If good nesting habitat is destroyed, then the number of birds returning to the tropics in winter will be reduced and a higher percentage of birds will be able to find high-quality winter habitat. Consequently, the average rate of survival through the winter should increase, boosting the percentage of birds that return to the breeding areas. Conversely, if winter habitat is destroyed or if there is a drought in the wintering range, fewer individuals will return to the north to breed in the spring. Survivors will have a better chance of finding good breeding sites, so the loss of adults in winter will be offset to some extent by higher average nesting success in summer. As a result, population changes depend on a complex interplay between the effects of summer and winter habitat changes. Of course, extensive destruction of either summer or winter habitat could lead to extinction, but more moderate environmental changes in either breeding or wintering areas would be buffered by reduced competition for good habitat at the other end of the migratory route.

Stopover Habitat

Migratory birds spend substantial time in both their breeding habitats and tropical winter habitats, and degradation of either could imperil them. They spend much less time in habitats along the migratory route, where they stop briefly to feed and replenish the fat deposits that fuel their long migratory flights. Migrants may become highly concentrated along shorelines, on the ends of peninsulas, or on offshore islands as they rest and feed before or immediately after crossing a large lake or stretch of ocean. Some of these stopover sites may be critically important to the survival of migratory birds, but research on their importance has just begun.[94]

Recently several groups of researchers have attempted to monitor the occurrence of birds in different habitats at stopover sites during spring or fall migration. The largest-scale study was completed on the Delmarva Peninsula in Virginia and Maryland and the Cape May Peninsula in New Jersey, where numerous volunteer birders counted birds at thousands of survey points. This study showed that although fall migrants were widely distributed on the two peninsulas, they were more abundant along the bay coasts and on barrier islands than in either the interior or along the ocean coasts.[95] Also, migrant abundance was highest in scrub-shrub habitat (ar-

FIG. 5.10. Unbroken deciduous forest in Shenandoah National Park, Virginia. Many species of forest birds depend on large expanses of forest. Photograph by Robert Askins.

eas dominated by low shrubs). Some species were primarily found in this habitat, while other species were concentrated in woodland.[96] Similarly, most migrants stopping on Horn Island, a barrier island off the coast of Mississippi, were concentrated in shrub-scrub habitat during spring migration.[97] Both the abundance of individuals and the number of species was higher in this habitat than in marsh-meadow, pine forest, or dune habitats. Although only 14 percent of the island was covered with shrub-scrub, 46 percent of the migrants were counted there. In the St. Croix River valley of Minnesota, spring migrants were more abundant in alder and willow thickets along the river than in floodplain forest, pine forest, conifer swamps, or other habitats.[98] During fall migration, however, migrants were more evenly distributed in the different habitats. A tendency for spring migrants to be concentrated near the water's edge was also discovered by David Ewert and Michael Hamas when they surveyed birds at different distances from the northern shore of Lake Huron in Michigan.[99] Spring migrants were much more common next to the lake, where they fed on aquatic midges, than in inland forests. These studies suggest that some sites may be particularly important to birds when they stop to feed and rest during migration. Scrub and woodland habitats along the shores of rivers, lakes, and bays may be particularly important for migrating songbirds, but more research is needed to determine whether this is a general pattern.

An interesting and unanswered question is whether small fragments of forests that are generally not good breeding habitat for forest birds might still play an important role as stopover habitat when these birds are migrating. Remnant patches of forest in urban or agricultural areas may serve the same role as wooded shelterbelts (planted by farmers as windbreaks) in the farmland of South Dakota; even small shelterbelts are used intensively as stopover sites by migrants.[100]

Conservation of Migratory Birds of the Eastern Deciduous Forest

After twenty years of increasingly intensive study of the ecology of migratory birds that nest in the eastern deciduous forest, we have found that they are vulnerable to habitat degradation and destruction in both their summer and winter ranges. Different populations of the same species live in habitats that vary greatly in quality (as measured by survival or reproductive rates) during both seasons, and it is clear that conversion of favorable habitat into poor habitat will eventually result in population declines. However, most of the forest-dwelling neotropical migrants in eastern North America have not shown overall, continentwide declines since the Breeding Bird Survey was initiated in 1966 (although many have declined at particular sites or in particular regions).[101] Populations of some species have actually increased. This may result from the regrowth of forest in many regions where farming has been abandoned, which may have more than compensated for forests that have been cleared and fragmented in rapidly growing urban areas. Significantly, many of the species showing steady, continentwide declines concentrate in wintering areas that have suffered particularly severe habitat destruction, such as lowland rain forest in southern Mexico or mid-elevation forest in the Andes. Protecting the habitats needed by these declining species should be the highest priority, but even species with increasing or relatively stable populations need attention. We now know that many of these species do not do well in temperate forests that have been fragmented into remnant patches, that they need specific habitats during the winter, and that they may depend on particular types of stopover sites during migration. We have the information needed to protect critical habitat before these birds become rare, and it is always easier to prevent a severe population decline than to pull a species back from the brink of extinction.

Work on the distribution and reproductive success of woodland birds has shown that it is not sufficient to study a particular forest and the areas that abut that forest. The fate of forest bird populations depends on the landscape patterns that are best viewed from an airplane or even from a satellite image. The health of a bird population may depend on the intricate patterns of forest cover, cropland, pasture, and housing developments over several counties or even several states. This research on migratory birds contributes to a more general realization that nature reserves and even large national parks are not really self-sufficient islands of natural habitat. Instead, they are influenced by events in the larger region in which they are embedded. Preserving the biological diversity and ecological functioning of natural areas is not merely a matter of establishing parks and patrolling their boundaries. It is also critical to work with land owners and land managers beyond the boundaries of a park or wildlife refuge. In the case of migratory birds, we must work with people in even more distant places, at the stopover sites where birds refuel during migration, and in the tropics where they reside for several months of the year. As a result of this new perception, conservation efforts have become more challenging but potentially much more effective.

Conservation of neotropical migrants will require coordinated efforts throughout the Western Hemisphere (including the Caribbean islands). Coordination of efforts to protect migrants is already under way with the Partners in Flight Program, which is run by a network of conservation groups, government conservation agencies, and ornithologists from numerous countries. Protection of large, unbroken expanses of breeding habitat; high-quality wintering habitat (which, for many species, consists of lowland tropical forest or mangrove swamp); and important migratory stopover sites have become leading goals of this program. The future of numerous species of thrushes, vireos, warblers, and other migrants depends on how these habitats are protected and managed.

CHAPTER 6

Industrial Forestry and the Prospects for Northern Birds

It is a country full of evergreen trees, of mossy silver birches and watery maples, the ground dotted with insipid small, red berries, and strewn with damp and moss-grown rocks,—a country diversified with innumerable lakes and rapid streams . . . the forest resounding at rare intervals with the note of the chickadee, the blue jay, and the woodpecker, the scream of the fish hawk and the eagle, the laugh of the loon, and the whistle of ducks along the solitary streams.

—HENRY DAVID THOREAU, *The Maine Woods*

I N contrast with the eastern deciduous forest, great expanses of the northern boreal forest of North America remained intact throughout the history of European settlement. A few species of trees, most of them conifers (spruce, pine, fir, and larch) dominate this forest, which stretches southward from the tundra edge to the region along the U.S.-Canada border. On the other side of the Bering Strait, a strikingly similar forest stretches across Eurasia, from eastern Siberia to Scandinavia and Scotland. The Old World version of boreal forest has the same general types of trees (usually a dominant pine, a dominant spruce, an aspen, and one or two common types of birches), but the particular species are different. Although both Eurasia and North America have distinctive groups of birds that are absent from the other continent, some species, such as the Red Crossbill, Bohemian Waxwing, Pine Grosbeak, and Three-toed Woodpecker, inhabit the coniferous forests of both continents.[1] Also, some of the same groups of birds (woodpeckers, chickadees, and kinglets) are well represented in the coniferous forests of Canada, Scandinavia, and Siberia.[2] The similarity of these northern forests and their birds has important implications because much of the spruce-fir and aspen forest of North America

PLATE 6. Red Crossbill in a white spruce on the edge of a clearcut. Short-rotation clearcutting reduces the number of cone-bearing trees needed by crossbills and other seed-eating birds of the boreal forest.

may soon be managed like the "industrial forests" of Sweden, Norway, and Finland. Northern European forestry is sustainable and efficient for wood production but costly in terms of the loss of biological diversity. Some of the most distinctive species of boreal birds have declined in Scandinavia because of intensive forest management, and the same is likely to occur in North America unless forest managers learn from the experience of northern Europe.

Ice Age Forest

The boreal coniferous forest has been called the "moose-spruce biome" or, more evocatively, the taiga. Unfortunately, the word taiga, which originally referred to the boreal spruce-pine-birch forest in general, has been used by some authors for the transition zone between boreal forest and open tundra, so the term is now ambiguous.[3]

The boreal forest spread across the northern hemisphere as the climate cooled with the beginning of the Pleistocene about 2 million years ago.[4] Colder weather favored the spruces, pines, and birches of the boreal forest but also brought the cycles of glaciation that almost eliminated many of these species. Repeatedly, continentwide glaciers pushed southward, covering almost all of the region now occupied by the northern forest. For example, all of the boreal region of Canada except for a small area in the northwest was covered with deep ice during the last glacial period.[5] The boreal forest survived farther to the south in the central plains and southeastern lowlands of the United States, and, in the Old World, in eastern Asia.[6] The distinctive species of Old and New World boreal forests probably evolved during these long periods of isolation. The warm interglacial periods, when the boreal forests formed an almost continuous band across the Northern Hemisphere, were brief compared with the periods when vast glaciers separated Asian and North American forests.

Only about 100 tree generations have passed since spruces, pines, and birches colonized the barren land left by melting glaciers in Canada.[7] Thus, all of the species of the boreal forest are recent colonists. Many of these species have wide distributions because their spread across the continent was unobstructed after the glaciers and melt-water lakes receded. The forests of the current interglacial period are somewhat different from those of earlier periods, however, because different species of trees migrate independently, expanding into ice-free lands at different rates and in response to different environmental conditions.[8] From the perspective of geological

FIG. 6.1. Boreal forest in Isle Royale National Park, Michigan. Photograph by Robert Askins.

time, the current boreal forest is relatively young, and it has always been in a constant state of flux. The birds of this environment must be flexible and resilient. Most are capable of living in a wide range of types of boreal forest. Despite this, some species of boreal forest birds tend to disappear from forests that are intensively managed for wood production. The cause is not the destruction of trees—which has always occurred, sometimes on a massive scale—in boreal forests. The problem for many bird species is that a diverse patchwork of different types of vegetation is replaced with a monotonous and simplified managed forest that is more similar to a tree plantation than to a natural forest.

Forest Mosaics

The original forests in many boreal regions of both North America and Eurasia were subject to frequent intense fires. In North America most of these fires are caused by lightning.[9] This apparently was true even before European settlement. Forest fires were frequent before Europeans arrived in North America, but they were not primarily due to burning by Indians. Indians who lived in the northern coniferous forest did not set fires during the summer fire season because these could not be controlled. The Cree of northern Alberta used fire to clear trails and hunting areas in swamps and along the margins of lakes, but they did this in early spring or late fall to avoid igniting the surrounding upland forest.[10]

The fine needles, resinous wood, and large number of dead branches and twigs make conifers exceptionally flammable.[11] Moreover, the conical array of branches on many conifers provides a ladder that transmits fire from the ground to the tree crowns. Crown fires then sweep across the landscape, often burning huge areas and killing most of the trees and understory plants. In northern Canada, single fires sometimes burn more than 250,000 acres (100,000 ha).[12] The forest usually rebounds quickly, however, even after most of the trees are killed. All of the common trees of the boreal forest have seeds that are dispersed by wind, which results in rapid recolonization of burns.[13] The number of seeds falling to the ground diminishes rapidly within a few hundred meters of the forest edge, but small, scattered islands of live trees remain in most large burns, providing a source of seeds for a new forest. Also, jack pine and black spruce have cones that are coated with a thick resin. The cones open when fire melts this resin, so seeds are released after a forest fire. Aspens and some types of shrubs form clones in which many trunks or stems are connected by a network of underground roots, so a grove or stand is actually a single plant. When the stems and leaves are burned by a fire, new stems quickly sprout from the underground root system.

Particular regions of the boreal forest can be characterized in terms of a fire cycle—the average length of time when forest can grow up in an area before another fire occurs. The frequency of fires can be calculated from a variety of sources: charcoal deposits in lake sediments, fire scars on trees, historical records of fires, and the distribution and area of forest stands of different ages in a region where natural fires have been the main source of disturbance.[14] If fires occur every ten years, the result is an aspen parkland (a grassland with scattered aspen trees). Fire cycles longer than 100 years permit the growth of closed conifer forests. Fire frequencies in most boreal forests average 50 to 200 years.

Although forest fires can be large, most affect a relatively small part of a regional forest. In the boreal forest of Alberta, for example, the average area of burns is about 1,000 acres (400 ha).[15] Consequently, a large expanse of boreal forest often has many distinctive patches of vegetation, each recovering from a different fire.[16] Recent burns are covered with low shrubby vegetation, while older burns are dominated by aspen and birch trees. Mixed woods of conifers, aspen, and birch grow in areas that have not burned for decades.[17] In the absence of fire, the aspen and birch eventually are replaced by spruce and fir, which grow up in the shade of these short-lived, broad-leaved trees. Aspen and birch may persist, however, if windstorms or outbreaks of insects such as spruce budworm create openings in the spruce-fir forest, resulting in a particularly diverse old forest.[18] Mixed forests of spruce, fir, aspen, and birch are common in parts of eastern Canada and European Russia that have never been subjected to intensive lumbering.

Land surveyors described the forests of northeastern Maine between 1793 and 1827, before the forest had been extensively cleared. Significantly, they reported several large burns, some covering tens of thousands of acres, as well as numerous areas where sections of forest had been blown down by storms. Using these survey reports, it is possible to estimate the mix of different successional stages that characterized the forest before Europeans arrived: 2 percent recently burned land, 14 percent birch-aspen forest, 25 percent young forest (75–150 years old), 32 percent mature forest (150–300 years), and 27 percent old-growth forest (more than 300 years old).[19] Thus, a wide variety of habitats was available for birds and other forest organisms.

Many birds depend on periodic fires to create their preferred habitats. A few species of birds benefit almost immediately when a fire burns through the forest and kills large numbers of trees. Black-backed, Three-toed, and Hairy woodpeckers concentrate, sometimes in large numbers, in recent burns where they feed on the abundant grubs that infest dead and dying trees.[20] Adults of some species of wood-boring beetles (such as the whitespotted sawyer) are attracted by the smoke of burning wood.[21] They lay their eggs on burned trees immediately after fire has swept through the forest, so the wood is soon filled with large numbers of large white grubs. Woodpeckers extract these efficiently by drilling into the wood.

Other species depend on periodic fires or other disturbances to create early successional habitats. For example, Alder Flycatcher, Chestnut-sided Warbler, Mourning Warbler, Common Yellowthroat, White-throated Sparrow, and Dark-eyed Junco are common in recent burns or clearcuts dominated by low shrubs, while Least Flycatcher, American Robin, Red-eyed Vireo, Black-and-white Warbler, Canada Warbler, American Redstart, and Rose-breasted Grosbeak are common in the birch or aspen woodlands that grow on burns and clearcuts after a few years.[22] A number of other species are primarily associated with spruce and fir forests that become established when there is no disturbance for a long period: these include Spruce Grouse, Pileated Woodpecker, Yellow-bellied Flycatcher, Gray Jay, Red-breasted Nuthatch, Boreal Chickadee, Brown Creeper, Golden-crowned

FIG. 6.2. Black-backed Woodpecker, a species that frequents recent burns. Photograph by Steven Pantle. Reproduced with permission from the Cornell Laboratory of Ornithology.

Kinglet, Cape May Warbler, Blackburnian Warbler, and Bay-breasted Warbler.[23] Elimination of any of these stages of vegetation from the regional landscape would cause some bird species to decline and, in some cases, to disappear.

Large-Scale Harvesting of the Boreal Forest

The boreal forests of Canada and the northern United States have become an increasingly important source of lumber and of wood pulp used for papermaking.[24] The relatively simple, homogeneous structure of the northern conifer and aspen forests, which have only one or two dominant species of trees, along with the relatively flat terrain, facilitates large-scale, mecha-

FIG. 6.3. Gray Jay, one of the species associated with older stands of spruce-fir forest. Photograph by Isidor Jeklin. Reproduced with permission from the Cornell Laboratory of Ornithology.

nized harvesting of trees. Mechanical harvesters grab a tree, cut it at the base, and strip off the limbs. A skidder then drags the stripped trunks to a road where they are loaded onto logging trucks. This system is most efficient when large blocks of forest can be harvested.[25] The forest is allowed to regenerate naturally or is replanted, but the new trees are cut when they are still relatively young. In Michigan, for example, aspen and jack pine stands are cut every 30 to 60 years.[26] Although this type of forest management can provide a stable source of wood and creates plenty of habitat for early successional birds, it eliminates the older, more structurally complex forest needed by other species of birds. In Lake County, Minnesota, for example, the percentage of red and white pine forest older than 120 years declined from 30 percent before European settlement to only 2 percent in the 1990s as a result of commercial logging.[27]

Older forests have many characteristics that are missing from the relatively young forests maintained by forest management.[28] After 120 years, the trees in red and white pine forests are large enough so that when one dies or falls, the gap is too large to be filled by the outward growth of the branches of neighboring giants.[29] Instead, the gap in the canopy is closed by saplings growing up from the forest floor. Before they are sealed by

young trees, these openings allow sunlight to reach the forest floor, resulting in the growth of an understory (which is typically missing in younger
stands of forest). Also, dead trees and branches are more abundant in these
older forests. All of these features are important to some species of birds,
and these species tend to disappear in the rapidly growing, structurally simple forests propagated by modern forestry.

One of the best studies of the impact of industrial forestry on North
American birds was recently completed by John Hagan and his colleagues at
the Manomet Observatory. They surveyed birds in clearcuts, regenerating
forest, and old forest in north central Maine.[30] Their study sites are in a transition zone between northern coniferous forest and northern hardwood forest, but many of the bird species associated with boreal forest occurred at
these sites. They found that some species obviously benefited from the creation of clearcuts: early successional species such as Common Yellowthroat, Chestnut-sided Warbler, Mourning Warbler, Lincoln's Sparrow,
and Song Sparrow were primarily concentrated in areas that had been
clearcut within the past five years. Other species, such as Black-capped
Chickadee and Black-and-white Warbler, were found across a broad range
of habitats, including recent clearcuts and old forest. Finally, species such
as Spruce Grouse, Three-toed Woodpecker, Pileated Woodpecker, Boreal
Chickadee, Gray Jay, Philadelphia Vireo, and Bay-breasted Warbler were primarily found in forest more than 60 years old. These species are threatened
by the short rotation harvesting that has increasingly dominated the private
industrial forest land of Maine. Between 1981 and 1992, the amount of
coniferous forest older than 60 years declined from 33 percent to 24 percent
of their extensive study area. Based on estimates of bird densities in different successional stages, the populations of many mature-forest species
probably declined by more than 15 percent in the region of their study since
1981 as a result of loss of mature forest.

Bird Declines in the Managed Forests of Northern Europe:
Lessons for North America

Intensive forestry has been practiced in Scandinavia and Finland for
decades, and North Americans have much too learn from these countries
about managing boreal forests. To some extent, clearcut logging in northern Europe simulates the large-scale destruction of trees once caused by forest fires. After clearcutting, the land is either planted with tree seedlings or

trees are allowed to grow back naturally. The new stand of trees is thinned several times, resulting in a vigorous young forest that can be harvested after several decades. This rigorously planned and managed forestry system has been criticized by biologists, however, because it generates a simplified, homogeneous forest that lacks the biological diversity of natural forests.

The forest industry in Finland exemplifies the efficiency and productivity of Nordic forestry. More than 75 percent of Finland is covered with forest, most of which is owned by farm families and other small landowners.[31] Private forest land is dominated by native species of pine, birch, and spruce that characterized the original boreal forest of northern Europe. Periodically trees are selectively harvested throughout a stand so that the remaining trees will grow more vigorously.[32] These thinning harvests, which provide a steady source of income to the landowners, create a more open forest. Eventually all of the remaining trees in a forest stand are cut to create favorable growing conditions for the seedlings that will give rise to a new forest.

Although some farmers use chain saws and tractors to harvest trees in the winter, most arrange for forest companies to remove trees using specialized and efficient equipment.[33] Typically, a mechanical harvester cuts trees at the base, strips off their branches, and chops them neatly into logs. The logs are immediately picked up by a "forwarder," a vehicle with a crane that reaches over the forward cab, grabs a log, and places it on a flat bed behind the cab. Working in tandem, these machines can quickly and surgically thin a stand, or harvest and carry away all of the trees in the stand.

This intensive system of forest management appears to be sustainable. Annual growth of wood exceeds the annual harvest in Finland,[34] and clearcuts are disc-trenched, seeded, and even fertilized to ensure the growth of a new forest of native spruce, pine, or birch. A profitable timber industry appears to be compatible with maintenance of a green and beautiful countryside.

The mechanized forestry of Finland generates a seminatural forest, but the demand for trees that are 60–100 years old results in a landscape dominated by young forest.[35] Thinning operations and removal of dead trees also contribute to the homogeneity of forests in Finland. Some species of birds do very well in these managed forests, but other species have shown steady population declines as most of the natural forest in northern Finland has been clearcut and replaced by young forest.

Ornithologists surveyed bird populations along hundreds of kilometers of transects in northern Finland between 1941 and 1977, a period when the amount of coniferous forest older than 120 years was reduced from 55 percent to 32 percent of total forest area.[36] Not surprisingly, many early successional bird species increased substantially while species typical of old forest declined. Most of the species that increased—species such as Great Tit, Willow Warbler, Song Thrush, and European Robin—are generalists that thrive in disturbed habitats. Many are widely distributed and common throughout western Europe. In contrast, many forest specialists declined, including Capercallie, Black Grouse, Black and Three-toed woodpeckers, Crested Tit, Willow Tit, Siberian Tit, Siberian Jay, Redstart, and Red and Parrot crossbills.[37] Nearly all of the distinctive species of the boreal forest were hurt by forest management. Many of these species—Siberian Jay, Siberian Tit, and Redstart—are substantially more abundant in old, uncut forest than in forests that are managed for timber harvesting in the same region.[38]

Many species that depend on old coniferous forest declined even more than would be expected from the reduction in the total area of old forest, suggesting that they may not be able to survive in remnant patches of old forest.[39] Finnish ornithologists were surprised to find that many of the species associated with old forest, such as Siberian Tit, Black Woodpecker, and Redstart, declined between 1915 and 1981–1983 at Törmävaara, a square-kilometer patch of old spruce forest surrounded by gravel pits and clearcuts.[40] During the same period, more generalized species, such as Song Thrush and Willow Warbler, increased. Although the spruce forest at Törmävaara did not change appreciably, many of the same bird species declined as in highly disturbed parts of northern Finland, suggesting that a larger area would be needed to sustain some of the old-forest specialists. In 1915, twenty bird species at Törmävaara were represented by only one to four pairs.[41] Populations this small could easily disappear after the site became isolated from other old forests. In contrast, in the Vuotso region of northern Finland, where extensive areas of old, unharvested forest are protected in a national park and a nature reserve, the populations of forest specialists did not change greatly between the 1940s and the early 1980s.[42]

Unlike migratory songbirds that decline in forest remnants in eastern North America, most of the songbirds that decline as boreal forests are reduced and fragmented by industrial forestry in Europe are resident species. Many, such as the jays and grouse, have large home ranges, a trait that probably makes them more vulnerable to forest fragmentation. Siberian Jays may require large expanses of forest because the young stay with their parents throughout the year.[43] Although they are small, Siberian Tits also have large breeding territories (35 acres, or 15 ha).[44] The Capercallie, the giant grouse of Eurasian coniferous forests, needs particularly large areas of old forest. This important game species is one of the most spectacular inhabitants of the boreal forest, and its habitat requirements have been studied intensively. Capercallies depend on coniferous forest older than 60 years.[45] During the winter they feed primarily on pine needles, and they tend to feed in the crowns of old trees with broken and irregular tops where the needles are particularly rich in protein.[46] In contrast to young, managed forests, older forests have more old pines, and more small openings caused by tree falls. These brushy openings provide Capercallies with the blueberries and other berries they need during the summer months. Capercallies also need forest openings for attracting mates. Like many other species of grouse, they have dancing grounds called leks where the males gather in spring to attract females. These leks generally are found in openings with good visibility. Males gather at the lek to defend small territories for the sole purpose of attracting females for mating. When females appear, they are courted with elaborate strutting displays accompanied by loud calls (double clicks leading to a loud "roll," and ending in a penetrating pop that sounds like a cork pulled out of a bottle).[47]

Traditionally, foresters have attempted to maintain Capercallie populations by protecting lek sites and special feeding trees, but the populations of this species have declined throughout northern Europe. When clearcuts cover more than 50 percent of the regional landscape, Capercallies are found only in remnant patches of old forest larger than 125 acres (50 ha).[48]

The effect of clearcuts on bird populations in remnant patches of boreal forest has not been studied as intensively in North America as in Europe. However, a large-scale experiment in a mixed quaking aspen, balsam poplar, and white spruce forest in central Alberta provided preliminary evidence that some species of birds decline in patches of old forest surrounded by clearcuts.[49] Remnant patches of different areas were isolated by harvesting the forest around them. Compared to control plots of the same size inside intact forests, plots in isolated patches of forest had lower numbers of several species of migratory birds, including Ruby-crowned Kinglet, Black-throated Green Warbler, Tennessee Warbler, and Rose-breasted Grosbeak. Only a few resident species, such as Red-breasted Nuthatch, were hurt by fragmentation of the forest, but the plots were not large enough to adequately sample woodpeckers, hawks, and other resident birds with large home ranges.

Although some species have declined in the managed forests of northern Europe because they require large areas of old forest, other species have declined primarily because of the scarcity of standing dead trees or snags. Hole-nesting birds depend on snags for nesting sites. Siberian, Crested, and Willow tits; Three-toed and Black woodpeckers; Redstarts; and a number of other species nest in cavities in dead wood.[50] The Willow Tit, Crested Tit, and woodpeckers construct their own cavities, while other species use old woodpecker holes or natural cavities.

Dead trees and branches are removed during thinning and clearcutting operations in industrial forests, leaving few nesting sites for cavity-nesting birds. In southern Finland, Great Spotted Woodpeckers, Three-toed Woodpeckers, Redstarts, Pied Flycatchers, and Treecreepers were all more abundant in nature reserves, where snags were not removed, than in woodlands managed for forestry.[51] Three-toed Woodpeckers, for example, were ten times more abundant in nature reserves than in managed forest. It therefore is not surprising that most of these cavity-nesting species have declined as intensively managed forests have replaced natural forests in Finland and other parts of northern Europe.[52] Moreover, a large number of other organisms that depend on dead wood, including numerous species of fungi, lichens, and beetles, have declined, and some are now considered threatened with extinction.[53]

Because of concerns about the loss of biological diversity, forest management is changing in Finland and other parts of northern Europe. An effort is now made to simulate some of the characteristics of unmanaged forest. To find an appropriate model for both forest structure and for the pattern of forest destruction and regeneration, Swedish and Finnish biologists have been studying the relatively untouched boreal forests in some parts of Russia. Natural pine forests in Russia have 33 times as many snags, 46 times as many fallen dead trees, and 8 times as many large trees as managed pine forests in Sweden.[54] In natural forests, storms and large fires periodically remove the overstory of trees, generating expanses of young vegetation. However, in wetter sites, particularly along rivers and streams, fire may be infrequent, and young birches and aspens grow in openings created by single treefalls.[55] The result is a complex forest dominated by spruce, but with many scattered broad-leaved trees of various ages. Natural complexity could be simulated, to some extent, by using clearcutting, perhaps with controlled burning, in upland forests, while using selective logging of individual trees along streams and rivers. In both types of sites, snags, fallen logs, and some large trees should be retained.

Crossbills and Other Nomadic Seed-Eaters

Some boreal-forest species probably would not survive long in a single reserve of mature forest, even a massive one, because they must continuously move from region to region in search of food. These species are highly dependent on particular food sources, such as conifer seeds or spruce budworms, that undergo boom and bust cycles in abundance. A particular forest may have a rich supply of such food during one year, but little or none the next, forcing these birds to move to another region. Nomadic movements, often across hundreds or even thousands of miles of boreal forest, have permitted some boreal species to evolve this highly specialized feeding behavior. Like the Passenger Pigeon, however, these species may be particularly vulnerable to extinction because they depend on great expanses of forest in several widely separated localities. A single forest reserve may be critically important for a population of nomadic birds, but it would never be sufficient. It constitutes only one of a series of widely separated forests that are needed in different years.

The highly nomadic species of crossbills of Finland, the Red (or Common) Crossbill and the Parrot Crossbill, have shown severe population declines in managed forests.[56] The susceptibility of these species to major changes in their habitat is not surprising because they are among the most highly specialized birds of the boreal forest. They feed almost exclusively on conifer seeds, and different species of crossbills specialize on different types of conifers. The seeds of most conifers are adapted for dispersal by the wind, not by birds, and they are protected from seed-eaters by the thick scales of the cone. Some seed-eaters, such as squirrels, bite through these scales to reach the seeds, but crossbills are able to obtain seeds much more quickly and efficiently. The upper and lower mandibles of their bills cross at the tip, permitting them to deftly extract seeds even from closed conifer cones. While a crossbill holds the cone steady with its foot, it bites its way between two scales, then moves the lower mandibles to the side (essentially "biting" sideways) to pry the scales apart. Its tongue then probes down to pull out the seed.[57] The seed is pushed into a groove in the palate on the roof of the mouth. The seed fits perfectly into this groove, where it can be held steady by the tongue while the seed coat is removed with the lower mandible.[58] The size of the bill and the width of the groove in the palate differ in different species of crossbills, depending on the types of cones and seeds that they normally use.

Although crossbills feed on spruce budworms and other insects and on the seeds and buds of a variety on nonconiferous plants,[59] they ultimately depend on conifer seeds for reproduction and winter survival. The young are primarily fed on conifer seeds that are regurgitated by their parents. Breeding, which is closely tied to the availability of these seeds, occurs at almost any time of the year. White-winged Crossbills typically nest during three seasons: late summer and early autumn, when tamarack and white spruce cones are maturing; January to March, when red and white spruce cones are filled with seeds; and March to June, when black spruce cones open, exposing the seeds.[60] Winter nests are larger than summer nests, providing better insulation, and the female covers the eggs and young nestlings almost continuously during cold weather.[61] During this period she is fed by the male.

Red Crossbills also may nest more than once during a year. Like many other temperate-zone songbirds, they have a seasonal cycle and normally nest in the summer, but unlike most other species, they can quickly come into reproductive condition in the winter if conifer seeds are available. In a series of well-designed experiments with captive birds, Thomas Hahn showed that progressively increasing day length in the spring causes the release of luteinizing hormone (which induces reproductive activity) and the growth of the testes in males.[62] As days become shorter in the fall, the birds go out of reproductive condition and start molting, even when food is abundant. By January, however, after molting has been completed, crossbills will come into reproductive condition if food is plentiful. Consequently, they can nest during both the longest and shortest days of the year.

Just as Passenger Pigeons needed to move between regions of oak and beech forest to find enough acorns or beechnuts to support a nesting colony, crossbills must shift from region to region within the boreal forest to find abundant crops of conifer seeds. This was originally shown by A. Reinikainen, who skied across a standard route in northern Finland each year between 1927 and 1937, counting crossbills and spruce cones. He found that crossbills were abundant in years with a good seed crop and rare in years with few spruce seeds.[63] An area that is favorable for nesting one year may not be used again for several years. Most conifers produce few

Color plates

PLATE I. Male Bobolink singing in grassland on a reclaimed strip mine (The Wilds) near Caldwell, Ohio.

PLATE 2. Yellow-breasted Chat singing from a smooth sumac in a thicket maintained along a powerline cut.

PLATE 3. Mountain Plover nest in a blacktail prairie dog town.

PLATE 4. Carolina Parakeets (two adults below an immature) feeding on cypress cones in a swamp along the Suwanee River in Florida.

PLATE 5. Hooded Warbler feeding two of its own nestlings and a larger Brown-headed Cowbird nestling in the understory of an eastern deciduous forest. A jack-in-the-pulpit is in bloom behind the nest.

PLATE 6. Red Crossbill in a white spruce on the edge of a clearcut. Short-rotation clearcutting reduces the number of cone-bearing trees needed by crossbills and other seed-eating birds of the boreal forest.

PLATE 7. Varied Thrush in an old-growth forest in Oregon.

PLATE 8. Phainopepla (female on left and male on right) feeding on berries of desert mistletoe in a dry wash in southern Arizona.

PLATE 9. Red-cockaded Woodpeckers at a nest cavity in a live pine. An immature male (partially behind tree) is helping the pair feed the nestlings.

PLATE 10. Male Ivory-billed Woodpecker feeding a wood-boring beetle grub to nestlings in sweetgum and oak bottomland forest in Louisiana.

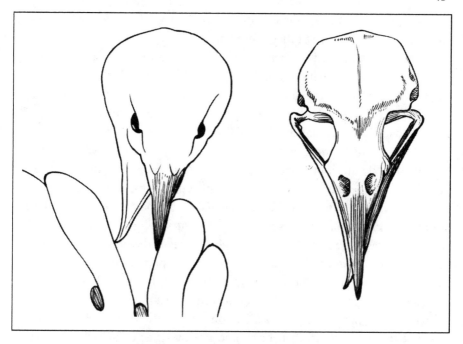

FIG. 6.4. Method used by White-winged Crossbill for removing seeds from a spruce cone. From Benkman, 1992, with permission from the artist, Daniel Otte.

seeds in the year or two following an abundant seed crop, so crossbills are forced to move to another region.[64] Consequently, even a large preserve of northern forest, such as the more than 2 million acres in the Boundary Waters Canoe Area in Minnesota and adjacent Quetico Provincial Park in Ontario, may be too small to sustain crossbill populations. Only a system of forests in different regions would guarantee enough food to support nesting each year.[65]

Today crossbills can still move readily across the great boreal forests of North America, from Nova Scotia to Alaska, in search of good cone crops. Industrial forestry will increasingly limit their options, however. Carefully planned logging will not eliminate the coniferous forest, which should grow back vigorously, but it will reduce the cone crop. A short rotation time that calls for cutting trees every 60–100 years will eventually yield a forest that produces relatively few cones. Much of the landscape will be covered by forest too young to produce cones, and even the oldest trees will produce fewer cones than the 100- to 200-year-old trees in a mature forest.[66] Increasingly, crossbills will depend on forests that are managed with longer rotation times or where timber is not harvested.

The two recognized species of crossbills in North America need different types of conifers, which complicates the effort to provide adequate habitat for them. White-winged Crossbills have relatively small bills, so they extract seeds from the small cones of spruce and tamarack most efficiently.[67] In contrast, many populations of Red Crossbills depend on pine seeds.[68] To complicate conservation planning even further, there are several distinct forms of Red Crossbills that have different food requirements. Traditionally, some of these types have been recognized as subspecies, but work by Jeffrey Groth indicates that they may actually be separate species.[69] These would be "cryptic species," species that are so similar in appearance that they could easily be mistaken for a single species if the only basis for comparison is museum specimens. However, if two populations do not interbreed when they live in the same region, then they are considered separate species even if they are virtually identical in appearance. Cryptic species of birds typically have different vocalizations, probably because vocal signals are critical for preventing hybridization between species. The different forms of crossbills have distinctly different flight calls, which may be important in courtship and mate recognition. Wild birds approach a tape recorder if it is playing a recording of flight calls like their own, but not if it is playing another type of flight call.[70] Also, flocks typically consist of birds that share the same type of call. Curtis Adkisson observed birds of three different call types migrating in separate flocks through the Colorado Rockies.[71] The status of these call types is still questionable, however, because biochemical analysis indicates that there are only slight genetic differences among the types.[72]

Groth found that in 24 pairs of Red Crossbills that were tape-recorded in the southern Appalachians, the male and female consistently shared the same type of flight call.[73] Moreover, individuals with one call type were consistently larger (and had larger bills) than individuals with the other call type; there was no overlap between the two. Although only birds with two call types nest in the southern Appalachians, four call types are found in the boreal forest of eastern Canada and six occur in the coniferous forests of the Pacific Northwest.[74]

The forms of Red Crossbills differ not only with respect to calls and body size but also in feeding behavior. In the Pacific Northwest, for example, each form depends on a different "key conifer," with one form primar-

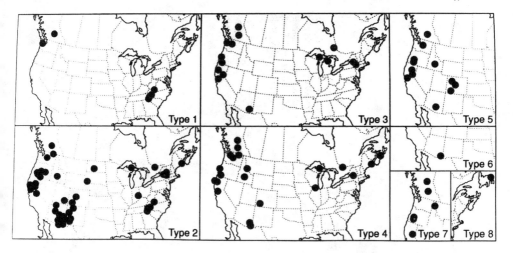

FIG. 6.5. Distribution of records of the different call types of Red Crossbills. From Groth, 1993, with permission from the University of California Press.

ily using western hemlock, another using Douglas-fir, and a third requiring ponderosa pine.[75] The situation is similar in the southern Appalachians, where the small-billed form of Red Crossbill primarily extracts seeds from white pine cones while the larger-billed form concentrates on the thicker, more heavily armored cones of table mountain pine.[76] The shape of the bill and palate of each form is optimally suited for extracting and husking seeds of its preferred conifer. When seed removal and husking were timed for captive crossbills, each form was most efficient at processing the seeds of its key conifer.

Birds with the four call types in the boreal forest of eastern Canada have not been carefully compared. If each of these forms is a cryptic species that depends on a different species of conifer, then conservation efforts may become complicated. Three of these forms are probably nomadic, which means that they move from region to region, tracking good cone crops of their preferred conifer. A good nesting site for one form may lack a sufficient number of the preferred conifer of another form, so each type of crossbill might require its own set of widely dispersed, cone-producing forests.

Other boreal seed-eating birds also move great distances in search of food. They not only move across the boreal forest, but periodically they move south of the boreal forest in "irruptions." These massive movements often occur synchronously for a number of seed-eaters—Red-breasted

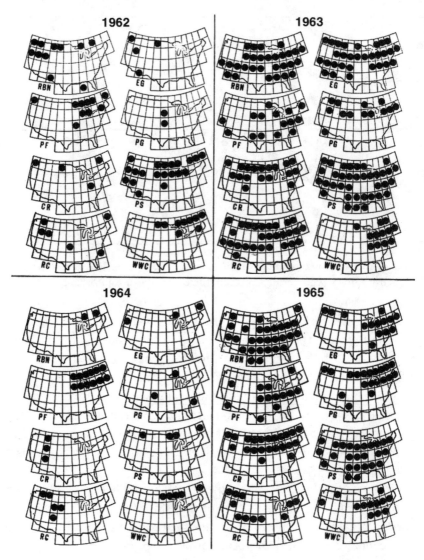

FIG. 6.6. Pattern of occurrence of eight seed-eating birds of the boreal forest between 1962 and 1965. Dots indicate higher-than-average abundance on Christmas Bird Counts. RBN = Red-breasted Nuthatch, EG = Evening Grosbeak, PF = Purple Finch, PG = Pine Grosbeak, CR = Common Redpoll, PS = Pine Siskin, RC = Red Crossbill, WWC = White-winged Crossbill. From Bock and Lepthien, 1976, with permission from the University of Chicago Press. © 1976 by The University of Chicago Press.

Nuthatch, Evening Grosbeak, Purple Finch, Pine Grosbeak, Common Red-poll, Pine Siskin, and White-winged Crossbill—in years when the spruce and birch trees of the boreal forest produce few seeds.[77] The seed crops of spruce and birch tend to occur synchronously across huge sections of the boreal forest. A banner year for these seeds results in high nesting success, but a good seed crop is usually followed by a year with few seeds. Consequently, during the following winter an inflated population of seed-eating birds is confronted with an environment with few seeds, and many birds must move south in search of food. The seed crops of pines are not synchronized with spruce and birch seed crops, however, so Red Crossbills, which depend primarily on pine seeds in eastern Canada, irrupt at different times from the other species. "Superflight" years, when there are truly massive irruptions of seed-eating birds, probably occur when the crop of pine cones happens to be low during a poor year for spruce and birch seeds.[78] Superflights occur irregularly and rarely (in 1935 and 1969, for instance), but normal irruptions occur every two or three years, coinciding with the cycle of seed production in spruce and fir. The drama of superflight years has given many people the impression that irruptions are extremely irregular and unpredictable, but data from Christmas Bird Counts show that they are surprisingly regular and frequent.[79]

Birds such as sparrows and goldfinches that feed on the seeds of herbaceous plants rather than on the seeds of trees either remain in the north as permanent residents or migrate regularly between specific nesting and wintering areas. Their food supply tends to be predictable and relatively constant from year to year, in contrast with the boom and bust cycle of tree seeds. Species that depend on tree seeds show dramatically irregular movements. They may change both their wintering areas and their breeding areas from year to year, so that the populations are not regional but continental. For example, of approximately 17,000 Evening Grosbeaks banded at a station in Pennsylvania, only 48 were recovered at the same site in subsequent winters. Another 451 were recovered across North America, in 17 states and four Canadian provinces.[80] Preservation of a nomadic species of this sort cannot depend solely on a single concentrated local effort.

Decline of the Newfoundland Crossbill

Nomadic crossbills are vulnerable because they depend on finding appropriate habitat in many widely spaced regions; one forest will not sustain them. Ironically, however, the crossbill populations that are in the greatest trouble are essentially sedentary crossbills, especially those living on islands, such as Hispaniola and Great Britain (Scotland). In North America, the most threatened type of crossbill is the sedentary Newfoundland subspecies of the Red Crossbill (*Loxia curvirostra percna*)[81] restricted to the island of Newfoundland.[82] Because this crossbill has a distinctive flight call, it may constitute a separate species.

Newfoundland Crossbills are in trouble because they depend on a single species of conifer, the black spruce.[83] Unlike most conifers in the boreal forest, black spruce produces cones dependably each year. Black spruce seeds could potentially sustain crossbills throughout the year, but on the mainland these seeds are unavailable in winter because the cones are closed and crossbills are unable to open them. On Newfoundland, however, Red Crossbills can extract seeds from closed cones because the resident type of Red Crossbill has a substantially larger bill than do other types in North America, and because black spruce cones have thinner scales on Newfoundland than they do on the mainland. Black spruce has evolved on Newfoundland in isolation since the last glacial period, and its cones may be less well protected because one of the major "predators" of spruce seeds, the red squirrel, was absent from the island. With no loss of seeds to the gnawing attacks of squirrels, black spruce evolved a less heavily armored cone that was accessible to crossbills even when closed. Unfortunately, however, red squirrels were introduced to the island by the Newfoundland Wildlife Service in the early 1960s to provide food for the endemic Newfoundland subspecies of the marten (*Martes americana atrata*), which had declined after the destruction of old, uncut forests.[84] Because the black spruce on the island have such thin cone scales, squirrels were able to harvest a high proportion of their seeds, leaving few seeds for the crossbills. Not surprisingly, the Red Crossbill population declined rapidly in the 1970s. This subspecies (which may actually be a species) soon may be extinct on Newfoundland, and the best hope for its survival is that it will remain on small offshore islands where there are no squirrels.

Two other relatively sedentary populations of crossbills that are restricted to islands, the Scottish Crossbill and the Hispaniolan subspecies of the White-winged Crossbill (*Loxia leucoptera megalaga*), are also endangered.[85] Both of these crossbills depend on pine seeds, and much of the pine forest in both Scotland and Hispaniola has been destroyed. Ultimately, it is the high degree of feeding specialization that makes both nomadic and sedentary populations of crossbills vulnerable. The decline of sedentary populations is obvious, but a serious decline in nomadic populations may go undetected because of the unpredictable and infrequent occurrence of birds at any particular site. Like Passenger Pigeons, nomadic crossbills could decline drastically before anyone noticed.

Spruce Budworm Warblers

Another group of nomadic species are the warblers that concentrate in spruce and fir forests that are heavily infested with spruce budworm. Like the nomadic seed-eaters, the budworm specialists could suffer if the boreal forest of North America is homogenized by forest management.

Budworm moths lay eggs on conifer needles in late summer.[86] The caterpillars spend the winter in silken cases on the twigs. In spring they emerge and attack the buds and emerging needles, often causing defoliation. Repeated defoliation can kill trees, decimating large areas of forest. Next to fire, outbreaks of budworms and other defoliating insects are the most important natural disturbance in the boreal forest. Although budworms eventually destroy the forest overstory, in the short term they provide a rich source of food for forest birds. A few species of birds actually seem to depend on spruce budworms. Heavy infestations of budworm flare up in different areas every year, so these budworm specialists must be nomadic, moving across the boreal forest from one outbreak to another.

In a hundred-year-old balsam fir forest in New Brunswick, spruce budworm populations increased from 1,000 larvae per acre in a non-outbreak year (only one larva for about every 8,000 leaves) to 8 million per acre during an outbreak year.[87] During the four years of the outbreak (1950–1954) the forest was severely defoliated, and almost all of the mature trees were killed. Many species of birds, even such birds as crossbills and woodpeckers, which normally do not feed on caterpillars, shifted their feeding behavior to take advantage of the superabundant budworms. However, the greatest response was shown by two species of warblers that are budworm specialists. The population of Bay-breasted Warblers increased from 10 to 120 territorial males per 100 acres (40 ha), and the Tennessee Warbler, which had been absent from the site before the outbreak, increased to 45 males per 100 acres. A third budworm specialist, the Cape May Warbler, did not respond to this particular budworm outbreak, although it had increased during earlier outbreaks. Remarkably, some of the common resident warbler species, such as Back-throated Green, Magnolia, and Yellow-rumped warblers, declined during the years when budworms were abundant, perhaps because of competition with the suddenly abundant spruce budworm specialists or because their normal insect prey declined as the budworms removed most of the fir needles in the forest. The same resident species also declined during an earlier outbreak in Ontario.[88]

Anthony Erskine of the Canadian Wildlife Service has completed the most thorough study of the distribution of birds of the Canadian boreal forest. As part of his study, he counted the number of territorial males in fourteen widely scattered spruce and fir stands across Canada. Three of the four sites with high spruce budworm densities also had exceptionally high densities of Tennessee, Cape May, and Bay-breasted warblers. The Cape May Warbler, in particular, was highly associated with budworm outbreaks, with densities of 49–158 males per square kilometer in three sites with budworms, and densities lower than 8 males per square kilometer at sites without outbreaks. The other species had high densities at some non-outbreak sites but were generally most abundant at sites with outbreaks. The Blackburnian Warbler, which was restricted to fir stands, was also most abundant at outbreak sites.

Judging by Erskine's results, none of these species completely depends on spruce budworm outbreaks. All are found in at least small numbers in conifer stands where there are few budworms. However, outbreaks may be important to the long-term survival of these species. This would be true if nesting success were low in the absence of budworms so that the warblers ultimately depend on finding budworm outbreaks to sustain their populations. Two of the spruce budworm specialists, the Cape May and Bay-breasted warblers, lay larger clutches of eggs than other warbler species in the same habitat, and they lay more eggs during budworm outbreak years than during non-outbreak years, indicating that they are adapted for raising young in an environment with exceptionally abundant food.[89]

Conservation of Boreal Forest Birds

Although much of the North American boreal forest has not been greatly modified by forest management,[90] this is changing rapidly because of the great demand for aspen and 60- to 100-year-old conifers for pulp production.[91] Intensive harvesting and forest planting now occurs in boreal forests in Ontario, Minnesota, and northern Maine. The result could be a simplified, homogeneous forest like the highly managed forests of Finland, Sweden, and Norway. Foresters in the Nordic countries have already tried this experiment on a massive scale, and they have witnessed the decline of some of the most distinctive species of animals of the northern forest. North Americans have much to learn both from their success in maintaining wood production and from their failure in sustaining a diversity of birds and other organisms that depend on particular features of a natural forest. Although the species in the boreal forests of North America differ from those in Europe, the general results of intensive forestry are likely to be the same. Some of the species that have declined in Europe have close ecological counterparts in North America; the Boreal Chickadee is similar to the Siberian Tit, the Gray Jay is in the same genus as the Siberian Jay, and the Three-toed Woodpecker is found on both continents.[92] Boreal forests in North America have a counterpart for the pine needle–eating Capercallie (the Spruce Grouse), as well as a counterpart for the large Black Woodpecker of Europe (the Pileated Woodpecker). Many North American species, like their European counterparts, depend on dead wood for nesting cavities. Also, judging by the experience in deciduous forests on the two continents, North American boreal forests may have many more species that are adversely affected by forest fragmentation, particularly among the wood warblers (Parulinae) and vireos (Vireonidae), groups that are important in North America but are absent from Europe.

The diversity of birds and other organisms of the boreal forest will quickly be eroded if the complex natural forest is converted to a monotonous young forest managed solely for wood production. Biologically diverse mixed forests of conifers and hardwoods are particularly threatened by forest management, which tends to favor pure hardwood or conifer stands.[93] Large reserves should be protected so that we have models of a functioning natural forest (something that has been almost completely lost in some parts of northern Europe). On timber lands, the negative impact of logging can be greatly reduced by retaining some of the internal complexity of the forest in the form of snags, fallen logs, some large trees, and a mixture of tree species, and by maintaining a mosaic of different forest types to support a wide range of birds, from the Mourning Warblers and White-throated Sparrows of shrubby second-growth to the Blackburnian Warblers and Boreal Chickadees of mature spruce-fir forest.

CHAPTER 7

Birds of the Western Mountain Slopes

East of the Rio Grande the junipers are small and widely interspersed among the green rosettes of the one-leaved piñon pines, with which the pale hills are spotted like an ocelot. The green of the junipers is slightly yellower than the pines, which may also be identified for the stranger by their tendency to run to a true tree form with an upright stem, and by the blue back and the white ellipse of the spread wings of the piñonero, the piñon jay, whose winter pasturage is extracted from small, globose cones.

—MARY AUSTIN, *The Land of Journeys' Ending*

T HE coniferous forests of the western United States are more similar to the boreal forests of Canada and Alaska than to deciduous forests at the same latitude farther east. While much of the eastern forest was cleared for agriculture, both the western and boreal forests have been used primarily for timber production. Even though timber is sometimes harvested with massive clearcuts, these forests are generally managed to perpetuate commercially valuable forest, not to create farmland or suburbs.[1] Instead of the archipelagoes of isolated forest patches found in many parts of the East and Midwest, western mountains have interconnecting networks of mature forest with embedded clearcuts and burns. Most of the ancient forest has disappeared, but there is still considerably more old growth than in the East.[2]

Many bird species that range across the forests of the Canadian flatlands also are found in similar forests of the Rocky Mountains and Pacific coast ranges. Black-backed Woodpeckers, Gray Jays, Brown Creepers, Red Crossbills, and numerous other species are found in similar habitats in both regions. Like the boreal forests, the forests of the western mountains are dominated by conifers growing in a patchwork of different successional stages. The patchwork was once burned into the landscape, but increasingly it is dominated by the rectangles and straight lines engineered during tim-

PLATE 7. Varied Thrush in an old-growth forest in Oregon.

ber harvesting. Some species of birds originally depended on fire to create their preferred habitat, which might be freshly killed and charred stands of trees, dense growths of young pine, or groves of tall aspens. Other species depend on the giant trees, shrubby openings, and dead wood found in an old forest.

Most of the features that set the western coniferous forests apart from the boreal forest are associated with altitude. The western mountains are high enough that changes in elevation can result in dramatic changes in the environment. Higher elevations typically have lower temperatures, more rain, greater snow depth, more solar radiation, and more wind.[3] It is not surprising, therefore, that the vegetation and bird life steadily change with increasing elevation.

In proceeding up a mountain slope from the grassland or sagebrush at the base of a mountain, one passes through a sequence of different forest types before reaching the treeless alpine meadows at the highest elevations. These forest types grade into one another, and each elevational zone has a mosaic of successional stages generated by fire, wind, and timber harvesting. Thus, birds of the western mountains have adapted to a landscape with astonishing diversity. In contrast with the relatively homogeneous deciduous forest that once stretched almost continuously for hundreds of miles across eastern North America, or the unbroken boreal forest that still covers great stretches of Canada and Alaska, forests of the western mountains have always been patchy. Particular types of forest are restricted to a particular elevational belt, and similar belts in adjacent mountain ranges may be separated by distinctly different types of vegetation. These habitats have been fragmented since at least the end of the last glacial period, so the creation of habitat fragments by people might not have the same impact as in eastern forests.[4]

In the southern Rocky Mountains, the lower slopes are covered with a pygmy conifer woodland dominated by scattered junipers and pinyon pines.[5] These are typically low trees with rounded crowns. Black-chinned Hummingbird, Western Scrub-Jay, Pinyon Jay, Plain Titmouse, Rock Wren, Blue-gray Gnatcatcher, and Gray Vireo reach their greatest abundance in this habitat.[6] Upslope, the pygmy conifer woodland grades into ponderosa pine woodland, an open, grassy habitat with scattered pines. In older stands, the straight, red-barked ponderosa pines have massive trunks. Williamson's Sapsucker, Pygmy Nuthatch, Western Bluebird, Olive Warbler, and Yellow-eyed Junco all reach their greatest abundance in this open

FIG. 7.1. Changes in vegetation with elevation in Rocky Mountain National Park, Colorado. Photograph by Robert Askins.

woodland.[7] Douglas-fir forests—which also may contain white fir, blue spruce, limber pine, and quaking aspen—predominate at still higher elevations.[8] Not many bird species are concentrated in this intermediate zone, but MacGillivray's Warbler and Western Tanager are more abundant there than in other zones.[9] Still higher, the forests are dominated by Engelmann spruce and subalpine fir, close relatives of the white spruce and balsam fir of the boreal forests farther north.[10] Not surprisingly, many of the same species of birds that range across the boreal coniferous forest are also found in this zone. These include Boreal Owl, Gray Jay, Ruby-crowned and Golden-crowned kinglets, Wilson's Warbler, and Pine and Evening grosbeaks.[11] Other species, such as Rufous Hummingbird, Calliope Hummingbird, and Williamson's Sapsucker, are closely associated with the spruce-fir forest of the mountains but do not range eastward across the lowland spruce-fir forests of Canada.

Some mountain ranges have additional forest types along the elevational gradient. Oak-pine forests cover the lower slopes of ranges in Arizona and Texas, quaking aspen or lodgepole pine stands grow on mid-elevation slopes where the forest has burned relatively recently, and stands of short, gnarled pines (bristlecone pine, whitebark pine, or limber pine) border the alpine meadows near timberline.[12] Adjacent vegetation zones grade into one another, and higher altitude forest tends to spread down to lower elevations in such moist, protected areas as valley floors. Thus different zones not only blend but also interdigitate. The complex landscapes resulting from the combination of disturbances and elevational change offers a wide range of habitats for birds.

Fire and the Western Forest

Next to differences in altitude, fire is probably the most important factor for inducing diversity in western forests. Most of the mountain forests of the West burn periodically and relatively predictably, but there is wide variation among forest types and regions in both the frequency and intensity of fires. Frequency and intensity are closely associated: where fires are infrequent, there is often an accumulation of dead wood that fuels intense fires. These fires burn hot, climb to the crowns of the trees, and leave behind a forest of charred snags. If fires are frequent they tend to be relatively cool ground fires that do not kill many trees. Thus, frequent fires produce a more stable environment. Forests with infrequent fires may appear stable over the short term, but they are prone to sudden and catastrophic change.

Open woodlands of ponderosa pine are sustained by frequent fires. Fire scars on trees indicate that in some regions the grassy understory in these woodlands burned as frequently as once each year.[13] Frequent fires prevent the grassy areas from being replaced by trees, but they are not hot enough to kill the old, thick-barked pines. Pine seedlings are killed, however, and successful growth of young pines occurs in infrequent pulses during a wet period when fire does not burn through the understory for several summers. When fires are suppressed or when the frequency of fires is reduced because livestock remove most of the flammable understory vegetation, trees grow more densely, converting the open woodland to a thick forest. Dead wood then collects on the forest floor, and when a fire eventually occurs, it burns intensely and kills the trees.[14]

The lodgepole pines of mid-elevational forests are adapted to fire in a distinctly different way from ponderosa pine.[15] Fires are infrequent in most of these forests and usually occur only after dead wood has built up. When a major fire erupts because of extremely dry conditions and high winds, it may sweep across huge areas of forest. Fire scars in the trunks of old trees in Yellowstone National Park indicate that large firestorms sweep across the lodgepole pine forests of this region approximately every 200 to 300 years, destroying much of the forest.[16] The fires that burned 720,000 acres (290,000 ha) of Yellowstone National Park in 1988 were an expected part of this cycle.

During the first 250 years after a stand of lodgepole pines starts growing, it is not very flammable. The understory is usually green throughout the summer, there is relatively little dead wood on the forest floor, and the understory and fallen wood are well separated from the treetops. After 250 years, however, the oldest lodgepole pines begin to die, and they are replaced by younger trees (including subalpine fir and Engelmann spruce). Now there is a high density of fallen wood and a "ladder" of small trees to carry the fire to the crowns of the old pines. The stage is set for a firestorm, with fires so intense that they create their own wind. This wind is generated by the air pressure gradient between the air near the fire and the cooler air farther away. Fire scars indicate that large proportions of the Yellowstone forest burned in the late 1600s and mid-1700s, so by 1988 much of the forest was old enough to be highly flammable.

Lodgepole pines are well adapted to environments with intense, tree-killing fires. Unlike the ponderosa pine, with its fire-resistant trunks, lodgepole pines are highly flammable. Ironically, this feature gives lodgepole pine an edge in competition with other tree species. They usually burn long before more shade-tolerant species, such as Douglas-fir, replace them. These competitors are burned along with lodgepole pine, and the fire creates a seedbed for lodgepole pine seedlings. Also, lodgepole pines have cones (similar to the cones of jack pine and black spruce in the boreal forest) that open after they are singed. Following a fire, lodgepole pine cones release massive numbers of seeds that initiate a new forest (and, incidentally, attract large numbers of seed-eating birds, such as Clark's Nutcracker, Cassin's Finch, Red Crossbill, and Pine Siskin[17]). Pine seedlings typically become established over several decades, resulting in a forest with trees of different sizes.

Fires occur even less frequently in the coastal coniferous forests of the Pacific Northwest than in the lodgepole pine forests of the Rocky Mountains. The moist forests that stretch along the coast and adjacent mountains from the Gulf of Alaska to northern California are dominated by such giant conifers as Douglas-fir, western hemlock, western redcedar, Sitka spruce, and coastal redwood.[18] Major fires and blowdowns are infrequent, so the forests may be ancient. At Mount Rainier National Park in Washington, for example, fires occur on average every 434 years.

The reduction of the frequency of wildfires was one of the major changes in western forests following European settlement. The impact of fire suppression is dramatically obvious when nineteenth and twentieth century photographs of the same sites are compared. George Gruell has compiled numerous pairs of photographs taken at the same localities in the northern Rocky Mountains between 1871 and 1982.[19] In the earlier photographs the forests are more open, and there is more young forest and grassland. Later views show the encroachment of trees and shrubs, and, especially, the expansion of dense coniferous forest. Although some species of birds undoubtedly benefited from these changes, others must have lost their preferred habitats.

Birds That Use Burns

The charred landscape following a major fire is the preferred habitat for some bird species. As in the boreal forests, Black-backed Woodpeckers are virtually restricted to recent burns. Also, Three-toed and Hairy woodpeckers are more abundant in recent burns than in other habitats.[20] These woodpeckers thrive in recent burns because dead trees provide good sites on which to excavate cavities for nesting and roosting, and they are often filled with the larvae of bark beetles that woodpeckers extract and eat.

Even species that do not feed on bark insects may depend on dead trees for nest sites. Cavity nesters such as Northern Flickers, House Wrens, and Mountain Bluebirds increase in abundance after a fire because of the greater availability of nest sites. Mountain Bluebirds are particularly closely associated with recent burns. Two flycatchers, the Western Wood-Pewee and Olive-sided Flycatcher, also reach exceptionally high densities in recent burns, perhaps due to the availability of open perches from which they can make short sallies to snap up flying insects.[21] Still other species, such as the Chipping Sparrow and Broad-tailed Hummingbird, colonize burns as a shrub layer grows up.

Other species depend on vegetation molded by ground fires that do not kill mature trees. For example, Purple Martins and Western Bluebirds are much more abundant in the open ponderosa pine, where fire occurs frequently, than in the dense pine forest that results when fires are suppressed. The same is true of Lark Sparrows and Chipping Sparrows, which may decline when fires are suppressed because of the loss of the grassy understory where they feed on grass seeds.[22]

FIG. 7.2. Photographs taken at the same site in Lewis and Clark National Forest, Montana, in 1909 (above) and 1980, showing the loss of open habitats and young forest resulting from suppression of fires. From Gruell, 1983. Reproduced from USDA Forest Service, Gen. Tech. Rep. INT-GTR-158. This is one of numerous pairs of photographs showing the decline in early successional habitats in the northern Rocky Mountains during the past 50 to 100 years (Gruell, 1982).

Comparison of the Effects of Fire and Logging

In much of the West, logging has replaced fire as the main agent of forest disturbance. Although clearcut harvesting creates early successional habitat, it does not simulate all of the ecological effects of fire.[23] In particular, it does not produce stands of standing dead trees to attract Black-backed Woodpeckers and other species that need dead wood for feeding or nesting. Traditional clearcut logging entails the removal of all dead trees, resulting in a decline in the large number of species that nest and roost in tree cavities. Current forestry practices call for the retention of some snags to provide shelter for both birds and mammals that use tree cavities, but this is not an adequate substitute for the numerous dead trees produced by an intense fire.

When Richard Hutto compiled information from more than 250 published and unpublished studies to compare the birds of 15 habitat types in the northern Rocky Mountains, he found that the composition of bird communities of burned forests were more similar to clearcuts than to any other vegetation type.[24] Despite this, he found substantial differences between the two. Black-backed Woodpeckers were found in most burns but were absent from clearcuts. Calliope Hummingbird, Three-toed Woodpecker, and Clark's Nutcracker were more frequent in burns, while Red-naped Sapsucker and Cedar Waxwing were more frequent in clearcuts. Species that use low shrubby vegetation, such as American Kestrel, Fox Sparrow, and Dark-eyed Junco, occur frequently in both burns and clearcuts. However, in some of the most heavily managed forests, particularly in Douglas-fir forest in the Pacific Northwest, the grass and shrub stages of succession are shortened following logging because herbicides and fertilizer are used to quickly establish a dense cover of young Douglas-fir.[25] This reduces the amount of habitat for birds that require more open habitats and that generally do well in less heavily managed clearcuts. Mountain Quail, Calliope Hummingbird, Wilson's Warbler, Lazuli Bunting, Fox Sparrow, Vesper Sparrow, and Savannah Sparrow are some of the species likely to decline where forest regeneration is accelerated in western Oregon because they primarily nest in the grass and shrub stages.[26]

Effect of Logging on Birds of Western Forests

Although the habitat requirements of forest birds generally have been much more intensively studied in eastern deciduous forests than in western forests, the effect of logging on bird populations has been more thoroughly studied in the West. Comparing the results of these studies is daunting because the studies were completed in distinctly different types of forest (from ponderosa pine woodlands in the Southwest to Douglas-fir forests in the Pacific Northwest) and at sites that have been harvested with a wide range of methods. Sallie Hejl and colleagues compiled information on the impact of logging on western birds not only from published studies, but also from graduate theses and unpublished reports.[27] Although there is considerable variation in how particular bird species respond in different places and to different types of logging, there are some patterns.[28] Three-toed Woodpeckers, Mountain Chickadees, Red-breasted Nuthatches, Pygmy Nuthatches, Brown Creepers, Golden-crowned Kinglets, Ruby-crowned Kinglets, Winter Wrens, Swainson's Thrushes, Varied Thrushes, and Townsend's Warblers were always less abundant in clearcuts than in uncut forests. Brown Creepers and Pygmy Nuthatches were also less abundant in partially cut forests. Other species increase with forest harvesting. Mountain Bluebirds and Song Sparrows were usually more abundant in shrubby clearcuts than in uncut forests, and Calliope Hummingbirds and Rock Wrens were always more abundant in partially cut forests than in uncut forests.

One of the most detailed studies of the impact of forest harvesting on birds was completed in ponderosa pine forest in Arizona.[29] An uncut plot was compared with a nearby harvested plot of the same size (35 acres, or 15 ha). Different types of timber harvesting had been employed on the harvested plot: moderate thinning of trees, strip cutting (creation of alternating strips of clearcut and forest), severe thinning, and clearcutting. Violet-green Swallows, Pygmy Nuthatches, Cordilleran Flycatchers, Red-faced Warblers, and Black-headed Grosbeaks had substantially higher densities in the uncut and moderately thinned plots than in any of the plots that had been severely cut. Other species, such as Western Wood-Pewee, Mountain Bluebird, and Spotted Towhee, were more abundant in the open habitats created by timber harvesting.

In a distinctly different type of western coniferous forest, the temperate rain forest of Prince of Wales Island in Southeast Alaska, there were also species of birds that were associated with older, uncut forest.[30] Golden-crowned Kinglet, Red-breasted Sapsucker, and Pacific-slope Flycatcher were all more abundant in ancient forest than in recently harvested areas, and Brown Creeper was detected only at sites in ancient forest.[31] Other species, such as Orange-crowned and Wilson's warblers, were most abun-

FIG. 7.3. Brown Creeper, a species that is usually found in extensive tracts of old forest in the western mountains. Photograph by O. S. Pettingill. Reproduced with permission from the Cornell Laboratory of Ornithology.

dant in the young forest that had grown on recent clearcuts. Both young and old-growth forests are dominated by western hemlock and Sitka spruce in this region, but the young forests are missing certain features, such as large trunks, large dead trees, and multilayered vegetation, that are needed by some species of birds.

Forest harvesting has a particularly severe effect on species that depend on ancient forests. Unlike other vegetation stages, old-growth forest is not replenished through regrowth. Although new forests grow up in clearcuts, they are usually cut long before they achieve old-growth characteristics, such as giant trees, large snags, fallen logs, and well-developed shrub patches where the canopy has been ripped open by the collapse of a large tree. No species of songbird appears to completely depend on ancient forests, but the Brown Creeper is consistently more abundant in uncut old-growth forests than in clearcuts or partially harvested forests. Other species, such as Golden-crowned Kinglet, Varied Thrush, and Townsend's Warbler, were more abundant in old-growth than in mature second-growth forests in most regions, but they showed the opposite pattern in other parts of the West.[32]

Logging not only reduces the total amount of ancient forest, it also creates a landscape in which older forest is restricted to islandlike patches surrounded by brush or young forest. In eastern deciduous forests, many species of songbirds decline or disappear in small, isolated patches of forest. In the forests of western mountains, however, most species of forest birds persist even in small relict patches of old forest. In a comparison of 46 patches of Douglas-fir forest in northern California, Kenneth Rosenberg and Martin Raphael found that only a few species of birds (Acorn Woodpecker, Pileated Woodpecker, and Winter Wren) were significantly more abundant in large forests than in forest remnants smaller than 48 acres (20 ha).[33] These remnant patches were surrounded by recent clearcuts. A similar study of birds in Douglas-fir forests in the Washington Cascades showed that the abundance of most species of birds was similar in small and large stands of old growth.[34] However, the smallest stand included in this study was 96 acres (40 ha). Even in these relatively large patches of ancient forest, the Winter Wren appeared to be sensitive to the amount of forest fragmentation. This species was less abundant in stands surrounded by recent clearcuts than in stands adjacent to mature (but not old-growth) forest. In subalpine forests dominated by Engelmann spruce and subalpine fir in the Snowy Range of Wyoming, the bird communities are surprisingly similar in continuous forest and in stands where numerous strips or patches of trees have been removed. However, one of the species most closely associated with big trees and old-growth conditions, the Brown Creeper, was completely missing from the sites where the continuity of the old forest had been disrupted.[35]

In remnant patches of old-growth forest in British Columbia, the number of species classified as old-growth specialists was similar for small and large sites, and most old-growth species were equally abundant in sites of different sizes. In this case, the "islands" of old growth were surrounded by a "sea" of young forest, so the birds on the island patches probably are not subject to the adverse influences from outside the forest edge that have been documented for woodlots surrounded by farmland or suburban streets in eastern North America. In fact, Brown-headed Cowbirds were absent from these forests, so nest parasitism was not a factor, and the abundance of ravens, crows, and jays (which are important nest predators) was not significantly different in large and small old-growth patches. Perhaps more important, birds of western mountain forests have evolved in a patchwork

landscape with many forest types. Fire, windstorms, and elevational differences create patches of different types of forest, so many species of western forest birds may have evolved the capability to live and reproduce successfully in relatively small patches of their preferred habitat.[36]

Owls of Old-growth Forest

Although few species of songbirds are closely associated with old-growth
forests, four species of owls appear to depend on this habitat.[37] Flammulated Owls tend to be found in mature or old-growth stands dominated by
ponderosa pine or aspen, while Boreal Owls are found in old stands of
spruce and fir in the northern Rocky Mountains. Spotted and Great Gray
owls also tend to be found in ancient forests. Of these species, the Spotted
Owl has been the most intensively studied because of its key role in legal
and political disputes about harvesting of ancient forests.[38] In the Pacific
Northwest, Spotted Owls are associated with tall coniferous forests that are
more than 250 years old.[39] Spotted Owls equipped with radio transmitters
spend most of their time in forests that have big trees, a complex sequence
of vegetation layers at different heights, and such a dense overstory that relatively little light reaches the forest floor. These old forests provide appropriate feeding, nesting and roosting sites for the owls. They nest high above
the ground in large cavities in trees or on natural platforms provided by broken treetops or intertwined debris on large branches, features that are typically associated with ancient trees. Also, Spotted Owls are sensitive to high
temperatures, and they typically spend the day roosting in the deep shade
of old, multilayered forest in stream bottoms.[40] During hot weather they
may open their talons to expose the pads of their feet, stand up to expose
their legs, and raise their breast and back feathers in order to increase heat
loss from their bodies.[41] They sleep at different levels in the forest on different days, apparently searching out the most favorable temperature.
Thus, old, multilayered forests may be important for daytime roosting as
well as for nesting.

Spotted Owls also hunt primarily in mature and old-growth forests, and
avoid open areas.[42] They feed on a variety of species, but in many forests
their main prey is either the dusky-footed woodrat or the northern flying
squirrel, both of which are typically most abundant in old forests. The large,
fallen trees that litter the floor of old forests provide good habitat for

woodrats and other rodents, and large dead trees provide the cavities and hollows needed by flying squirrels. Also, in contrast to young forests, old-growth forest may provide better perches for hunting owls because of the availability of horizontal branches at many levels in the multilayered vegetation.

Although Spotted Owls occur in second-growth forests (especially second-growth stands with relict clusters of ancient trees), they are substantially more abundant, and are much more likely to nest, in undisturbed areas of old forest.[43] They are threatened not only by direct loss of large amounts of this habitat, but also by the increasing isolation of pairs or small clusters of pairs in relict stands of ancient forest. As more and more of the suitable habitat for Spotted Owls is restricted to islandlike patches surrounded by young, regenerating forest, the number of juvenile owls finding nesting territories may plummet, leading to an irreversible decline in the owl population. Mathematical models based on documented mortality and reproductive rates of Spotted Owls and on information about dispersal distances of young owls indicate that if the frequency of successful dispersal between isolated patches of ancient forest becomes too low, then the population could collapse.[44] Thus, Spotted Owls could disappear from large regions even though there is enough suitable habitat to sustain numerous territorial pairs.

A Seabird That Depends on Ancient Forests

The most unlikely old-growth species is the Marbled Murrelet, a small seabird that nests on the branches of massive conifers in coastal forests from northern California to Alaska. The nesting behavior of this species was an enigma during most of this century. Marbled Murrelets were often seen far inland during the breeding season, but no nests were discovered.[45] As early as 1905 they were observed flying over forests, however. At the foot of Mount Baker in Washington, W. L. Dawson wrote that "having risen before daybreak for an early bird walk, on the morning of May 11, 1905, I heard voices from an invisible party of Marbled Murrelets high in the air as they proceeded down the valley as though to repair to the sea for the day's fishing."[46] Other observers saw birds carrying fish over inland forests, or found eggs, nestlings, or stunned adults with brood patches in the debris left after old-growth forests were logged. Still, no nests were found.

In 1961 the first nest of a Marbled Murrelet was discovered on a lichen-covered branch of a relatively small larch tree in Siberia.[47] The first North American nest was not discovered until 1974, when a tree surgeon cutting dangerous branches from trees in a campground in Big Basin Redwoods State Park in California found a chick squatting on a large, flat branch of an ancient Douglas-fir.[48] The downy chick was sitting in a bowl-like depression on the top of the limb, 150 feet (45 m) above the ground. Like the nests of many seabirds, the murrelet nest was surrounded by a ring of droppings and reeked of fish.

Nearly all of the nests discovered since then have also been on wide branches high above the ground in ancient forest. Marbled Murrelets nest on the ground in treeless or almost treeless areas of northern Alaska, but only a small proportion of the population lives in such areas.[49] The only ground nest found in a forest was nestled in the interlacing roots of a large hemlock tree in southeastern Alaska, but it was perched at the top of a 36-foot (11-meter) cliff, effectively suspended above the forest floor.[50]

The odd (for a seabird) nesting site of the Marbled Murrelet explains its unusual breeding season plumage. Like related species of alcids (auks), Marbled Murrelets are blackish above and white below during the winter. But unlike other species, they molt to a mottled brown plumage, above and below, during the breeding season. Their coloration is effective camouflage for an incubating bird surrounded by the reddish brown bark of a Douglas-fir or hemlock. The related Kittlitz's Murrelet also nests at solitary inland sites rather than in large colonies. It nests on the ground, however, in loose rock on talus slopes, and its breeding plumage is mottled light gray, which matches this background.[51]

Marbled Murrelets are usually found in forests with large trees that are older than 200 years.[52] They nest in sites dominated by Sitka spruce and western hemlock in British Columbia, and coastal redwood in California.[53] Nests are usually located on a large horizontal platform (such as a wide, flat branch) with sufficient moss to accommodate a nest depression, and with cover overhead and easy accessibility for landing birds coming in from the side.[54] On islands in Prince William Sound, Alaska, murrelets nest in trees that are considerably smaller than nest trees farther south in British Columbia or California, but these are still old trees with flat, moss-carpeted branches.[55]

Territorial chases have been observed in murrelet nesting areas, indicating that good nesting sites may be in short supply. Also, birds visit the nest-

FIG. 7.4. Location of a Marbled Murrelet nest in a 200-foot (61-m) high Douglas
Fir in Big Basin Redwoods State Park in California. From Singer et al., 1991, with
permission from *Condor*. Photo by S. Singer and M. Nixon.

ing areas during the winter months, long before the breeding season.[56] Visits to the nesting areas in winter could be important for locating mates, but they might also reflect the importance of defending good nest sites. Territorial chases have occasionally been seen in nesting areas in the fall and winter.

Another indication of the importance of nesting areas is the distribution of Marbled Murrelets in coastal waters during the breeding season.[57] They usually feed in protected waters (bays, fjords, or passageways between islands) near ancient coniferous forest, and they have disappeared in regions of Oregon and California where nearly all of the old-growth forest has been harvested.[58] Most nest sites are within 40 miles (64 km) of the coast, so relatively easily harvested coastal forests are critically important for these seabirds.[59]

Marbled Murrelets are also vulnerable to changes in their coastal feeding areas. Although they sometimes feed on fish in freshwater lakes near the coast, they primarily feed in shallow bays and estuaries where their favored prey, sand lance and surf smelt, are common.[60] Large numbers of Marbled Murrelets have been caught in gill nets in intensively fished coastal waters of British Columbia and Alaska.[61] Also, the concentration of these murrelets near the coast makes them vulnerable to oil spills. The Exxon Valdez spill killed about 8,400 Marbled Murrelets (3 percent of the population of this species in Alaska).[62]

Like many other habitat specialists, the Marbled Murrelet has suffered population declines, and since 1990 it has been listed as a threatened species in British Columbia, Washington, Oregon, and California.[63] Historical records and survey data indicate severe declines in Alaska, British Columbia, Washington, Oregon, and California.[64] Also, there is evidence that reproductive rates are too low to sustain the population. In late summer, only 1–3 percent of the birds in coastal feeding areas have the distinctive juvenal plumage, which is a much lower percentage than in the populations of other well-studied alcids.[65] Using data on the proportion of juveniles in several murrelet populations and estimates of annual mortality, Steven Beissinger developed a model that predicts that the Marbled Murrelet population is declining at a rate of at least 4–6 percent each year.[66] More direct evidence for a recent population decline comes from Christmas Bird Counts for coastal Alaska, which indicate that the population dropped in this region by about 50 percent between 1972 and 1991.[67] Moreover, systematic boat surveys in Clayoquot Sound off Vancouver Island, British Columbia, showed that the murrelet population declined by 40 percent between 1982 and 1993.[68] Substantial amounts of old-growth forest have been harvested in the area around this sound, particularly at low elevations where murrelets nest.

The Marbled Murrelet has specialized requirements both for feeding and nesting, and its population is probably declining for several reasons, but its dependence on old-growth forest makes it particularly vulnerable. Intensive clearcutting of old-growth forest along the coast has directly eliminated nesting habitat in many regions. Clearcutting may also indirectly degrade remaining stands of old-growth habitat by creating forest edge that attracts such predators as Common Ravens and Steller's Jays. Murrelets nesting near clearcuts and other openings may lose their eggs or young to these nest predators.[69] Of the 32 nests that were monitored by researchers at numerous locations along the Pacific Coast between 1974 and 1993, 72 percent were unsuccessful, and the leading cause of nest failure was predation of eggs or chicks. Nest predation may be particularly frequent near campgrounds and picnic areas that attract ravens, crows, and other predators.[70] This problem is especially serious in California, where many of the ancient redwood forests are in state or national parks.

The potential impact of nest predators on Marbled Murrelet was recently tested in an ingenious study in coastal Washington and Oregon.[71] Because real murrelet nests are so difficult to find, the researchers used artificial nests. These were located in typical Marbled Murrelet nesting sites: close to the trunk on large, moss-covered branches protected by overhanging foliage. Plastic eggs that were painted to resemble murrelet eggs, and mounted (stuffed) chicks or live domestic pigeon chicks were placed in these artificial nests. The nests were monitored with video cameras triggered by motion detectors. A high proportion of the nests were visited by predators: 75 percent at the Washington sites and 77 percent at the Oregon sites. Predation rates were higher near campgrounds and towns primarily because of the higher density of American Crows and Steller's Jays at these sites. These results, and the results of observations of real nests, indicate that old-growth areas remote from human activities should be protected because they provide safer nesting sites. Also, crow and jay populations near campgrounds and other centers of human activity might be reduced by cutting off their access to garbage and other sources of food provided by people.

A Habitat Planted by Birds: The Pinyon-Juniper Woodland

Recent excavations at San Josecito Cave in Nuevo León, in the mountains of northern Mexico, yielded bones of extinct horses and ground sloths, and an odd assortment of birds, including both wetland and grassland species.[72] Apparently both wetlands and savanna occurred in this region in the late Pleistocene, 27,000 to 45,000 years ago. Pinyon-juniper woodland must have grown nearby as well, because bones of a Pinyon Jay were discovered at the site. Modern Pinyon Jays are seldom found far from pinyon pines.

The area around San Josecito Cave is now covered with dry desert scrub, and the nearest Pinyon Jays are about a thousand miles to the north, in the pinyon-juniper woodlands of New Mexico. During the Pleistocene, these woodlands apparently moved north and south, and up and down the slopes of mountains, with the cyclical retreat and advance of glaciers. Pinyon pines responded to climatic changes substantially more rapidly than other tree species did, moving northward soon after the ice disappeared.[73] Despite their large, heavy seeds, pinyon pines colonize new areas more quickly than pines that have lightweight, winged seeds that float on the wind. The answer to this enigma is that efficient, long-range seed dispersal is achieved with the help of seed-eating birds, especially Pinyon Jays.

Pinyon Jays depend on the seeds of pinyon pines both to survive the winter and to feed their young. Pinyon pines produce large, nutlike seeds that are easy to harvest and husk. Most species of pines have cones that are protected by spiny, fibrous scales that open only enough so that the wind shakes out the flat, winglike seeds. In contrast, the scales of pinyon pine cones are not armored with spines, and they open far down into a horizontal position, conspicuously displaying the large seeds on upright cones as if they were being served on platters. The round seeds are held in place on the cone scale by membranous tissue until they are plucked off by a seed-eating animal.[74] These seeds have a thin husk, so they are not only accessible, but they are also easily opened and eaten. The seed (and the embryonic plant inside the seed) is ground up and digested when it is eaten by an animal, in contrast to the seeds in the berries of junipers and other plants, which can pass through a bird's digestive tract unharmed. It can germinate only if it escapes being eaten, yet it appears to be well adapted to attract seed-eaters. How could such a defenseless and delicious seed evolve?

Pinyon pine seeds are not dispersed on the wind or in the guts of animals, but instead depend on the peculiar food-storing behavior of Pinyon

FIG. 7.5. Pinyon Jay harvesting seeds from the cones of a pinyon pine. Reprinted from Marzluff and Balda, 1992, by permission of the publisher, Academic Press Limited, London.

Jays and other corvids (species in the Family Corvidae, which includes crows, jays, and nutcrackers). Unlike the squirrels of the Southwest, which tend to concentrate seeds in large, underground storage chambers where they are unlikely to germinate, pinyon jays bury single seeds or small groups of seeds. They frequently bury seeds in open areas, such as recent burns, where sunlight reaches the ground during the winter, melting down the snow pack so that seeds are easier to retrieve. Jays jab a hole into the ground, insert seeds, and cover the site with soil or a leaf.[75] Many of the seeds are eventually unearthed and eaten by the jays that hid them, but some are left in the ground, where they germinate. As a result, openings in the pinyon-ju-

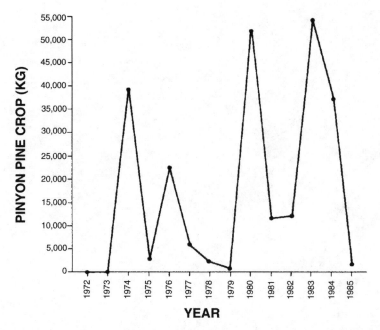

FIG. 7.6. Annual variation in the harvest of pinyon pine seeds by people at Flagstaff, Arizona. Reprinted from Marzluff and Balda, 1992, by permission of the publisher, Academic Press Limited, London.

niper woodland are quickly recolonized by pinyon pines. Flocks of Pinyon Jays may fly up to 7 miles (12 km) to bury seeds in a favorable open site,[76] so it is not surprising that pinyon pines spread rapidly northward as the climate changed at the end of the last glacial period.

Pinyon pines have a dramatic adaptation that ensures that many seeds are stored rather than eaten: seeds are produced synchronously by nearly all trees in a region during particular years, while hardly any seeds are produced in other years. The unpredictability of cone crops in any particular woodland prevents populations of seed-eating insects and rodents from building up. It also ensures that Pinyon Jays and other species that store seeds will have more seeds than they can eat at one time, increasing the chance that most seeds will be buried for later use.

The best evidence for the extreme fluctuations in pinyon pine seed production comes from records of the total weight of seeds harvested commercially for human consumption (as "pine nuts") in Arizona. Virtually no seeds are harvested during some years, while tens of thousands of kilograms are harvested in other years.[77] Massive crops of pinyon seeds occur when trees throughout a large region produce cones in synchrony. Intermediate crops occur when nearly all of the trees in large stretches of woodland produce cones, while other large woodlands in the same region have few or no cones.[78] Pinyon Jays live in permanent flocks of 100–200 birds that respond flexibly to these changes in pinyon seed abundance. Normally these flocks restrict their activities to a large home range that overlaps the home ranges of other flocks. When pinyon cones are scarce in the home range, however, the flocks may become nomadic, ranging over hundreds of miles in their search for pinyon seeds.[79]

When Pinyon Jays find a stand of pinyon pines that are covered with cones, they first eat their fill and then begin holding seeds in their throats for later storage. Pinyon Jays have an expandable esophagus that can accommodate up to 38 seeds.[80] Even if the cones have not yet opened, these jays can hammer through the cone scales to extract the seeds.[81] Each seed is tested by being held or rattled inside the bill, and seeds that are not heavy enough are rejected.[82] Also, the jays discriminate between aborted, inedible seeds (which are yellow) and edible seeds (which are dark brown). Only brown seeds are gathered. After seeds are harvested for a period, the flock begins to assemble in response to special calls ("kaws") given by some of the birds. The flock will then fly to one of the several communal food storage areas in the flock's home range. The birds walk parallel to one another in a roughly straight line, burying single seeds or a small cluster of seeds in the soil beneath the leaf litter. This behavior ensures that the seeds are widely scattered and hidden from rodents and other seed-eaters. Also, if each bird remembers the starting point of its walk, the seeds are easy to find several months later.[83] Not all seeds are recovered, however, and the scattered survivors germinate and grow into the trees of a new pinyon woodland.

Adaptations of Pinyon Jays to Pinyon Pines

Pinyon pines are critically important to Pinyon Jays. Between November and February, as much as 90 percent of the jays' diet consists of pinyon pine seeds, and their behavior and physiology reflects the importance of this food source. For example, while most species of North American birds reproduce on a seasonal schedule in response to the longer days of spring, Pinyon Jays also may nest in late summer or even in the middle of winter if a large crop of pinyon seeds is available to provide food for both the adults and nestlings. Captive birds come into breeding condition (as indicated by an increase in the size of the gonads) in response to longer days in the spring.[84] Gonadal development occurs more quickly, however, if numerous pinyon pine seeds or green cones are placed in their cage. In late summer Pinyon Jays will reproduce in the presence of abundant pinyon seeds even though the days are becoming shorter. This flexible response means that these jays can nest whenever pinyon seeds are abundant. Interestingly, they do not respond in the same way if other types of food are abundant. Also, in the northern part of the Pinyon Jays' range, where winters are colder, nesting does not occur if there is heavy snowfall even when pinyon pine cones are abundant.[85]

Pinyon Jays are primed to nest quickly in response to a massive crop of pine seeds. Pairs mate for life and travel together, remaining near one another as they perch even in fall and winter. If they encounter pinyon pines laden with cones, the male and female rapidly cooperate in building a nest with an outer shell of sticks and a bulky lining of grass.[86] The nest is thicker than the nests of most jays, so it provides good insulation for the eggs and young when nesting occurs in late winter or early spring.[87] The eggs are warmed and protected by the incubating female, which can stay on the nest almost constantly because she is fed by her mate.[88]

Pinyon Jays nest in loose colonies.[89] Dozens of pairs of birds build their nests, incubate eggs, and raise their young in close synchrony. They also fly together to retrieve pinyon seeds from the communal storage areas and to collect insects to feed the young, and they cooperate in driving predators from the colony. The young gather into large creches after leaving the nest. A creche may contain 20 to 60 young jays, all of which beg noisily for food. Parents leave the creche and return to feed the young in a cohesive flock.[90]

The behavior of Pinyon Jays is closely tied to the production of seeds by pinyon pines, but they are not completely dependent on this species. North of the range of the pinyon pine, they gather the seeds of ponderosa pine.[91] Conversely, the pinyon pine depends on Pinyon Jays for seed dispersal over much of its range, but other species (particularly Steller's Jays and Western Scrub-Jays) also gather and bury pinyon pine seeds. The Pinyon Jay is the most effective disperser of these seeds, however.

Pinyon Jay populations declined on Breeding Bird Survey routes in the western United States between 1980 and 1994.[92] If this trend continues and

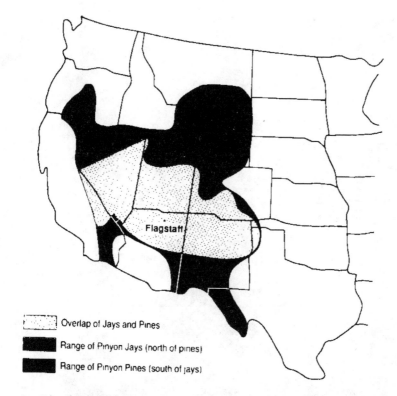

Overlap of Jays and Pines

Range of Pinyon Jays (north of pines)

Range of Pinyon Pines (south of jays)

FIG. 7.7. Distribution of Pinyon Jay and pinyon pine. Reprinted from Marzluff and Balda, 1992, by permission of the publisher Academic Press Limited, London.

Pinyon Jays become rare, then the rate at which pinyon pine seeds germinate would be lower. Aldo Leopold wrote that removing Pinyon Jays from the juniper foothills of the Southwest would result in the "ecological death" of that natural community.[93] Leopold was alluding to the esthetic loss of one of the distinctive features of the pinyon-juniper woodland, but he also may have sensed that the loss of the jay would ultimately change the entire woodland. Pinyon pines, and the animals that depend upon them, would slowly decline throughout much of the Southwest.

Another Pine Seed Specialist: The Clark's Nutcracker

Nutcrackers sometimes descend from the high mountain forests where they normally live to feed on pinyon pine seeds. In many respects, they are even more specialized for feeding on pine seeds than are Pinyon Jays. During the

FIG. 7.8. Clark's Nutcracker feeding on the seeds of limber pine. Photograph by Ronald M. Lanner.

summer and fall they live in the stunted forests of whitebark and limber pines that grow immediately below timberline. Like pinyon pine, these high-altitude pines produce large, edible seeds that are dispersed by animals. Stored seeds sustain nutcrackers during the winter and provide food for their young in the early spring.[94] In the Sierra Nevada in California, nutcrackers spend the autumn storing whitebark pine seeds, then descend to lower-elevation forests, where they feed on the large, winged seeds of Jeffrey pine during the winter and early spring.[95] The nestlings are fed with stored Jeffrey pine seeds during the early spring, but after they fledge, they follow their parents up to the subalpine forests, where they are fed the whitebark pine seeds stored during the preceding autumn. This regular altitudinal migration is timed for harvesting ripe pine seeds at different elevations and storing them for periods when food is scarce.

FIG. 7.9. Interior of whitebark pine cone. Note the large seeds, which are harvested by Clark's Nutcrackers after they remove the fragile cone scales. From *Made for Each Other: A Symbiosis of Birds and Pines* by Ronald M. Lanner. Copyright © 1996 by Ronald M. Lanner. Used by permission of Oxford University Press, Inc.

To release their seeds, the cones of whitebark pines must be opened by an animal. In a ripe cone, the scales open enough to reveal the seeds, but not enough for the seeds to fall out when a cone is shaken by the wind. However, nutcrackers can quickly snip off the scales of a cone with their powerful bills and then remove the seeds. They store these in a flexible pouch on the floor of the mouth below the tongue.[96] This pouch can hold as many as 82 whitebark pine seeds. When it is filled, nutcrackers may fly several miles to an area where the seeds are stored. Typically, one to six seeds are buried in the ground at each storage location.[97] Seed caches are well dispersed, and they are often dug in open areas such as recent burns, so unrecovered seeds have a good chance of germinating and growing.

The importance of nutcrackers in planting whitebark pine is clear from the structure of the forest. Whitebark pines typically grow in discrete clumps of three to seven trees, each clump derived from a nutcracker cache that was never tapped.[98] Similarly, limber pines grow in tight clumps on ridgetops in Utah where nutcrackers store limber pine seeds.[99]

Like Pinyon Jays, Clark's Nutcrackers are adept at finding the thousands of seeds they have buried. Typically, they fly directly to a cache site and retrieve the seeds without any apparent searching.[100] Remarkably, nutcrackers can burrow through the snow or peck through ice to reach stored seeds.[101] Stephen Vander Wall investigated seed storage by Clark's Nutcrackers by creating artificial storage areas in a large flight cage. By monitoring the behavior of color-banded birds, he showed that individuals usually find only the seeds that they had stored, not the seeds stored by other birds, indicating that they use memory and not smell, soil disturbance, or some other sensory clue to find seeds.[102] In a key experiment he permitted two nutcrackers to bury seeds in the flight cage, which was equipped with a large number of landmarks, such as shrubs, rocks, and logs. He then moved all of the landmarks in one half of the flight cage a distance of 20 centimeters to the right. The birds were able to retrieve stored seeds effectively in the half of the flight cage where the landmarks had not moved, but they found few seeds in the half of the enclosure where the landmarks had been shifted. They consistently searched for seeds at a point to the right of the actual storage site, positioning themselves at the distance from obvious landmarks that would have been correct if the landmarks had not been moved. Remarkably, they remembered the positions of storage sites for several months. Based on her observations of nutcrackers in the Sierra Nevada in California, Diana Tomback estimated that each nutcracker buried whitebark pine seeds at about 7,700 separate cache sites, so they must have an impressive capacity to remember landmarks.[103]

Whitebark pine is declining in most parts of the West for a variety of reasons.[104] Ground fires create the open conditions needed for the germination and growth of whitebark pine seedlings, so fire control has resulted in the slow replacement of whitebark pine by trees, such as fir and spruce, that can grow in deep shade. An indirect consequence of fire control is the outbreak of mountain pine beetles that frequently occur in lodgepole pine forests that have been protected from fire. The outbreaks spread upslope and decimate whitebark pine stands. The greatest threat to whitebark pines, however, is white pine blister rust, an introduced fungus that can kill more than 90 percent of the whitebark pines in a stand. Without active intervention, the mountain-top whitebark pine forests may disappear from the West, and with them we will lose the intricate, mutually dependent relationship between whitebark pines and nutcrackers. Although they feed on other types of food, Clark's Nutcrackers may be in trouble if their most important source of food disappears. There is no evidence for a recent decline in this species, however; their populations showed a significant increase on Breeding Bird Survey routes in the western United States between 1980 and 1994.[105]

Western Red Crossbills and the Challenge of Preserving Nomadic Birds

The different call types of Red Crossbill (which may represent different species that are almost identical in appearance) may also be vulnerable because each type depends on large seed crops of a particular species of conifer (see Chapter 6). While four call types are found in the boreal forest of eastern Canada, even more call types have been identified in the western mountains. Craig Benkman studied four of the six call types found in the Pacific Northwest.[106] Each call type has a bill that is the correct size and shape to extract and husk the seeds of a particular species of conifer. The preferred conifers are western hemlock for call type 3, lodgepole pine for call type 5, Douglas-fir for call type 4, and ponderosa pine for call type 2. A less well-studied call type, type 1, may depend on Sitka spruce in the Pacific Northwest. Some of these call types could be in trouble because they depend on older forests (which produce substantially more cones than young forests) and because they may depend on forests in different regions to sustain them in different years. For example, Douglas-fir and ponderosa pine do not produce good seed crops at a particular locality each year, so the crossbill types associated with these conifers must continually move across great distances to find cone-bearing stands of their preferred conifer.

Like the nomadic, acorn- and beechnut-eating Passenger Pigeon in the nineteenth century, the nomadic birds that track conifer seed crops in the western mountains may be more vulnerable than their population sizes and wide distributions might suggest. Clark's Nutcrackers, Pinyon Jays, and Red Crossbills all move either seasonally or nomadically to different elevations in the mountains and to different mountain ranges in search of their preferred types of seeds. To survive and reproduce, each species may need several widely dispersed forests dominated by particular kinds of conifers. A few small preserves will probably not sustain their populations.

Population declines are more difficult to detect in nomadic birds because local declines may be caused by movement to another region rather than a reduction in abundance. Hence, there is a danger that we may not realize that these species are in trouble until they are on the verge of disappearing. We should pay special attention to trends in the abundance of nomadic species in the Breeding Bird Survey results because these provide a continental rather than local perspective on population changes.

Conservation of Forest Birds in the Western Mountains

Breeding Bird Surveys indicate that populations of most species of birds in western forests have not decreased since 1966.[107] Even songbirds that are primarily found in unlogged forests, such as Brown Creeper and Pygmy Nuthatch, have not declined significantly across the entire region. Of the species that decline when the forest is completely or partially cut, only the Golden-crowned Kinglet appeared on a list of species that showed a significant population reduction on Breeding Bird Survey routes throughout the western United States and Canada.[108] Many western forest birds are distributed across a wide range of forest types and successional stages, so they appear to be resistant to forest harvesting and other human activities. Moreover, many of the migratory birds that nest in western forests use disturbed and early successional habitats in the mountains of western Mexico, where they spend the winter.[109] Most of these species do not appear to depend on a specific type of winter habitat that is disappearing, although some do. For example, in western Mexico, American Redstarts are primarily found in mangroves, and Ovenbirds and Least Flycatchers are concentrated in tropical deciduous forest.

There are three groups of species that could potentially be in trouble in the near future, however, because they use specific habitats that could disappear if they are not considered in land management planning. One group consists of species, such as the Black-backed Woodpecker, that cannot use clearcuts as a substitute for burns. These species will probably find adequate habitat if we continue the policy of allowing most natural fires to burn in wilderness areas and national parks. Another group consists of species that require forests that are hundreds of years old. These species will be represented by vulnerable, widely scattered, relict populations if large areas of old-growth are not protected from clearcut logging. Among the old-growth specialists, some species are particularly vulnerable: the Spotted Owl and some other owl species because they need large territories, the Marbled

Murrelet because of its special requirement for nest sites in coastal forests, and the Winter Wren and Brown Creeper because they are concentrated in large expanses of unharvested, old-growth forests. Finally there are some species that potentially may be at risk as much as the old-growth specialists because they require large areas of one particular type of habitat in widely scattered locations. These include the nomadic, seed-eating species that depend on one or a few types of trees for food. They must move over large areas to track "seed crops," and they could not persist in a single large preserve because the trees in one locality would not produce an adequate number of seeds every year. The Pinyon Jay, Clark's Nutcracker, and some types of Red Crossbill are vulnerable because they need to move from seed crop to seed crop in different coniferous woodlands or forests.[110]

CHAPTER 8

Declining Birds of
Southwestern Floodplains

Below them, in the midst of that wavy ocean of sand, was a green thread of verdure and a running stream. This ribbon in the desert seemed no wider than a man could throw a stone,—and it was greener than anything Latour had ever seen, even in his own greenest corner of the Old World. But for the quivering of the hide on his mare's neck and shoulders, he might have thought this a vision, a delusion of thirst.

Running water, clover fields, cottonwoods, acacias, little adobe houses with brilliant gardens, a boy driving a flock of white goats toward the stream,—that was what the young Bishop saw.

—WILLA CATHER, *Death Comes for the Archbishop*

A FTER finding the Greater Roadrunners, Cactus Wrens, and Black-throated Sparrows of the desert upland, birders visiting the Southwest usually concentrate their efforts in the narrow, green strips of woodland and shrub along rivers and creeks. Despite their arid setting, these floodplain or riparian woodlands have a rich diversity of birds.[1] Breeding bird censuses show that they have a density of songbirds as high as in any habitat in temperate North America.[2] The majority of southwestern bird species nest in floodplain woodlands, and many species are almost restricted to this habitat.[3] For example, along the Verde River in Arizona, over 50 percent of the bird species depend on the Fremont cottonwoods of the floodplain for nesting, and along the Gila River of New Mexico, 49 percent of the breeding bird species occur primarily or exclusively in riparian habitats.[4]

More than 95 percent of the streamside habitats in the western United States have been destroyed or degraded, however, so the remaining patches of intact habitat are critically important.[5] Species that depend on floodplain woodlands have declined as their habitats have been destroyed. For exam-

---

PLATE 8. Phainopepla (female on left and male on right) feeding on berries of desert mistletoe in a dry wash in southern Arizona.

FIG. 8.1. Streamside woodland in the Gila Wilderness Area, New Mexico. Photograph by Carmen Burch.

ple, Yellow-billed Cuckoo, Vermilion Flycatcher, and Verdin all showed statistically significant population declines on BBS routes in the Southwest (U.S. Fish and Wildlife Service Region 2) between 1966 and 1994.[6] Another floodplain bird, the southwestern subspecies of the Willow Flycatcher (*Empidonax traillii extimus*), has become so rare that it was classified as an endangered subspecies by the U.S. Fish and Wildlife Service in 1995. In many regions, so much habitat has been lost that recovery of floodplain bird populations will require replanting and nurturing of native woodland vegetation. Without such efforts, the arid lowlands of the West will continue to lose much of their bird diversity.

A Landscape Shaped by Water

In the absence of human intervention, the floodplain woodlands of the deserts are molded by natural disturbances as much as the coniferous forest in the nearby mountains. In this case, however, the primary disturbance is flood, not fire. The streamside woodlands change constantly, and sometimes dramatically, as single trees or entire stands of trees are swept away by spring floods. The floods create steplike terraces, each with a distinctive type of vegetation, and they lay down the moist and fertile silt that serves as a seedbed for new woodland. Superimposed on the abrupt changes caused by flooding are the steady changes caused by the slow, wavelike movement of the stream across the landscape over a period of decades or centuries. At each curve in a river or creek, the woodland on the outside of the arc is steadily undercut and toppled, while new soil is deposited on the inside, providing a seedbed for new woodland. Consequently, old and regenerating woodlands always grow together along the stream. Numerous species of birds are adapted to this continuously shifting landscape, and some species are specifically associated with one of the many distinctive types of floodplain vegetation.

Most boundaries in nature are gradual and subtle, but the boundary between riparian woodland and upland desert can be abrupt, usually spanning less than a meter.[7] On the moist side of this boundary, roots of trees and shrubs can reach down to the water table associated with the stream.[8] Although moisture is the key factor, soil fertility is also important for the growth of tall trees and dense clusters of shrubs along the stream. Flood waters lay down a mixture of soil types (from coarse gravel to fine clay) that is porous enough to retain water and aerate roots.[9] This soil is also rich in

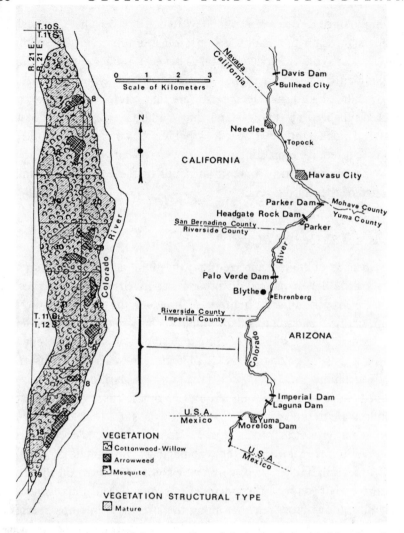

FIG. 8.2. Reconstruction of the vegetation of the lower Colorado River based on surveyor notes from 1879. From Ohmart et al., 1977. Reproduced from USDA Forest Service, Gen. Tech. Rep. RM-43.

organic matter that, as it decomposes to humus, provides nutrients needed for plant growth and acts as a sponge to hold moisture. This rich soil is periodically replenished by floods.

Each spring, snowmelt in the mountains fills the creeks and rivers of the Southwest, which then fan out across the lower part of the floodplains. The area along a stream that is flooded each year is built up and flattened by

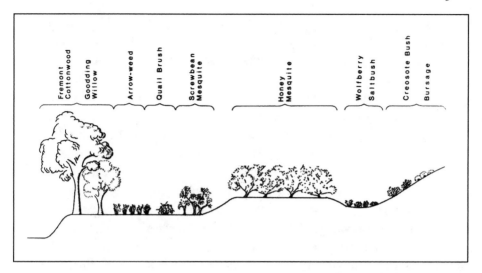

FIG. 8.3. Cross-section of a drainage along a major river or creek, showing the characteristic vegetation of the first and second terraces. Ohmart and Anderson, 1982. From *Reference Handbook on the Deserts of North America*, edited by G. L. Bender. © 1982 by Gordon L. Bender. Reproduced by permission of Greenwood Publishing Group, Westport, Connecticut.

the sediments that the flood waters leave behind. This becomes the first terrace,[10] where sandy soil and the proximity of water favor the growth of tall Fremont cottonwoods and Goodding willows.[11] The outer edge of the first terrace is eroded by the flood waters, forming an abrupt, 3-to-6-foot-high (1–2 m) boundary with the second terrace.[12] The second, and, in some cases, third terraces of western rivers are created by occasional heavy floods that bring moisture and rich alluvial soil to the higher reaches of the floodplains. These higher terraces support honey mesquites and other trees and shrubs that can survive in drier conditions than do cottonwoods and willows. These species still require access to the groundwater near the stream, but they can reach farther down into the soil with long taproots. Different sets of birds live in the tall, multilayered cottonwood-willow woodland of the first terrace and the open honey mesquite woodlands of the second terrace.

Birds of the Floodplain Terraces

The cottonwood and willow woodlands are the tallest and most intricately layered habitats in the desert lowlands. The forest canopy may be 70 feet (24

m) above the ground.[13] Numerous species nest in both the shrub layer and the dense canopy of this habitat. For example, in the lower Colorado River Valley, Yellow-billed Cuckoo, Willow Flycatcher, Vermilion Flycatcher, Yellow Warbler, Yellow-breasted Chat, and Summer Tanager are largely restricted to this habitat. Cavity-nesting birds, such as Elf Owl, Gilded Flicker, Gila Woodpecker, and Brown-crested Flycatcher, also reach their highest densities in floodplain cottonwoods and willows because these large trees provide more potential nest sites. In the Sonoran Desert, however, the giant saguaro cactus also provides nest sites, so many of these cavity-nesting species are common in the upland desert as well as in the streamside woodlands. Gilded Flickers and Gila Woodpeckers excavate cavities in living cacti, and these cavities are later used by other species.

A different set of species dominates the honey mesquite woodlands of the second river terrace. Here one finds Crissal Thrashers, Verdins, Black-tailed Gnatcatchers, Ash-throated Flycatchers, and Lucy's Warblers. The branches of honey mesquites usually support dense clumps of parasitic desert mistletoe that produce numerous bright red berries. These berries are a stable source of food throughout the winter, from November until March, and they are especially abundant in midwinter. They attract a variety of fruit-eating birds, included Phainopeplas, Cedar Waxwings, American Robins, Mountain and Western bluebirds, and Sage Thrashers.

The most distinctive species feeding on mistletoe berries is the Phainopepla, the only representative of the silky flycatcher family (Ptilogonatidae) in the United States. Males are black except for white, flashing wing patches. They have an elegant crest and sleek, shiny plumage. Females are similar, but with gray in place of black.

The Phainopepla is specifically adapted for feeding on mistletoe berries.[14] Whereas the gizzards of most birds are used to break open and grind hard food, the thin-walled gizzard of the Phainopepla is specialized for "skinning" mistletoe berries and extruding them into the intestine. The gizzard is only slightly wider than a single berry, and the skin or exocarp of a berry is easily squeezed off because there is a "slip zone" between the skin and the sticky pulp underneath. As "skinned" berries are extruded into the small intestine, they stick together in a train of 8–16 berries. Trains are separated by a stack of empty, flattened skins. The pulp is digested as the string of berries passes through the intestine, but the seeds are not destroyed. Quick removal of the indigestible outer covering of the berry by the gizzard makes the bird more efficient at obtaining nutrition. This also

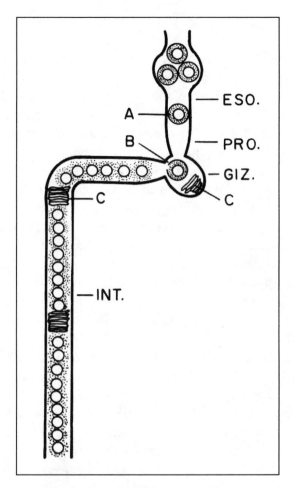

FIG. 8.4. Processing of mistletoe berries in the digestive tract of a Phainopepla. A = whole berry, B = berry with split skin or exocarp, C = packet of skins, ESO. = esophagus, PRO. = proventriculus, GIZ. = gizzard, INT. = intestine. From Walsberg, 1975, with permission from *Condor*.

benefits the plant because the seeds are not ground up but are excreted in long, sticky strings that adhere to the branches of trees, where they may germinate and sink their roots into a new host. In the Sonoran Desert of Arizona, mistletoe fruit is eaten by a variety of birds other than Phainopeplas, especially by Gila Woodpeckers and Northern Mockingbirds, but Phainopeplas feed on this fruit more consistently and spread the seeds to new host plants more dependably than any other species of fruit-eater.[15]

Thus, the desert mistletoe and the Phainopepla have evolved to depend on one another.

Not surprisingly, the life history of Phainopeplas centers on the distribution and fruiting time of mistletoe. In the lower Colorado River Valley, Phainopeplas concentrate in the honey mesquite woodlands and dry washes where mistletoe is most common.[16] They arrive in the valley in the autumn at about the same time as other winter-resident songbirds, and they begin feeding on mistletoe fruit before it is ripe. The abundance of Phainopeplas peaks in midwinter, but after March many birds leave the region as berry concentrations decline. Others remain through the end of June, subsisting on mistletoe berries while they nest, and frequently hiding their nests in dense, globular clumps of mistletoe.[17] In May, the abundance of mistletoe berries declines again, and breeding Phainopeplas shift to feeding on wolfberries. After the breeding season, they leave the region, returning only at the end of the summer. The importance of mistletoe berries was demonstrated during the winters of 1974–75 and 1978–79, when freezing weather destroyed nearly all of these berries and few Phainopeplas successfully bred in the lower Colorado River Valley.[18]

Phainopeplas that nest in the Colorado Desert of California also depend on mistletoe berries during the early spring breeding season.[19] Here they concentrate in dry washes with a fringe of palo verde, catclaw acacia, mesquite, desert willow, and other desert trees, many of which support clumps of desert mistletoe. Breeding occurs here in March and April, when the abundance of both mistletoe berries and flying insects is high. The breeding season is short, with only five weeks between the completion of egg laying and the fledging of the young. Adults subsist mainly on berries, and they feed berries to the young.[20] They also may make quick sally flights from perches to catch flying insects to provision the young.[21]

Soon after Phainopeplas leave the Colorado Desert, they arrive in the floodplain woodlands of chaparral-covered mountains along the coast of southern California. These are probably the same birds that nest in the desert in March and April.[22] If so, then these birds have a second breeding season in a distinctly different habitat. They nest in coastal woodlands between May and July. Instead of feeding on concentrations of long-lasting mistletoe berries, they primarily eat the berries of buckthorn, which is widely scattered in the chaparral above the floodplains. Their nests are hidden high in the oaks and California sycamores of the floodplains.

Descriptions of the behavior of Phainopeplas in the southwestern deserts and in the coastal woodlands of California are so different that they could refer to two different species.[23] In the desert, male and female Phainopeplas defend separate winter territories, ensuring that each individual has a supply of mistletoe berries.[24] In March, however, males and females on adjacent territories unite their territories or females leave their winter area to join a male on his territory. The large breeding territory provides food for both the adults and young. Males advertise for mates with display flights in which they circle 75–300 feet (25–100 m) above the ground. Males on adjacent territories may display high above their territories at the same time, and they are sometimes joined by females. The male builds the nest and uses it as the site for another courtship display, which involves holding nest material and spreading the wings to expose the white wing patches.

In the oak woodlands in coastal canyons, Phainopeplas defend only the area around the nest rather than a large territory.[25] Pairs defend a single tree or even half of a tree from other pairs, and they often nest close together in small colonies of six to eleven pairs. In contrast to the breeding territories in the desert, coastal territories do not include the food supply for the breeding season. Unlike the highly concentrated, long-lasting berries of desert mistletoe, the fruit of buckthorn is too widely scattered and too ephemeral (lasting only a short time on a particular bush) to be effectively defended through the breeding season. Instead, birds often forage together, probably following each other to concentrations of berries or flying insects. Also, the courtship behavior in coastal habitats is distinctly different from what is observed in the desert. Perhaps because of reduced visibility in oak-sycamore woods and the small size of the territory, courtship flights are not used, and the main courtship display is the nest display. Thus, Phainopeplas adjust their territorial and courtship behavior to fit two distinctly different habitats. The same individuals apparently change their behavior as they migrate from one habitat to the other.

If a single population of Phainopeplas depends on floodplain habitat in both coastal California and the interior desert of the Southwest, then it will be vulnerable to habitat degradation in either region. Although Phainopeplas did not show an overall decline on Breeding Bird Survey routes in the Southwest between 1966 and 1994, they declined substantially in many regions in Arizona.[26] Phainopeplas disperse the fruit of mistletoe and many species of shrubs on which other species depend, so a reduction in the population of Phainopeplas could eventually lead to the decline of many plants and fruit-eating animals. If not a "keystone species," the Phainopepla is at least a key species in the riparian woodlands of the Southwest.

Winter Residents and Migrants in Transit

Floodplain habitats are not only important for Phainopeplas and other birds that need them for nesting, but also for a variety of migratory birds that use woodlands during migration or winter. Many of these species nest in the forests of the western mountains or the lowlands of western Canada. Most species stop in riparian woodlands only briefly to refuel as they migrate between their northern breeding areas and their wintering areas in western Mexico or farther south.[27] For example, Willow Flycatchers stop during migration in riparian vegetation (especially stands of willow) along the Rio Grande in New Mexico.[28] Fifty percent of these flycatchers had depleted their fat stores by the time they arrived at the river, showing that the riverside vegetation was essential for feeding and rebuilding fat supplies before the next leg of the migratory journey.

A few species spend the winter along southwestern rivers and streams as long as the weather remains relatively mild. For example, Ruby-crowned Kinglets, Orange-crowned Warblers, and Yellow-rumped Warblers are common in the remnant patches of floodplain woodland in the lower Colorado River Valley during warm winters, but they are much less common during cold winters. Similarly, Yellow-rumped Warblers remain in the riparian woodlands of southeastern Arizona during most winters, but they disappear in midwinter if the weather turns cold and insect densities decline. Presumably they continue traveling south to Mexico or east to the Gulf Coast during harsh winters, a possibility that is strongly suggested by the unusual tendency of captive birds to show nocturnal migratory restlessness well after the normal migratory period.[29]

Some winter residents require specific vegetation types on floodplains, so destruction of these habitats could affect populations of species that nest hundreds of miles to the north. For example, Orange-crowned Warblers and White-crowned Sparrows tend to concentrate in quail bush and saltbush, shrubs that are most abundant in the honey mesquite woodland.[30] In contrast, Sage Sparrows primarily spend the winter in thick patches of inkweed bushes.

Floodplains in Mountain Canyons

In southeastern Arizona, creeks flow out of the Huachuca Mountains and across the desert grassland. In the lowlands, these creeks are fringed with cottonwoods that host many of the same species found along the lower Colorado River: Yellow-billed Cuckoo, Vermilion Flycatcher, Yellow Warbler, and Yellow-breasted Chat. Farther upstream, where the creeks flow through steep-walled canyons in the mountains, the floodplain woods are dominated by Arizona sycamores and big-tooth maples, and the slopes above the floodplain are covered with oaks. These canyon woods have an exceptionally distinctive bird fauna, with species such as the Magnificent Hummingbird, Elegant Trogon, Strickland's Woodpecker, Dusky-capped Flycatcher, and Painted Redstart that are largely restricted to canyon bottoms in Mexico and adjacent areas of Arizona and New Mexico.[31]

The willow, aspen, and birch woods that border canyon streams in the Toiyabe Mountains of Nevada also support bird species that are not found in the surrounding uplands, but, unlike many of the Arizona canyon species, these birds are widely distributed in similar habitat throughout the Rocky Mountains: Broad-tailed Hummingbirds, Cordilleran Flycatchers, Warbling Vireos, and Lazuli Buntings.[32]

Like the lowland floodplain woodlands, the canyon bottoms support bird species that do not live in the surrounding uplands, and they are narrow habitats that are exceedingly vulnerable to disturbance. Road building and house construction pose a special threat to some of these canyon-bottom woodlands.

Loss of Woodlands Along the Colorado River

The floodplain woodlands of the arid West are among the most threatened habitats in North America. In the 1600s, an estimated 4,800 acres (2,000 ha) of the floodplain of the lower Colorado was covered with Fremont cottonwood, but only about 240 acres (100 ha) remained in 1990. The loss of floodplain woodlands along the lower Colorado River Valley was particularly severe, but the history of other western watersheds is similar. For example, only 29 acres (12 ha) of the once extensive mature cottonwood-willow woodland remain along the lower Rio Grande.[33]

In the nineteenth century, nearly all of the large cottonwoods and willows along the lower Colorado River were cut to provide wood to fuel steamboats.[34] This was a transient disturbance, however. The quick-growing trees of the floodplain recovered in much the way that they recover from a catastrophic flood or other natural disturbances. A more serious threat was initiated with the construction of Laguna Dam in 1907, the first of a series of dams on the Colorado River. With the construction of the massive Hoover Dam in 1936, the river was largely brought under human control, and regular seasonal flooding of the riverbanks below the dam ceased. Dams eventually resulted, either directly or indirectly, in the destruction of nearly all natural floodplain habitat along the lower Colorado. Some wood-lands were inundated by dams, and the water level of the reservoirs fluctu-ated too frequently to permit a stable woodland to grow along the reservoir banks. The water supply of the reservoirs was available for irrigation, so much of the woodland below the reservoirs was cleared for agriculture. Without seasonal flooding, backwaters and oxbow lakes dried out, killing the adjacent streamside vegetation. Also, seasonal floods no longer created silt banks that serve as a seedbed for new woodland. Along the undammed upper Gila River, young stands of cottonwood and willow are restricted to recently formed sand bars.[35] Sand bars coalesce over time, so saplings are typically found on clusters of sand bars. Typically these clusters fuse with each other and with the riverbank before the cottonwoods and willows ma-ture. The sand bars are fortified and protected from damaging floods by seepwillow, which grows on most sand bars. Seedlings are absent from ma-ture stands, indicating that sand bars are essential for reproduction. Below the large dams of the Colorado River, however, the water runs clear and there are no seasonal floods, so few sand bars form and few cottonwood and willow seedlings germinate.

To compound these problems, much of the natural floodplain vegeta-tion along the lower Colorado River has been replaced by saltcedar, an in-troduced shrub that grows in dense, homogeneous stands.[36] Water that sits in reservoirs or flows across fields becomes salty because evaporation oc-curs quickly in the dry, hot southwestern climate, leaving salts behind. Reservoir and irrigation water drain into rivers, increasing the salt content of the river water and the salinity of the soil along the banks. Unlike most native plants, saltcedar thrives in saline soil and quickly colonizes areas where the native vegetation has been removed. Once established, it tends to spread. Saltcedar loses its leaves in winter, and dense stands of this shrub usually burn after dead leaves have accumulated on the ground for ten years. Saltcedar quickly grows back after a fire, but many native trees and shrubs are killed. The result is a steady increase in the proportion of saltcedar until almost pure stands of this plant dominate the banks of rivers and creeks.

Few species of birds live in uniform stands of saltcedar.[37] Saltcedar can provide good nesting habitat for White-winged Doves, but it has a much lower diversity of birds than native floodplain woodlands.[38] This is particularly true during winter, when saltcedar is leafless. Blue Grosbeaks and Lucy's Warblers are among the few birds that regularly nest in pure stands of saltcedar along the lower Colorado River. Remarkably, some of the species that are restricted to native woodlands in the lower Colorado River are also found in pure saltcedar in the river valleys of the Pecos River and Rio Grande. Yellow-billed Cuckoo, Bell's Vireo, and Summer Tanagers are on the verge of disappearing along the lower Colorado River, where they are restricted to remnant patches of cottonwood-willow woodland, but are doing much better along the Pecos River and Rio Grande, where they use the abundant saltcedar.[39]

Threats to the Riparian Woodlands

All of the major rivers in the southwestern desert are intensively managed with systems of dams.[40] In many cases, the river channel has been straightened and stabilized with a lining of rocks to speed the flow of water and prevent the natural meandering that occurs in rivers that flow through flat floodplains.[41] Moreover, the floodplains (particularly above the second levee) are often cleared for agriculture. A major tributary of the Colorado River, the Gila River, is now dry for most of its length because of the construction of Coolidge Dam.[42] All of the floodplain woodlands have died along this river except trees supported by the discharge of sewage effluent from Phoenix or seepage from large irrigation projects. Similarly, the lower Salt River has dried up as a result of dam construction, and the large reservoirs along this river flooded large areas of floodplain woodland.[43]

Riparian woodland has also been lost in the lowlands of California. Along the rivers of the Sacramento Valley, cottonwood-willow woodlands support a diverse community of birds, including many species found in similar habitats in the interior desert. Red-shouldered Hawk, Yellow-billed Cuckoo, Willow Flycatcher, Bell's Vireo, Yellow-breasted Chat, and Blue Grosbeak are restricted to this habitat.[44] Many of these species are declining because of destruction of the floodplain woodlands, which were reduced from 775,000 acres (315,000 ha) to 12,000 acres (4,900 ha) by the

1970s. The Yellow-billed Cuckoo population, in particular, has dropped to a perilously low level.

In some regions in Arizona and New Mexico, riparian vegetation has been cleared along rivers to reduce the amount of water lost through evaporation from plant leaves. Along the Pecos River in New Mexico, for example, 50,000 acres (20,000 ha) of streamside vegetation was cleared specifically to reduce water loss from plants.[45] Not surprisingly, removal of trees to control evaporation results in a substantial reduction in the number of breeding birds.[46]

A more insidious threat to floodplain habitats is the lowering of the water table brought on by the pumping of hundreds of wells near developed areas. In Arizona, lowering of the aquifer killed the honey mesquite woodlands at San Xavier, south of Tucson, and pumps in rapidly expanding Fort Huachuca threaten the protected riparian woodland along the upper San Pedro River.[47]

Effect of Grazing on Floodplain Woodlands

The floodplains in the Southwest must have been subject to grazing and browsing by large mammals during the Pleistocene, before most large mammals in North America became extinct. A glimpse of this lost fauna is provided by the remains in an ancient California Condor nesting site in Sandblast Cave, which is located high on a vertical cliff face in the Grand Canyon.[48] Here, 10,000 to 13,000 years ago, condors brought food to their young. The cave deposits contain the skeletons of condor fledglings and bone fragments from their food, which included a variety of such large mammals as horses, bison, mammoths, and camels. Presumably at least some of these large herbivores were feeding in the cottonwood and willow woodlands along the Colorado River, and the birds of these woodlands must have lived in a habitat modified by large grazers and browsers.

Although floodplain woodlands may have originally hosted a set of grazers and browsers, they clearly did not evolve with the intensity of grazing that now occurs in many western rangelands. In arid areas, cattle spend a disproportionate amount of their time in floodplain habitats, which provide shade and water as well as food. They spend 5 to 30 times longer in floodplain habitats than in surrounding upland habitats.[49] The continual pressure of grazing cattle can convert a biologically rich woodland into an eroded scrubland. Grazing is one of the greatest long-term threats to the remaining patches of floodplain woodland.[50]

Cattle feed upon and trample the herbaceous understory, destroying cottonwood seedlings and ultimately preventing the regeneration of the woodland overstory.[51] Particularly when cattle are permitted to use floodplains throughout the summer growing season, the woodland may slowly change to a scrubland dominated by annual grasses and shrubs[52] or, in some cases, saltcedar, which can grow into a dense tangle that cattle do not enter.[53] When ranges are overstocked, cattle may eliminate nearly 100 percent of the understory in riparian woodlands, removing all palatable vegetation lower than about 5 feet (1.5 m).[54] Typically only old trees remain along heavily grazed floodplains, and, as they die, a woodland that once supported a diversity of streamside birds is replaced with an almost bare, deeply eroded landscape used by few species of birds.

Removal of ground cover and trampling by cattle result in severe erosion. The little rain that comes to southwestern deserts usually occurs in downpours. Soil washes freely from the banks of heavily grazed banks, widening the stream channel and converting small tributaries into deep, steep-walled arroyos. Water rushes more quickly and forcefully into the stream channel, cutting the channel down to bedrock or large cobble. The stream bed may sink so deep that the roots of streamside trees and shrubs are stranded above the water table, causing them to slowly die. They are replaced with upland species, such as juniper and sage.[55] Moreover, the rapid flow of water off eroded banks produces intense floods that may rip out riparian woodland downstream from a heavily grazed area.

On smaller streams and creeks, the elimination of beaver by trappers may have contributed to the deep erosion of stream channels.[56] Beaver dams create ponds that reduce channel erosion by slowing the velocity of the stream while increasing the deposition of sediment.

Heavy grazing can ultimately destroy the woodland on which many floodplain birds depend, but the short-term effects of grazing on birds are more subtle. Victoria Saab and her colleagues summarized the results of nine studies of the effects of grazing on birds in floodplain areas from Oregon to Nevada and Colorado.[57] Most of the study areas were cottonwood-willow woodlands, but some were dominated by aspen or mixtures of cottonwood and pine. Of the 68 species that were assessed in at least two of these studies, 46 percent were less abundant in relatively heavily grazed woodlands. Common Yellowthroats and Willow Flycatchers consistently declined in more heavily grazed plots. Not surprisingly, species that forage and nest in the canopy generally were not affected by grazing (at least in the short term), while many species that depend on the understory declined. Some species, such as Killdeer and American Robin, were more abundant in grazed areas, but these species do not depend on floodplain woodlands.

The sensitivity of floodplain specialists to grazing is illustrated by a study of winter-grazed and summer-grazed willow shrublands along the Illinois River in northern Colorado.[58] Three species that were found only in shrub-willow habitats in this region (Willow Flycatcher, Lincoln's Sparrow, and White-crowned Sparrow) were missing from the summer pastures, where the vegetation is severely thinned by grazing, but were common in the denser willow clumps of the winter-grazed pasture. More generalized species such as Yellow Warbler, Savannah Sparrow, and Song Sparrow were common in both types of pastures but were also found in a variety of other floodplain habitats in the region.

Restoring Floodplain Habitats

Restoration of heavily damaged or destroyed vegetation can be costly and time-consuming, but it is often the only option available for preserving specialized species that depend on an endangered habitat. Restoration efforts have been directed at habitats where only small relict patches have been saved from destruction, such as tallgrass prairies or oak savannas in the Midwest, or where most of the habitat has been heavily modified by human activity, which is the case with New England salt marshes. In most areas of the Southwest, only relict patches of floodplain woodland remain, and most of these are heavily modified by grazing and flood control. Many bird species that are essentially restricted to riparian woodlands in the Southwest will disappear from the region without large-scale restoration efforts.

If a floodplain woodland has been intensively grazed and mature trees are still present, then restoration is straightforward. A few years after the removal of livestock, the shrub layer recovers and young trees begin growing. For example, both the understory and the bird life recovered four years after grazing was eliminated from a section of the San Pedro River in Arizona.[59] Bell's Vireo, Yellow Warbler, Common Yellowthroat, Yellow-breasted Chat, Summer Tanager, and Song Sparrow all at least doubled in density. Similarly, along Mahogany Creek in Nevada, a dense understory and numerous young cottonwoods and willows returned only ten years after grazing ceased.[60]

Recovery of badly overgrazed floodplains takes much longer if grazing

FIG. 8.5. Top, cattle grazing along San Pedro River in 1985. Below, the same site in 1995, eight and a half years after a restoration plan was instituted by the Bureau of Land Management. In addition to removal of cattle from the riparian wood-land, the plan also eliminated off-road vehicles, sand and gravel operations, and agricultural pumping in abandoned agricultural fields. Photographs from BLM files.

is carefully managed rather than eliminated. The impact of livestock can be reduced by grazing only during the late fall or winter or briefly during the spring, ensuring that some cover remains to reduce erosion.[61] Summer grazing tends to reduce the density of the vegetation, which depresses the abundance of many species of birds.[62]

In some desert streamside habitats, even brief grazing by cattle can cause serious damage, and protection of the riparian woodland requires exclusion of cattle. Based on his efforts to protect streamside woodlands in Coronado National Forest in Arizona, Charles Ames concluded that even a few days of grazing can eliminate tree seedlings, reducing the rate of reproduction of the dominant floodplain trees.[63] According to Ames, "It's like having the milk cow get in the garden for one night." In most parts of the Southwest, the remaining patches of cottonwood-willow and honey mesquite woodlands have become so rare that they should be completely protected from livestock grazing.[64]

Where floodplain woodland has been destroyed, so that there is no local source of seeds for native trees and shrubs, then removal and exclusion of livestock is not sufficient to restore the natural habitat. A carefully planned and expensive program of active restoration is then required.

Some of the most successful efforts at restoration of habitats for floodplain birds are under way in the lower Colorado River Valley, which has suffered almost complete loss of natural floodplain habitats.[65] The goals for restoration were set after a long-term study of the habitat requirements of floodplain birds. Under the leadership of Robert Ohmart and Bertin Anderson, intensive, year-round research on birds along the river and its tributaries between 1973 and 1977 prepared the way for efforts to re-create natural habitats in the late 1970s. Planting of native species was carefully planned, beginning with an assessment of the soil type (sandy or clay), soil salinity, and depth of the water table. Early attempts to reestablish native plants provided a number of lessons. When the soil is saline, native trees will not grow well, but native shrubs such as quail bush, inkweed, and wolfberry (all of which provide food for floodplain birds) may thrive. Also, restoration sites must be carefully prepared before seedlings are planted. For example, one site was covered with introduced saltcedar. During the first year of treatment, the saltcedar were bulldozed and burned, and then the roots were pulled out mechanically.[66] Despite this systematic assault, saltcedar resprouted within a year, and the sprouts had to be chopped down. Native quail bush was then planted, and it spread enough to prevent saltcedar from reoccupying the site. Mesquite, cottonwood, and willow trees were then planted successfully.

Planted tree seedlings survive better and grow faster if the site is prepared by tilling and if the plants are watered.[67] Drip irrigation, in which water is dripped at the base of each tree through a system of small plastic tubes, was used effectively in the lower Colorado River project. In another restoration project, on the floodplain of the Rio Grande south of Albuquerque, New Mexico, stem cuttings from large cottonwood and willow saplings were planted by drilling holes about 7–12 feet (2–3 m) deep, down to the water table.[68] These stems sprouted and grew well except at sites where they were cut down by beaver or killed by flooding that lasted for more than three weeks. These problems show that restoration sites must be selected carefully.

Plots in the lower Colorado River project were planted specifically to create the complexity and diversity of vegetation needed by floodplain birds.[69] As predicted, particular types of plants attracted birds that had been absent or rare at the site. Quail bush was used by Crissal Thrashers, Verdins, and wintering Blue-gray Gnatcatchers and Orange-crowned Warblers, and inkweed attracted wintering Sage Sparrows. Once the cottonwoods were several years old, several were girdled to create dead trees for cavity-nesting birds. Ladder-backed and Gila woodpeckers soon excavated nesting holes in these snags. Brown-crested and Ash-throated flycatchers were also attracted to the site once nest cavities were available.

One of the goals was to provide habitat for Yellow-billed Cuckoos and Summer Tanagers, which had almost disappeared from the region. As predicted, both species appeared at restored sites after trees attained a height of 27 feet (9 m). However, neither has established a permanent breeding population.[70] The 72-acre (30-ha) site may be too small, and too isolated from other floodplain woodlands, to support these species.

Those trying to restore floodplain woods might benefit from learning more about the ancient farming traditions of the Southwest. Native American and Spanish American farmers planted cottonwoods and mesquites to control the effects of seasonal floods.[71] In the Sonoran Desert of Mexico, farmers still plant stem cuttings of cottonwoods and willows between their floodplain fields and the stream channel. After these grow into trees, the farmers build a fence of interwoven branches between the trunks, creating a permeable barrier that reduces the force of flood waters. The slower-moving water causes less soil erosion and drops a thicker layer of nutrient-rich

FIG. 8.6. Restoration of riparian woodland along the Colorado River at a site south of Palo Verde, California. Trees were planted in holes augered in the sand of this dredge spoil site, and they were irrigated with drip lines. Top, the restoration site in 1979, soon after trees were planted. Below, the same site in 1980; opposite, the site in 1986. Photographs by Robert Ohmart.

FIG. 8.6. (continued)

sediment on the fields. Many "natural" cottonwood stands in the south-western United States may have started out as field-edge plantings of this sort.

When floodplain woodlands are planted, they can be designed to provide habitat for specific species of birds. However, the process is laborious and expensive. The effort required to replicate this habitat dramatizes the importance of protecting the intact remnants of floodplain woodlands in the Southwest.[72] These centers of biological diversity are exceedingly difficult to replace, and the challenge of restoration becomes greater as more and more of the original habitat is degraded or destroyed. Although grazing has degraded many of these woodlands, the greatest immediate threats to their survival are the damming of rivers, the diversion of water for irrigation and urban areas, and the lowering of water tables.

Conservation of Birds in Desert Streamsides

Intact streamside woodlands have become rare in the Southwest, so they should be protected. Floodplains that still undergo a natural flooding cycle

FIG. 8.7. Aerial view of streamside woodland along the San Pedro River in Arizona. Photograph by Adriel Heisey.

are particularly important because floods create the conditions needed for the growth of tree and shrub seedlings. Most floodplain habitats require more than protection, however: they also require active management and restoration. This may be as simple as reducing or eliminating livestock grazing, or as complex as planting and nurturing all of the dominant native plants of the woodland. These efforts are clearly worthwhile: without them, we are in danger of losing much of the natural diversity of the Southwest, including some of the most distinctive and beautiful birds of the region.

CHAPTER 9

Red-cockaded Woodpeckers and the Longleaf Pine Woodland

This plain is mostly a forest of the great long-leaved pine (P. palustris Linn.), the earth covered with grass, interspersed with an infinite variety of herbaceous plants, and embellished with extensive savannas, always green, sparkling with ponds of water.

—WILLIAM BARTRAM, *Travels through North and South Carolina, Georgia, East and West Florida* (1791)

B EFORE European settlement, forests of tall, straight longleaf pine covered much of the coastal plain from southeastern Virginia south to central Florida and west to East Texas. They grew on sandy, dry soil, interspersed with the hardwood forests growing in river valleys and in uplands with moist, well-drained loams.[1] In most of the Southeast, longleaf pines grew in an open woodland or savanna, with ancient pines scattered regularly across an expanse of low wiregrass or bluestem. The canopy of these savannas was pure longleaf pine or longleaf pine mixed with other fire-resistant trees, such as loblolly and shortleaf pines. The "bilayered" structure of this woodland—low grass and tall, ancient pines, with little or no shrub layer—was its most characteristic feature.[2] In 1540, Hernando De Soto and his army crossed long, monotonous stretches of this type of pine savanna in Georgia; immense pines were scattered across an otherwise open grassland.[3]

Despite its structural simplicity, this woodland supports a remarkable diversity of understory plants.[4] It also supports a mixture of open-country and woodland birds, including the highly specialized Bachman's Sparrow and Red-cockaded Woodpecker. Unfortunately, however, only tiny remnants of this ecosystem remain, and some of its most distinctive species have become

PLATE 9. Red-cockaded Woodpeckers at a nest cavity in a live pine. An immature male (partially behind tree) is helping the pair feed the nestlings.

FIG. 9.1. The Wade Tract, an old-growth, longleaf pine forest in Thomas County, Georgia. This forest is privately owned but is managed by the Tall Timbers Research Station. Photograph by Todd Engstrom.

rare. To see open expanses of longleaf forest one must travel to such sites as Okefenokee National Wildlife Refuge in Georgia, Apalachicola National Forest in northern Florida, and the Vernon Ranger District of the Kisatchie National Forest in Louisiana. Beginning with the first English settlements in Virginia in the early 1600s, longleaf pine was exploited for tar, pitch, rosin, and turpentine, which were collectively called naval stores because they were used in building and maintaining ships.[5] Tar and pitch were extracted by burning pine wood in kilns. The bark of living trees was cut to collect resin that was distilled to produce turpentine and rosin. Scoring the bark did not kill the trees, but it left scars that made the trees more susceptible to fire. Many longleaf pine woodlands were also cleared for agriculture, and most of the trees that were not killed for naval stores or agricultural clearing were harvested between 1870 and 1930, when a commercial logging boom occurred in the Southeast. Nearly all of the remaining ancient longleaf pines were cut during this period. Even worse, it soon became apparent that the cutover areas grew back into other types of forest. Longleaf pine seldom regenerated after ancient stands were removed.

On the basis of historical analysis of the distribution of woodlands dominated by longleaf pine, Cecil Frost estimates that only about 3 percent of the original 92 million acres (37 million ha) of the longleaf pine savanna are intact. This makes the longleaf pine woodland one of the most critically endangered ecosystems in North America.[6] Not surprisingly, a large number of species of plants and animals associated with the longleaf pine-wire-grass ecosystem are now rare.[7] This includes 187 species of plants[8] and 2 species of birds: the Red-cockaded Woodpecker, which is listed as endangered, and the Bachman's Sparrow, which is proposed for federal listing.

To save these species from extinction and to maintain the longleaf pine woodland as a functioning ecosystem, it is not sufficient to set boundaries or even tall fences around relict stands of ancient longleaf pine. Unless we understand and duplicate the ecological processes that generated the old-growth longleaf pine woodland, the ecosystem will soon disappear even in the few places where it has been protected.

Loss of Fire and the Decline of the Pine Woodland

The openness of this ecosystem depends on fire, and fire suppression is the greatest current threat to the remaining longleaf pine woodlands. Without periodic ground fires, the open ground cover (typically wiregrass or bluestem) is slowly replaced by a dense understory of young trees, such as slash pine, loblolly pine, and various species of oaks and other hardwoods.[9] Eventually the open woodland grows into a dense forest. Fire is still suppressed in an estimated 74 percent of the remaining longleaf pine woodlands, so the encroachment of more fire-sensitive trees is a major threat.

Originally these pine woodlands burned every one to five years.[10] The flat coastal plain had few natural barriers that would stop a fire from spreading, so ground fires could sweep across huge areas, burning the wiregrass understory and killing the seedlings of most trees.[11] Longleaf pine seedlings and saplings often survived these fires, however. Young pines typically remain in a short "grass stage" for years until they survive a fire; then they grow upward rapidly, becoming large enough to easily survive the next ground fire. This spurt of growth is sustained by an extensive root system laid down during the grass stage.

FIG. 9.2. Ground fire in longleaf pine forest in Georgia. Photograph by Todd Eng-strom.

Longleaf pine seedlings have a number of adaptations that increase their chances of surviving a fire. The terminal bud (the growing point of the young stem) is protected by a dense layer of needles.[12] When these burn, the water in the needles vaporizes upward, deflecting the heat of the fire away from the bud. Also, the bud itself is protected by silvery down, and the stem is insulated with a thick, corky layer.[13] Finally, the seedling has an excep-tionally large taproot that stores the food needed for a rapid spurt of growth after a fire.

Most of the seedlings of other trees are killed by the ground fires that trigger the growth of longleaf pine seedlings. In many regions, wiregrass is the main fuel for ground fires, so ultimately the regeneration of longleaf pine depends on wiregrass. It is no coincidence that these two species are often found together. The open, two-layered structure of the longleaf pine forest is a direct product of frequent ground fires.

After European settlement, agricultural clearings and roads created new barriers to the spread of ground fires, but the longleaf woodlands were maintained by artificial burning.[14] Farmers set fire to the wiregrass in early spring to "green up" the forage for cattle.[15] As cattle-raising declined in the second half of the nineteenth century, however, this practice became less fre-

FIG. 9.3. Longleaf pine sapling after a fire. The long bud is new growth that oc-curred after the fire. Photograph by Todd Engstrom.

quent. After 1910, artificial burning was replaced with suppression of wild-fires. Preventing and suppressing forest fires became a priority for foresters. Ironically, one of the goals of fire suppression was to permit long-leaf pine regeneration. When old-growth longleaf pine was harvested, other species of trees grew in its place. In the late 1800s, foresters attributed the low rate of survival of longleaf pine seedlings to the fires set by farmers, so the elim-ination of fire was the obvious solution.[16] Regular burning in early spring did, in fact, kill most pine seedlings, which never became large enough to survive a fire.[17] However, Roland Harper was one of the first botanists to rec-ognize that longleaf pine ultimately depends on fire.[18] He noted that every longleaf pine forest shows charred bark and other evidence of recent fires.

With the recognition that the longleaf pine ecosystem depends on fire, many stands (including extensive stands in some national forests) were managed with controlled burning. Recently, however, it has become clear that it is not sufficient to simply burn the understory. It is also important to simulate the type of fires that burned across the coastal plains before human settlement. Originally these were caused by lightning, which is much more frequent in summer than in winter. In Florida, lightning strikes are most frequent in July and August, but lightning-caused fires are most frequent in May, when the vegetation is normally much drier.[19] Foresters have traditionally burned the ground cover in longleaf pine woods in fall or winter, when predictable winds and relatively moist, cool conditions make fires easier to control.[20] There was also a fear that fires set during the growing season were more likely to damage longleaf pine seedlings. However, in a carefully planned experiment, different plots in longleaf pine woodland in St. Marks National Wildlife Refuge in Florida were burned at eight different times, at six-week intervals throughout the year.[21] The season of burning had little effect on mortality or growth of young longleaf pine seedlings. Fires set during the early growing season killed more young oaks than fires set at other times of the year, however, so they are more likely to produce a pine woodland with a grassy ground cover and little oak. Also, although the season of burning does not affect the density of the ground cover, it does affect flowering and seed production. Wiregrass and other common grasses of the longleaf pine ecosystem produce many more flowers after fires in the growing season (particularly the early growing season) than after winter fires.[22] This may have little immediate impact because these grasses spread primarily by vegetative reproduction, but it may have long-term effects. Without good seed production, grasses may not quickly recolonize spots where the grass has been killed by particularly severe burns.[23]

The frequency of fires is also important. Annual burning will kill longleaf seedlings before they have grown large enough to be fire resistant.[24] If ground fires occur every two to five years, then young pines will survive but most young oaks will be killed, resulting in an open pine woodland.[25] If the interval between fires is longer, then young oaks have a high survival rate and the site is likely to become an oak woodland. At the time of European settlement, longleaf pine woodlands were probably restricted to areas where lightning fires occurred every few years, or where Native Americans frequently burned the woodlands.[26] Lightning fires can no longer be sustained in this ecosystem because the spread of fires is severely limited by roads, fields, and fire suppression. Consequently, we can maintain intact longleaf pine ecosystems only by using prescribed burns that simulate the timing and frequency of lightning fires.

When ground fires disappear from the pine savanna, many of the most distinctive birds disappear. When fire was excluded from a second-growth pine savanna in Florida, Eastern Kingbirds, Loggerhead Shrikes, Blue Grosbeaks, and Bachman's Sparrows disappeared within five years.[27] The rapid growth of young water oaks and other hardwoods between the pines soon made the habitat unsuitable for these open-country birds. They were replaced by such woodland species as Red-eyed Vireo, Wood Thrush, and Hooded Warbler. Similarly, pine savanna in the Ouachita Mountains of Arkansas supported many species that were absent or substantially less common in pine woodland that had been protected from fires and consequently had been invaded by numerous hardwood trees. When pine savannas were restored by removing hardwood trees and periodically burning the understory, they supported fewer forest-interior birds than unrestored stands, but more Northern Bobwhites, Brown-headed Nuthatches, Chipping Sparrows, Indigo Buntings, Prairie Warblers, and Pine Warblers.[28] Many of these species are declining, so it is important to maintain open pine savanna to sustain their populations. Two other species, the Red-cockaded Woodpecker and the Bachman's Sparrow, were also restricted to restored pine savanna, but they were too infrequent in the Arkansas study areas to permit firm conclusions about their habitat needs. Numerous other studies show, however, that these two species disappear when fire is removed from the pine savanna.

The Bachman's Sparrow (which was once called the Pine Woods Sparrow) is especially closely associated with the grassy expanses of wiregrass or bluestem in mature pine savannas.[29] It primarily feeds on seeds on the ground, and its nesting territories are usually in areas with a dense ground cover of grass and herbs, with relatively few shrubs.[30] Given the disappearance of pine savanna from most of the Southeast, it is fortunate that Bachman's Sparrows also use other grassy habitats, such as old fields, recent clearcuts, powerline rights-of-way, and the open grasslands (or glades) of the Missouri Ozarks.[31] However, pine savanna may be their preferred habitat.

When different habitats in Francis Marion National Forest in South Carolina were surveyed in 1988, Bachman's Sparrows were primarily found in relatively old stands of longleaf pine. These stands were managed with prescribed burns, which were repeated every three to five years.[32] Bachman's

Sparrows were infrequent in clearcuts. In September 1989, however, the national forest suffered a direct hit from Hurricane Hugo, which blew down large expanses of mature longleaf pine woodland.[33] The downed trees and the disturbance of the ground cover by equipment used in salvage logging apparently made these areas unsuitable for Bachman's Sparrows, and surveys in 1990 showed that they were absent from the damaged pine stands but were found in clearcuts where they had previously been absent. One section of the forest had little hurricane damage, however, and here the sparrows were found in mature longleaf pine and not in clearcuts.

In parts of the South where no longleaf pine savanna remains, Bachman's Sparrows can be found in clearcuts and other open habitats, but it is usually not known whether their rates of nest success and survival are comparable to those of Bachman's Sparrows nesting in pine savannas. One study in Arkansas showed that nest success was similar for birds nesting in recent clearcuts (less than four years old) and in shortleaf and loblolly pine savanna.[34] It is interesting, however, that when given a choice between clearcuts and mature longleaf pine savannas in Francis Marion National Forest, they nest in the savannas rather than in clearcuts. Moreover, clearcuts provide appropriate habitat only for a few years, until trees and shrubs become too dense, while pine savannas can provide a permanent habitat if they burn periodically.[35] Not surprisingly, the Bachman's Sparrow population declined substantially between the 1930s and the 1960s, a period when timber harvesting and fire control converted many open pine savannas to dense forests.[36]

Cavities and Fire: The Ecology of the Red-cockaded Woodpecker

The Red-cockaded Woodpecker is one of the most specialized species of birds in North America. It is closely associated with woodlands with an open understory and mature pines. A survey of woodpeckers in different types of forest (bottomland hardwood forest, mixed pine-hardwood forest and longleaf pine savanna) in eastern Texas showed that Red-cockaded Woodpecker was the most specialized of the eight species; it was found only in longleaf pine savanna, while all of the other woodpecker species were found in a variety of habitats.[37] It is not surprising, therefore, that the Red-cockaded Woodpecker has steadily declined with the loss of the pine savanna.[38]

Red-cockaded Woodpeckers tend to disappear from sites that are protected from fire. These birds primarily feed on wood roaches, carpenter ants, wood-boring beetles, and other bark insects, so, unlike the Bachman's Sparrow, they do not depend directly on an open, grassy understory for food.[39] However, they still abandon areas where an understory of hardwood trees has grown up around their nest trees, probably because their nests become more vulnerable to predators.[40] Open pine woodlands contains relatively few nest predators, but rat snakes are a major threat to woodpecker eggs and nestlings. Rat snakes can move straight up a pine trunk by latching their belly scales onto the rough, flaky bark.[41] The gray rat snake, a subspecies that is restricted to the Southeast, is an especially agile tree climber. To protect their cavities from these snakes, Red-cockaded Woodpeckers continually scale off loose bark to smooth the area around the cavity entrance, and they drill small holes around the entrance. Thick resin flows from these "resin wells" and coats a broad area.[42] The resin not only makes the tree even smoother and less easily climbed, but it also causes the scales of the snake to stick together and may even be toxic to snakes. In experimental trials, snakes writhed violently after they tried to crawl up logs coated with this resin.[43] By nesting on live pines and drilling resin wells, the Red-cockaded Woodpecker effectively protects its nest from predation by rat snakes, but this protection ultimately depends on ground fires that burn away the understory shrubs and saplings, leaving the cavity tree isolated. In the absence of fire, hardwood shrubs and trees provide snakes with new routes to the nest, making the nest cavity too vulnerable.

Red-cockaded Woodpeckers depend on the old-growth characteristics of longleaf pine stands because they need big, living trees for nesting.[44] The small trunks in a young, regenerating forest are too thin for construction of the roost and nest cavities that Red-cockaded Woodpeckers need. Also, unlike other North American woodpeckers that usually construct cavities in the relatively soft wood of snags, Red-cockaded Woodpeckers invariably nest in live trees, perhaps because snags do not long survive the frequent ground fires in pine woodlands.[45] Although living loblolly and shortleaf pines are used for nest sites, longleaf pines are superior because they produce more copious quantities of resin over a longer period of time.[46]

Only longleaf pines older than 80 years are typically large enough to accommodate a nest or roost cavity.[47] These cavities are essential for reproduction and are important in protecting the woodpeckers from predators and severe weather during the night. Because they are excavated in relatively hard, live wood, however, the cavities are difficult to construct. First an entrance tunnel is pecked through the resistant living sapwood, and then the

FIG. 9.4. Nesting area for Red-cockaded Woodpeckers in Kisatchie National Forest, Louisiana. Photograph by Robert Askins.

cavity is dug out of the dead heartwood. Many cavities are constructed in heartwood infected with red-heart fungus, which softens the wood, and this fungus is most common in older trees.[48] It requires at least ten months for a Red-cockaded Woodpecker to construct a cavity, and it usually takes much longer. In Texas, excavation of new cavities in longleaf pines requires an average of 6.3 years.[49] After the cavity is completed, however, it may be used for as long as 20 years.[50]

When they do not have a choice, Red-cockaded Woodpeckers build cavities in trees younger than 80 years old. Robert Hooper found that in Osceola and Ocala national forests in Florida, where few old trees remain, most cavities are located in young pines.[51] However, most of these trees either had decayed heartwood or exceptionally fast growth rates (and hence had large trunks for their age). Moreover, the density of Red-cockaded Woodpeckers was much lower in these Florida forests than in Francis Marion National Forest in South Carolina, where old trees were much more common and 94 percent of the cavities were in trees older than 80 years. There were three woodpecker groups per 1,000 acres (8 per 1,000 ha) in Francis Marion, compared with only about 0.6 per acre (1.5 per 1,000 ha) in the Florida sites. Hooper recommends that longleaf pines older than 95 years should be retained as nest and roost sites, and that clusters of younger trees that are slated to reach this age should be spared during timber harvesting.

The effort required to create cavities, and the consequent scarcity of cavities, molds the social behavior of Red-cockaded Woodpeckers.[52] Unlike most other species of woodpeckers, which tolerate only their mates and dependent young in their territories, Red-cockaded Woodpeckers often live in groups of several adults. Each group consists of a breeding pair and several nonbreeding adults (usually males) that are their grown offspring. Each adult has a separate roost cavity in a cluster of adjacent trees. The nonbreeding adults are "helpers"; they help incubate the eggs, feed the young, and defend the large territory used by the group. They also actively defend cavities from other cavity-dwelling animals, such as squirrels and other species of woodpeckers.

New groups seldom form because of the need for a cluster of cavities. This requires pines that are large enough to accommodate cavities as well as a lengthy period of woodpecker labor to excavate the cavities. Consequently, most young birds become breeders not by establishing a new territory with new cavities, but by replacing a breeder that dies on an established territory. A helper may inherit its parent's territory, or it may move out of its original territory in search of vacancies for breeders on other territories. Jeffrey Walters studied more than 200 groups of Red-cockaded Woodpeckers in the North Carolina Sandhills, and during eight years no new clusters of cavities were established.[53] Because young birds almost always become breeders on established territories, the number of groups may remain remarkably stable unless cavity trees are destroyed, which typically results in a rapid population decline. Cavity trees are the key to understanding the behavior of Red-cockaded Woodpeckers, and are an essential consideration in preventing their extinction.

Not surprisingly, Red-cockaded Woodpecker populations declined sharply when the old-growth longleaf pine woodlands were lumbered. They disappeared from many of the counties of the Southeast.[54] The remaining populations are small and scattered, with only 11 populations consisting of more than 100 breeding pairs.[55] These groups survived largely because many diseased and defective trees that could not be sold for lumber were left after most of the pines were harvested. Aerial photographs from the 1930s show that scattered large trees survived in many longleaf pine woodlands, and some of these probably housed Red-cockaded Woodpecker cavities.[56] More intensive forestry, with short rotation cycles for cutting and with removal of commercially useless trees, now threatens these relict old trees and the woodpeckers that depend on them.

The scattered old pines that Red-cockaded Woodpeckers need are also threatened by natural events. Old longleaf pines have always been toppled by tornadoes and hurricanes and killed by southern pine beetle infestations, but now these events can make a territory or even an entire section of a forest unsuitable for Red-cockaded Woodpeckers. When a cluster of old cavity trees is destroyed, there may be no alternative sites for nesting and roosting.

Natural disturbances can have a major impact on cavity trees. For example, during the autumn of 1985, two hurricanes hit the Wade Tract, an old-growth longleaf pine stand in southern Georgia.[57] Of the 33 cavity trees at this site, 24 percent were destroyed. Most snapped off at the level of the cavity. The impact of Hurricane Hugo on the Red-cockaded Woodpecker population in Francis Marion National Forest in South Carolina was even more devastating. Before 1989, when the hurricane struck, this was the second largest population of Red-cockaded Woodpeckers, with more than 1,000 individuals.[58] About 87 percent of the estimated 1765 active cavity trees and most of the pines on which the woodpeckers depend for food were destroyed, and about 63 percent of the woodpeckers were killed by the storm.[59]

Where Red-cockaded Woodpeckers nest in loblolly and shortleaf pine rather than longleaf pine, the southern pine beetle can be a major cause of death for cavity trees. In the pine woodlands in two national forests in eastern Texas, this insect was the main cause of the death of cavity trees. Steady mortality of single trees occurs in non-outbreak years, and major infestations of pine beetles can swiftly eliminate large numbers of nest and roost sites.[60] For example, more than 350 cavity trees were killed in this way in one section of Sam Houston National Forest between 1983 and 1985.[61] Longleaf pines are substantially more resistant to attack by southern pine beetles, probably because they release large amounts of resin into the tunnels that beetles attempt to dig through the bark.[62] Thus, longleaf pines not only provide safer nest sites because of copious resin production, but they are also less likely to succumb to pine beetles for the same reason.

Although storms and pine beetles are natural causes of death for pines,

FIG. 9.5. Top, mature longleaf pine woodland in Francis Marion National Forest, South Carolina, before Hurricane Hugo. Below, mature longleaf pines after Hurricane Hugo. Photographs courtesy of USDA Forest Service.

they may become more serious as a result of forest management. For example, trees near clearcuts are particularly vulnerable to wind damage,[63] as are stands of loblolly pine planted at sites that were originally occupied by longleaf pine.[64] Also, pine beetle outbreaks are more frequent in dense stands of young pine that have been protected from fire.[65] Ironically, in forests in Texas where the hardwoods have been removed around clusters of cavity trees to create an open pine woodland, a high proportion of the cavity trees have been killed by pine beetles.[66] Beetles may be especially attracted to these islands of open pine woods in a sea of dense mixed pine and hardwood forest.

In many parts of the South, the majority of cavity trees and potential cavity trees are isolated large trees left behind after timber harvesting. As these relicts of the original forest are steadily and inevitably lost, Red-cockaded Woodpecker populations decline. In time (and with proper management), other trees will become old enough to accommodate cavities, but in the near term many Red-cockaded Woodpecker populations will be in trouble. In the 1980s, Carole Copeyon, who was a graduate student at North Carolina State University, addressed this problem by inventing a method for creating artificial cavities.[67] She drilled cavities in the heartwood of living pines with a power drill.[68] The access hole for the drill is plugged with wood and sealed with wood filler, and a smaller entrance hole is drilled approximately horizontally (with a slight upward slope for drainage) into the cavity. Loose bark around the cavity entrance is flaked off, and artificial resin wells are drilled into the bark around the entrance so sap will spread out, creating a barrier to snakes. Another method is used for trees that are too small to drill a cavity into the heartwood. A prefabricated insert (a block of wood containing a cavity) is placed in the tree.[69] A rectangular hole is sawed out of the tree and replaced with the insert, which is fastened in place with small wooden wedges and sealed with wood filler. Resin wells are chiseled in the bark around the insert, and a steel restrictor is placed around the entrance hole so the hole cannot be expanded by other types of woodpeckers. Surprisingly, the inserts, which are 10 inches (25 cm) high and 6 inches (15 cm) deep, do not seem to harm the tree.

Woodpeckers readily accept these new cavities even on sites that were previously not occupied.[70] By 1993, 64 percent of the occupied cavity trees in Angelina National Forest in Texas were artificial,[71] and artificial cavities housed many of the woodpeckers left homeless after Hurricane Hugo hit Francis Marion National Forest.[72] In 1992, 61 percent of the 332 woodpecker groups in this forest were using artificial cavities for nesting.[73] This is an expensive program, but it can sustain woodpecker populations until pines grow old enough to be used by woodpeckers for natural cavity sites.

Ultimately, Red-cockaded Woodpeckers can be sustained in managed forests if enough old trees are preserved as nest and roost sites. Selective cutting, rather than clearcutting, could be designed to retain an adequate

FIG. 9.6. Installation of insert designed for use as a nesting and roosting cavity by Red-cockaded Woodpeckers. Photograph courtesy of USDA Forest Service.

density of old trees so that woodpecker groups do not depend on a few scattered relict trees.[74] Alternatively, Red-cockaded Woodpeckers could be maintained in woodlands managed with clearcutting if the rotations between harvests were longer (at least 100 years for longleaf pine), or if remnant stands of old pines were retained, or, preferably, if these two approaches were combined.[75] Prescribed burning is used in pine woods managed for pulp or wood production, so forests that are properly managed for both woodpeckers and timber harvesting could potentially provide better habitat than unmanaged pine woods that are not burned and soon become too crowded with trees to support Red-cockaded Woodpeckers and other species that are characteristic of the longleaf pine savanna.[76]

The potential for recovery of Red-cockaded Woodpeckers in the pine woods of the Southeast is illustrated by Francis Marion National Forest, which was almost stripped of pines in 1936 when the national forest was established.[77] Before Hurricane Hugo destroyed most of the mature pines, this area sustained a dense and increasing Red-cockaded Woodpecker population in old second-growth pines and relict pines that were not harvested. If this species had depended completely on ancient stands of forest, then it would probably have already become extinct because there are only a few remaining stands of truly old-growth longleaf pines, and most of these are small.[78] However, even though Red-cockaded Woodpeckers can use old second-growth pines, they will not remain long in forests where pines are uniformly harvested before they are large enough to accommodate a woodpecker cavity.

Islands of Pine Savanna

After continuous expanses of longleaf pine woodland were broken up into islandlike patches surrounded by roads, farmland, and pine plantations, the nature of the woodland changed. Fires are now less frequent because they can no longer spread freely. In some regions, wild hogs move from surrounding areas into the open pine woods, where they dig up longleaf pine seedlings to eat the taproot.[79] Even where hogs are not a problem and the structure of the woodland is maintained with prescribed burning, however, the birds that depend on the open pine woodland may be affected by habitat fragmentation. This is particularly true for the two most specialized pine savanna species, the Red-cockaded Woodpecker and the Bachman's Sparrow.

Red-cockaded Woodpeckers are largely restricted to open pine woodland, so their populations are prone to isolation on islandlike patches of favorable habitat. These patches may be so isolated from one another that successful dispersal between them is unlikely. Most of the long-distance dispersal in this species involves females because most yearling females leave their parents' territory, while a large proportion of yearling males remain with their parents as helpers. Also, if a male helper inherits his father's territory, his mother will disperse, probably to avoid breeding with her son. When a breeding female dies, she must be replaced by a female from another group, and this may be a problem for woodpecker groups on isolated territories. Although some birds disperse as far as 54 miles (90 km), the average dispersal distance for yearlings is only about 3 miles (5 km), so a breeding male on an isolated territory may have a long wait before a female finds him.[80] However, it was recently discovered that banded Red-cockaded Woodpeckers disperse more frequently than expected between different national forests in eastern Texas, often crossing many miles of open farmland.[81]

Another problem is that Red-cockaded Woodpeckers groups have large territories, averaging 213 acres (85 ha), but with a wide range in size (35–512 acres, or 14–213 ha).[82] Even in Francis Marion National Forest, where Red-cockaded Woodpecker densities are exceptionally high, the average size of the home ranges of woodpecker groups is about 200 acres (80 ha).[83] Because each group includes only one breeding pair, the number of breeding adults is low even in a relatively large tract of forest. Populations in remnant patches of pine woodland are therefore not only isolated, but also small. Small populations could be endangered by a decline in genetic variability unless they are reached periodically by birds dispersing from other populations.[84]

Red-cockaded Woodpeckers primarily feed on wood roaches, ants, termites, southern pine beetles, and bark beetles that are harvested from the bark of living pines, especially large pines.[85] Harvesting timber from the home ranges of Red-cockaded Woodpeckers therefore directly reduces their main source of food. However, at one study site, removal of as much as 37 percent of the pines on woodpecker home ranges had no effect on their survival or nesting success. In national forests in eastern Texas, however, the impact of timber harvesting depended on the size of the local woodpecker population.[86] In large populations, the size of woodpecker groups was not affected by clearcutting of up to 16 percent of the nearby woodland. In isolated patches of pine woodland with small woodpecker populations, similar levels of timber harvesting resulted in smaller groups with more unmated males and more pairs without helpers. Breaking up the woodland with clearcuts could reduce the frequency of dispersal between territories, especially in small populations where there are few sources for immigrating birds. As a result, unmated males may wait longer for potential mates to find them, delaying reproduction and the growth of the group. The number of inactive cavity clusters also increased, which could also be due to delays in dispersal of new birds to a territory after breeding birds die.

Because of the importance of successful dispersal for retaining breeding pairs on remote territories, some isolated populations may have to be sustained by artificially increasing the frequency of dispersal. Preliminary experiments indicate that this approach could work. Females are trapped at their roost cavities and transported to a territory where the breeding female has died.[87] A female is placed in an empty cavity in the new territory after dark and kept there until daylight, when she is released and may join the new group as a breeding female. Juvenile females have been transplanted successfully from Apalachicola National Forest in Florida, which has the largest population of Red-cockaded Woodpeckers, to other national forests throughout the Southeast.[88] Of the 18 females that were transported to new locations, 11 became established in breeding groups, and thus bolstered declining populations. New pairs have also been established successfully by releasing a young male and a young female from different territories on a vacant territory with cavity trees.[89] This is the key for establishing new breeding groups, but it still is not as dependable as translocating young females into territories with unmated males.[90]

Unlike Red-cockaded Woodpeckers, Bachman's Sparrows are found in a variety of habitats other than pine savanna, so their populations are not usually as isolated. In many regions of the Southeast, Bachman's Sparrows are found primarily in recent clearcuts and other areas with low, second-growth vegetation because the mature pine savannas where they once lived are gone. Young, regenerating pine stands become unsuitable for nesting after five years, however, so even large clearcuts provide only transient habitat. Mathematical simulations of population changes in hypothetical landscapes showed that if Bachman's Sparrows are restricted to clearcuts, the population may fluctuate wildly. This greatly increases the chance that the local population will disappear. However, if patches of permanent nesting habitat (representing mature pine savanna) are incorporated into the model, then the population becomes much more stable.[91] These long-term "sources" of young birds stabilize the population much more than would be expected on the basis of their total area, suggesting that even small areas of pine savanna may be important to the regional population of Bachman's Sparrow.

Conservation of Birds of the Longleaf Pine Woodland

Saving Red-cockaded Woodpeckers, Bachman's Sparrows, and other distinctive species of the southeastern pine woodlands requires that we understand a natural system of surprising complexity. A description of the habitat requirements of each species is insufficient; this would not reveal how the population replicates itself from generation to generation or how the habitat is sustained over the long term. Declaring a preserve for a known population of Red-cockaded Woodpeckers or Bachman's Sparrows will ultimately be useless if these populations are too small and isolated to sustain themselves, or if open savanna is slowly replaced with dense pine woods or oak forest.

Unlike prairies and other open grasslands, the longleaf pine savanna has an ancient component—massive longleaf pine trees—that does not recover quickly after a major disturbance, such as clearcut logging or a powerful hurricane. The two layers of the savanna function in distinctly different ways: the grass layer acts like an early successional habitat that will rapidly disappear in the absence of disturbance, while the tree canopy acts like an old-growth habitat that requires decades, if not centuries, to recover after it is destroyed. Sustaining this ecosystem therefore represents a special challenge: either too much protection or too much disruption will eliminate its special features and the species that depend on those features.

The few remaining stands of old longleaf pines should have the highest priority for protection; they not only constitute good habitat but also provide our best source of information on how this ecosystem functions. Many longleaf pine savannas have been heavily modified by fire suppression and timber harvesting, but these can be restored by simulating the natural cycle of burning and by permitting some of the remaining pines to grow to an old age. This can be done not only on nature preserves, but also on commercial forest land. A combination of prescribed burning of the understory, selective cutting to retain an overstory with some old trees, and longer rotation times for harvesting timber can create the conditions needed by the birds and other organisms of the longleaf pine savanna. Where longleaf pine has already been replaced with hardwoods or other types of pines, it can be replanted after the land has been cleared and burned. Periodic burning can then be used to slowly recreate the austere beauty and biological complexity of the pine savanna, and provide new habitat for the Red-cockaded Woodpeckers, Brown-headed Nuthatches, and Bachman's Sparrows.

CHAPTER 10

Landscape Ecology
The Key to Bird Conservation

That is where the Ivory-bill fits; a denizen of the tall trees in thick forests and swamps, he belongs in the same place as Wild Turkeys, bear, alligators, and other residents of our southern swamp wilderness. The most will be accomplished by preserving an area of virgin and primitive forest that will be a suitable habitat for the Ivory-billed Woodpecker and other forms of life, the whole a permanent monument of native trees, plants, birds and other animals inhabiting that wilderness area.

—JAMES T. TANNER, *The Ivory-billed Woodpecker*

I N the late 1930s, James Tanner completed an exceptionally thorough study of the Ivory-billed Woodpecker for the National Audubon Society.[1] He traveled 45,000 miles, visiting 145 sites where there had been reports of the species or where there appeared to be appropriate habitat. He also completed an intensive study of the feeding behavior, habitat requirements, and social behavior of ivorybills in the Singer Tract in Louisiana, which held one of the largest remaining populations. As a result of his study, he was able to make specific recommendations for saving the Ivory-billed Woodpecker from extinction: he listed the key sites that should be protected, estimated the minimum territory size needed to sustain a pair (2.5–3 square miles), and described the precise type of habitat needed by this species.[2] His main recommendation was that large areas of old-growth bottomland forest should be preserved. Recognizing the practical problem of purchasing and protecting forests with such valuable timber, he recommended sustained yield forestry that would preserve the habitat features needed by Ivory-billed Woodpeckers. His plan anticipated some of the later ideas of conservation biology: a completely protected reserve would be surrounded by a zone where trees would be selectively harvested. In the latter area, some of the old trees suitable for ivorybill feeding

---

PLATE 10. Male Ivory-billed Woodpecker feeding a wood-boring beetle grub to nestlings in sweetgum and oak bottomland forest in Louisiana.

would be left, and trees would be removed at a rate that would permit new trees to replace harvested trees so that the forest would never be completely cleared. Woodlands not used by ivorybills could be harvested by clearcutting.

Tanner's plan was based on sound scientific information and incorporated a pragmatic approach to economic demands, but it did not affect public policy. The Singer Tract, where he had made a detailed study of three pairs of ivorybills, was completely logged soon after his project ended. The Santee Swamp in South Carolina, where there were confirmed reports of Ivory-billed Woodpeckers and unconfirmed reports of Carolina Parakeets in the 1930s, was cleared and flooded for a hydroelectric project.[3] Two of the best remaining examples of floodplain forest were lost, and with them much of the hope for saving the Ivory-billed Woodpecker. This was foolish not only in terms of protecting the natural heritage of the country, but also in purely economic terms. The Singer Tract would have earned much more for surrounding communities from natural history tourism and sustained timber-harvesting than from the one-time sale of trees.

The loss of the ivorybill and other species of the floodplain forest was due to public policy as much as to the lack of foresight of the lumber companies. Tax policy in the southern states favored quick removal of all marketable timber because the assessments on land with standing timber resulted in high annual taxes.[4] Moreover, there was little concern about the endangered species of the southern bottomland forest during the period of active logging. Many scientists and naturalists were most interested in obtaining the skins and eggs of declining species for their collections, and remnant populations were sometimes decimated by collectors. Frank M. Chapman's assessment that the Carolina Parakeet was not able "to withstand contact with civilization"[5] was characteristic of the fatalism of many scientists in the late nineteenth and early twentieth centuries. Tanner's systematic search for the information needed to preserve Ivory-billed Woodpeckers represented a more optimistic and constructive response, but even this was ineffective without public policies that emphasized saving species threatened with extinction. With a few exceptions, such as the vigorous effort on Martha's Vineyard to save the last Heath Hen population, and the establishment of Audubon Sanctuaries to protect egrets and other long-legged waders, a true commitment to preventing extinction did not develop until the 1950s. By that time it was probably too late to save the Ivory-billed Woodpecker.

FIG. 10.1. Nestling Ivory-billed Woodpecker banded by James Tanner during his study of this species in 1938. Photo by James T. Tanner. From Tanner, 1942, © National Audubon Society. Reprinted by permission.

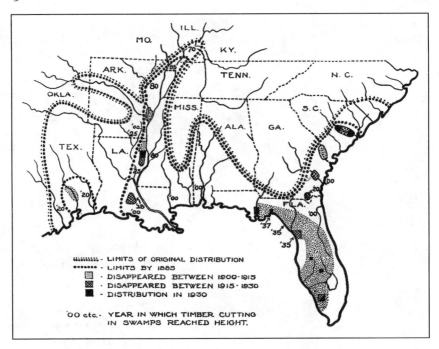

FIG. 10.2. Map showing the history of range contractions in the Ivory-billed Woodpecker. From Tanner, 1942, © National Audubon Society. Reprinted by permission.

The Causes of Population Declines and Extinction

Many North American landscapes have been so radically modified since European settlement that it is surprising that such a small percentage of terrestrial birds have become extinct. The relatively short list of extinct and endangered species suggests that North American land birds are generally adaptable and resilient. After all, these species are the survivors of massive landscape changes during the Pleistocene, when the distribution of plants and animals shifted and reshifted across hundreds of miles with the advance and retreat of continental glaciers.

As human influence spreads directly and indirectly across even the wildest of protected natural areas, however, many bird species face environmental changes that overwhelm their ability to adapt and persist. Birds are threatened both by direct destruction of their habitats and by more subtle changes that degrade habitats, making them unsuitable for particular species. Direct destruction affects wetlands when they are drained or filled, and old-growth forests that will require centuries to recover after they are cut down. Among landbirds, the old-growth specialists have been in most severe danger of extinction since European settlement because of the eradication of their habitats from most of temperate North America. The Ivory-billed Woodpecker is already gone, and the Spotted Owl is in trouble.

Other bird species can use young, regenerating forests, but they are intolerant of the fragmentation of the forest into small patches surrounded by open habitats. Similarly, many grassland specialists tend to disappear from small patches of grassland with unsuitable surroundings. Species disappear from small habitat patches because they have large home ranges and cannot find enough food in a small habitat area, or because they are intolerant of the increased rates of nest predation and cowbird parasitism that frequently characterize small patches. Habitat fragmentation has probably contributed to recent population declines in Upland Sandpiper, Grasshopper Sparrow, Cerulean Warbler, and Spotted Owl. Small size and isolation of habitat patches is more likely to affect bird populations in landscapes where more than 70 percent of the habitat has been destroyed; at this level of habitat destruction, many species decline more rapidly than would be expected simply on the basis of direct habitat loss.[6]

Some species are vulnerable because they must move from region to region in search of sporadically abundant food. These species require a series of widely separated habitats; they would quickly disappear from a lone preserve even if it had ideal habitat. This group includes species that depend on outbreaks of spruce budworm to nest successfully; species such as the Phainopepla that must move to find concentrations of berries; and species that depend on "mast crops" of nuts or conifer seeds. The need for great expanses of acorn- and beech-producing forest in several widely separated regions may have doomed the Passenger Pigeon. Without careful planning, other species that feed on the mast crops of particular types of trees—Pinyon Jays, Clark's Nutcrackers, and the various types of crossbills—could eventually be in trouble.

One of the least widely recognized but perhaps most pervasive causes of population declines in terrestrial birds in North America is human interference with natural agents of environmental disturbance. In contrast to old-growth specialists, which depend on the vegetation becoming complex, multilayered, and relatively stable for a long period, some species depend on the periodic destruction and simplification of the dominant vegetation. Frequent fires created the habitat for the characteristic birds of coastal grasslands, tallgrass prairies, and open ponderosa pine stands. Grazing by prairie dogs and bison generated the cropped grassland needed by McCown's Longspurs and Mountain Plovers in the shortgrass prairies, while frequent floods sustained the streamside woods where Vermilion Flycatchers and Bell's Vireos live in the Southwest. Similarly, windstorms and beavers were responsible for the shrubby openings where White-eyed Vireos and Chestnut-sided Warblers originally nested in the eastern forest. When these disturbances are removed—or when their frequency or intensity are reduced—then the habitat will slowly change until it is unsuitable for these species. To a casual observer, the landscape will look natural, even wild, but much of its biological diversity will be lost. As the ponderosa pine savanna slowly becomes a dense forest; as snowberry bushes spread across a prairie; or as the last expanse of shrubby vegetation in a region grows into forest, the bird species associated with these habitats disappear. The interruption of natural disturbances may have doomed the Bachman's Warbler and now threatens the Mountain Plover, Black-capped Vireo, Golden-winged Warbler, and Kirtland's Warbler.

Other factors may also come into play in driving populations down to dangerously low levels. Introduced organisms can spread explosively, replacing or eliminating the vegetation needed by habitat specialists. Saltcedar has replaced natural streamside vegetation along many southwestern rivers, forming a habitat used by relatively few species of birds. Balsam woolly adelgid, an introduced insect, has killed most of the balsam fir at upper elevations in the Great Smoky Mountains of Tennessee and North Carolina, greatly reducing the habitat of Red-breasted Nuthatch, Golden-crowned Kinglet, Swainson's Thrush, and the small-billed, southeastern form of the Red Crossbill.[7] The potential impact of introduced organisms, which may be pathogens, defoliating insects, predators, or competitors, deserves much more attention. The World Resources Institute estimates that 39 percent of the animal species that became extinct after 1600 disappeared because of introduced species, compared with 36 percent because of direct destruction of habitat and 23 percent because of hunting.[8] Although most of the species that disappeared because of alien species lived on oceanic islands where they had evolved in relative isolation from most pathogens, predators, and competitors, introduced species can also cause severe declines in species living on continents.

Birds that migrate to the tropics may decline because they lose their winter habitats. Cerulean Warblers may be in trouble not only because they disappear from forest fragments in their North American breeding areas, but also because of the loss of the mountain forests where they spend the winter in the Andes. Dickcissels are threatened because they are eradicated by farmers whose fields they raid in the wintering areas in Venezuela. Upland Sandpipers depend not only on extensive grasslands in North America for nesting, but also on the grasslands of Argentina, where they spend most of the year.

Conservation of birds depends on a clear understanding of both their habitat requirements and how their habitats are sustained. It is not enough to know that a species requires spruce-fir forest. The configuration and structure of the forest may be important: some species need large expanses of uninterrupted forest, while others require old-growth forest. The timing and pattern of disturbances may also be important: some species depend on fires to create a continuous supply of openings or snags. Another consideration is the proximity of other forests: dispersal between sites can help sustain populations. Although we must understand what happens at the scale of a single territory during a breeding season, we cannot fully understand the ecology of birds without zooming out from the territory of a particular individual bird to larger and larger scales, examining the context of a territory in the local and regional landscape, and considering how the landscape shifts over years, decades, and even centuries. For example, Ovenbirds need a forest floor with an open understory where they can search the leaf litter for insects, but they have substantially better nesting success in the interior of large forests than in small woodlots. The largest forests apparently produce a surplus of young that can sustain populations at other, sometimes distant sites. This type of analysis typifies the new field of landscape ecology, and it is the key to understanding and ultimately restoring natural ecosystems.

Lost Landscapes

Like the Ivory-billed Woodpecker, most of the declining bird species in North America are associated with "lost landscapes," habitats that have been almost completely destroyed or that have been dramatically transformed. To a surprising extent, modern conservation efforts depend on scraps of information about these landscapes from seventeenth- and eighteenth-century traveler's tales, and from the journals of early settlers and the first naturalists. Even vague descriptions of the pattern of vegetation in a region can help show how it was structured and how it functioned. However, relict patches of relatively untransformed habitat are a much better source of information about how these habitats work and how they can be sustained. Their destruction represents an irretrievable loss of information, like the burning of the library at Alexandria or the bulldozing of a major archeological site. For this reason, areas such as the Konza Prairie of Kansas, the floodplain woodlands of the San Pedro River of Arizona, the old-growth longleaf pines of the Wade Tract in Georgia, and the ancient cove forests in Great Smoky Mountains National Park should have the highest priority for protection and proper management.

For other ecosystems, large areas of relatively undisturbed habitat remain, but important ecological processes have been disrupted. Expanses of these ecosystems that are still subject to natural processes are especially important as a source of information. Prairies that are still cropped by prairie dogs and bison, forest watersheds that are still dammed by beavers, and wilderness areas that are still burned by lightning-triggered fires are as valuable as old-growth forests in terms of what they can tell us about how ancient ecosystems functioned, and about how birds have evolved to depend on particular ecological processes.

Some ecosystems are so completely transformed that there are no relict pieces large enough to show us how the original landscape worked. The towering, closed forests of the Ohio River Valley and the high grass stretching to the horizon on Illinois prairies are difficult to imagine, much less understand, because nearly all of the evidence was obliterated during the 1800s. Similarly, the forests of northern Scandinavia have been so completely transformed by efficient industrial forestry that Scandinavian ecologists travel to similar forests in Siberia to study the ecological roles of wildfires and old dead trees, key factors that have virtually disappeared in their homelands.

Sometimes understanding how ancient ecosystems functioned depends on the kind of careful detective work typical of archeology and paleontology. The fire scars on old lodgepole pines in Yellowstone National Park, the layers of fallen logs that date hurricanes and intense thunderstorms over the past few centuries in a New England hardwood forest, and the pollen and particles of charcoal in lake sediments can help us re-create lost landscapes. This record is usually easier to read in relatively undisturbed habitats.

Restoring Ecosystems

Once we understand how ancient ecosystems functioned, we can carefully reintroduce important ecological processes. In some cases this is relatively easy: wildfires can be allowed to burn in remote wilderness areas, beavers can be reintroduced to a national park, or tidal flow can be restored to a salt marsh by installing larger culverts under a coastal road. In more heavily settled areas, wildfires cannot be permitted, so prescribed burning must be used as a substitute. If beavers are permitted at all, the location and size of their ponds must be controlled. In the most extreme cases, we may need to artificially replicate ecological processes that have disappeared from a regional landscape. Cattle, sheep, or mowers may be used in place of fires or migrating herds of bison to sustain open grassland, and snags can be created in managed forests by girdling trees that are left after harvesting to simulate one of the effects of an intense forest fire.

Sometimes ecosystems must be rebuilt from the ground up. The floodplain woods along the Lower Colorado River were almost completely destroyed, and there was little hope that they would return naturally because even the seed sources were gone. After dense stands of introduced saltcedar were cut and burned, the soil was tilled and native mesquites, cottonwoods and shrubs were planted and nurtured with irrigation. They grew to form a woodland that supports many of the bird species found in natural floodplain woods.

In the future we may need to reconstruct many of the earth's ecosystems. This will require a clear understanding of how these habitats were molded by natural forces before people transformed the landscape. It will also demand an understanding of the requirements of particular species, many of which depend on specific types of vegetation, nesting sites, and food. Some ecological processes, and some species, cannot be sustained in small patches of habitat, so true restoration may require a large, continuous area. This kind of effort may not be applied to most ecosystems for a long time, but in the interim we must be careful not to lose the basic building blocks—the species of organisms—that are needed to rebuild lost ecosystems. This can be done in relict patches of natural habitat, in partially re-

stored habitat, or even in wholly artificial habitats that provide the basic ecological requirements of threatened species.

Landscape Ecology: The Key to Preserving Diversity

The underlying goal of most conservation efforts is to maintain fully functioning natural ecosystems with all of their constituent native species. However, conservation efforts that benefit one species will inevitably harm other
species. Creating grassland habitat for Grasshopper Sparrows may destroy
or degrade the woodland habitat of Cerulean Warblers. Both species have
declined in North America, and both species need large areas of their preferred habitats. Moreover, even a relatively large population in a large conservation area is likely to disappear if it is completely isolated from other
populations of the same species. Only coordinated planning of conservation efforts across a region will insure that these species will survive.

The greatest revelation—and the greatest shock—of landscape ecology
is that even relatively large nature preserves lose many of their most distinctive species if the surrounding landscape is completely transformed. Island-like preserves often lose species rapidly. For example, since 1932, the
215-acre (86-ha) tropical forest in the Bogor Botanical Garden in Java lost
nearly all of its characteristic forest bird species as surrounding woodlands
were cut down.[9] Of 22 species of forest specialists, 20 were extinct by 1985
and two others were represented by less than five pairs. The forest is now
dominated by the generalized species that also live in the cleared countryside around the preserve. Similarly, old-growth bird species disappeared in
the protected forest at Törmävaara in Finland, while species that are common in the surrounding managed forests increased. In small nature preserves in eastern North America, forest specialists declined and were replaced by more generalized birds that can also be found along wooded
residential streets and in suburban backyards. In each case, these small
patches of forest have lost some of their most distinctive birds.

Preserving these specialized species requires large, interconnected expanses of natural habitat. Not only must these areas be large enough to accommodate species which use large home ranges or which are sensitive to
the interruption of their habitat by developed land and abrupt edges, but
they must also be large enough to support a full range of successional
stages, from newly disturbed areas where the vegetation is regenerating, to
long undisturbed sites where the vegetation is relatively stable. They must

be large enough to sustain the patchwork of habitats caused by ecological processes that work on a grand scale: forest and grass fires, major storms, or extensive outbreaks of defoliating insects.

Even large, interconnected preserves must be part of a regional, and ultimately continental, system of preserves if we are to sustain nomadic species that must seek food in different regions during different years, and the migratory species that must move seasonally from summer to winter habitats. Learning where these species move—and which habitats they use along the way—will be critical to sustaining their populations.

Cooperative Management of Natural Landscapes

With increasing human populations and expanding land development, it is often difficult to protect even small, isolated nature preserves, much less huge tracts of wilderness with free-ranging large mammals and uncontrolled wildfires. How can we recreate the fabric of a natural landscape that has been cut and unraveled across most of its surface? Will we ultimately preserve only the generalized species that can live in small islands of natural habitat, losing the larger and more specialized species? This prospect has forced conservationists to look for new approaches. Like Tanner, who realized that Ivory-billed Woodpeckers require too much economically valuable land to be sustained on strict preserves where economic activity is excluded, many conservationists propose working with landowners and public land managers who use the land for farming, timber-harvesting, ranching or other economic activities. There are only a few places in the world where national parks and wilderness areas will be large enough to sustain an ecological landscape with all of its specialized, mobile species and large-scale processes. In most places, we must try to find ways to maintain ecological processes and the species that depend on them across landscapes that are used by people for economic purposes.

Part of the solution is to establish preserves that are close to other protected areas. Increasingly, different conservation organizations and government agencies are working together to cluster their preserves, link them with corridors, and coordinate their management. Few preserves or even national parks are large enough to sustain all of the populations and ecological processes that characterized North American landscapes before European settlement. Systems of preserves can come much closer to this goal. Although Yellowstone National Park is large, it is considerably more valuable because it is embedded in the Greater Yellowstone Ecosystem, which also encompasses Grand Teton National Park, six national forests, three national wildlife refuges, and other state and federal conservation land.[10] This system is large enough to permit bison herds to roam relatively freely and to allow many fires to burn themselves out.

When natural areas are restricted to relatively small, scattered areas, then it is often important to manage different areas in different ways—for example, a hands-off approach in every small preserve in a region might benefit woodland species while leading to the loss of grassland and shrubland species. Even if natural disturbances are not suppressed, they may be too infrequent to provide enough habitat for some early successional species simply because of the small area of remaining natural habitat. Letting "nature take its course" to reestablish a presumed "balance of nature" can lead to a severe loss in biological diversity. The resulting natural areas may be much more homogeneous than the diverse landscapes that preceded European settlement. To prevent this, managers of different natural areas need to coordinate their efforts, with some managers providing habitat for species associated with mature forest or other relatively undisturbed habitats, while others create early successional habitats that require frequent natural or artificial disturbances.

The Greater Yellowstone Ecosystem is primarily publicly owned, so the land can be managed not only for economic activities (such as tourism, forestry, and livestock grazing), but also for conservation. In most parts of North America, however, the majority of the land is privately owned, and effective conservation will depend on the efforts of landowners. Fortunately, most farmers, ranchers, and other landowners are interested in preserving the natural beauty of their land, and many people have a special interest in preserving birds and other wildlife. Although they may also need to derive income from the land, this is often compatible with maintaining regional biological diversity.

A good example of this approach—where many people in a community cooperate to preserve the natural landscape—is the effort orchestrated by The Nature Conservancy in the Texas Hill Country northwest of San Antonio and Austin.[11] The goal is to maintain functioning natural habitats on the Edwards Plateau. This is a distinctive region of rolling limestone hills covered with a stunted woodland of Ashe juniper, Virginia live oak, and Lacey oak.[12] Lacey oak is essentially confined to the Edwards Plateau, as is the Golden-cheeked Warbler, a federally endangered species.[13] Golden-

FIG. 10.3. Golden-cheeked Warbler, an endangered species restricted to the Edwards Plateau of Texas. Photograph by Greg W. Lasley.

cheeked Warblers are usually found in cedar brakes, which are stands of Ashe juniper. They depend on junipers for loose strips of bark that are used in nest construction, but only junipers older than 50 years have sufficient quantities of loose bark.[14] Ashe juniper is tolerant of ground fires, and fire suppression can lead to its replacement by other tree species and a consequent decline in Golden-cheeked Warblers.[15] Also, clearing of woodland results in a direct loss of Golden-cheeked Warbler habitat, and may also reduce populations in the remaining, intact woodland if it is heavily fragmented by openings. Golden-cheeked warblers are typically absent from woodland patches smaller than 125 acres (50 ha). The Texas Hill Country is adjacent to some of the most rapidly growing cities in Texas, so extensive areas of juniper woodland has been cleared or fragmented.

Another endangered bird species, the Black-capped Vireo, is also found in the Texas Hill Country. In contrast with the Golden-cheeked Warbler, however, it does not depend on mature juniper-oak woodland. Instead, Black-capped Vireos nest in dense, low oak thickets.[16] The vireos are most likely to occur in vegetation that is 6–9 feet (2–3 m) tall, with 35–55 percent shrub cover.[17] Open space between the shrubs permits an apron of vegetation to grow to the ground, and this is a consistent characteristic of

FIG. 10.4. Habitat of Golden-cheeked Warbler in Travis County, Texas. Photograph by Greg W. Lasley.

Black-capped Vireo territories. This type of shrubland grows in areas with thin, poor soil, or where the woodland overstory has been destroyed by a fire or other disturbance. Black-capped Vireos begin to disappear as taller junipers encroach on these shrublands. Consequently, careful regional planning is needed to provide both the relatively large areas of old juniper-oak woodland for the Golden-cheeked Warbler, and recently disturbed shrubland with few or no tall junipers for the Black-capped Vireo.

The Nature Conservancy's plan for the Texas Hill Country not only recognizes the habitat needs of different species, but also the economic needs of the people who own and use most of the land. Like Tanner's 1942 proposal to save the Ivory-billed Woodpecker, the Nature Conservancy plan combines core preserves and buffer areas.[18] When possible, core preserves, such as the Barton Creek Habitat Preserve and the Dolan Falls Ranch, are strictly managed for biological diversity. Core preserves are surrounded by buffer areas, such as ranches, where some areas of natural vegetation can be sustained. Although the buffer areas are used for cattle-grazing and other activities, they still protect large areas of relatively undisturbed woodland. The natural landscape on many of these ranches has been protected with conservation easements. Also, in some cases, ranchers derive income from

FIG. 10.5. Black-capped Vireo, an endangered species restricted during the breed-ing season to shrubby vegetation in Texas and Oklahoma. Photograph by Greg W. Lasley.

hunters and, increasingly, from birders and other naturalists who visit the region to see the Golden-cheeked Warbler and the other distinctive species of the Edwards Plateau. Thus, many ranchers in the region have an eco-nomic incentive (as well as a personal commitment) to saving the natural landscapes of the Texas Hill Country. Also, in the rapidly expanding Austin metropolitan area, a conservation plan calls for the protection of a system of preserves that would protect the largest expanses of continuous wood-land habitat, while permitting development in more fragmented wood-lands. The emphasis is on protecting sustainable populations of all species in the region rather than attempting to protect every relict population, re-gardless of its long-term prospects.

Many of the most interesting and ecologically important species in North America probably cannot be sustained in islandlike preserves. The

FIG. 10.6. Habitat of Black-capped Vireo in Kickapoo Cavern State Park, Texas. Photograph by Mark Lockwood.

large natural landscapes that they depend upon are controlled by numerous landowners, many of whom depend on their land for their livelihood. Only cooperation among landowners will preserve these species and the ecological processes on which they ultimately depend.

Beyond Birds: Preserving Biological Diversity

Throughout this book I have tried to illustrate the basic principles of landscape ecology with studies of particular species of birds. A similar book could be written using examples for amphibians, mammals, or butterflies, but birds are especially well studied, and many conservation efforts are directed at saving particular species of birds or particular bird communities. In the process of saving or restoring the habitats of birds, we often save and restore the habitats of many other threatened species. The longleaf pine savanna that supports a healthy population of Red-cockaded Woodpeckers will probably also support many species of rare plants that are endemic to this habitat,[19] along with southeastern pocket gophers, gopher tortoises, and indigo snakes. Although beaver ponds may be important to birds, they are even more important to bog turtles and some species of dragonflies.

Shrublands in New Hampshire provide habitat not only for Chestnut-sided Warblers, but also for the declining New England cottontail. Ultimately the goal should not be just to save particular species of birds, but to save the whole array of species associated with the different habitats of a region. Some of these habitats are special because they are ancient, with tall, majestic trees, while other habitats must constantly be renewed as the dominant vegetation burns or is blown down or is cropped by grazers. Without both types of habitats, we will not only lose species; we will also lose much of the intricacy and beauty of natural systems.

Afterword

Much has changed in the two years since this book was first published. Research on the landscape ecology of birds has proceeded on many fronts, and ideas generated by this research have molded conservation policy more rapidly than most researchers would have predicted. Programs like Partners in Flight and the Western Hemisphere Shorebird Reserve Network have opened communication between researchers and land managers, making everyone more responsive to new ideas. Research has molded land management and land management has molded research in unprecedented and unexpected ways.

As a member of the board of trustees of the Connecticut Chapter of the Nature Conservancy, I've participated directly in this process. The Nature Conservancy has begun a program to identify large blocks of forest and other natural habitat that should have a high priority for conservation in each of the Northeast's different ecological regions. I encountered this process after the criteria for identifying these blocks had been established and was pleased to see that they resonated nicely with major themes in this book. The goal is to locate forest blocks with roadless areas large enough to sustain forest-interior bird species, many of which disappear from small patches of forest surrounded by edge habitats. These blocks should also be large enough for salamander and frog species that spend most of their lives in upland forest but often need to move great distances to vernal pools to lay eggs, as well as for large mammals and birds that use extensive home ranges. Populations of black bears, bobcats (*Lynx rufus*), and Northern Goshawks (*Accipiter gentilis*) should be able to survive over the long term in these forests. Equally important, the forest blocks should be large enough and in good-enough condition so that a single catastrophic disturbance, such as a windstorm or fire, will not destroy them. The importance of this consideration is exemplified by the history of Cathedral Pines, an ancient white pine stand that was one of the only old-growth woods in Connecticut. The Nature Conservancy carefully protected and managed this site, but a rare New England tornado blew the ancient pines down in a few minutes. If this preserve had been 4,000 acres instead of 42 acres (17 ha), then the tornado would not have destroyed the old forest but would merely have created a biologically rich patch of young woodland in its midst. This was the pattern when a tornado flattened 99 percent of the canopy trees in 1,000 acres (400 ha) of the 4,000 acres of old-growth hemlock-beech forest in the Tionestra Scenic and Research Natural Areas in Pennsylvania's Allegheny National Forest.[1] If a block of woodland is large enough, it can become a mosaic of habitat types, from low thicket to tall forest, that supports a wide variety of species.

The Nature Conservancy's planning process also considers how each local population of a species needs to be tied into a system of local populations—a metapopulation—to survive over the long term. At times each local population will decline and then will be rescued by immigrants from other populations, and at other times it will send out emigrants, usually young birds, that serve as rescuers. If this process is prevented, then one local population after another may disappear. Hence, the next step in the planning process will be to identify "greenways" of natural habitat between the large forest blocks to connect different populations of the same species and facilitate immigration.

This is only one example of many efforts across North America to manage large landscapes. It is a clear departure from the traditional effort to protect nature preserves, which are often small samples of especially beautiful or diverse or pristine natural habitat. The shift from protecting preserves to protecting landscapes is difficult, however. In most cases, a single nonprofit organization or government agency cannot afford to buy or manage an area large enough to do this effectively. Many organizations and agencies must work together, and work with private property owners, to sustain large areas of roadless natural habitat.

Two studies in the southern Appalachian Mountains illustrate the importance of preserving large expanses of natural habitat. Between 1944 and 1946, Albert Ganier and Alfred Clesch surveyed birds in hardwood and conifer forests at high elevations in the Unicoi Mountains, which straddle the Tennessee–North Carolina border. In the late 1990s, Christopher Haney, David Lee, and Mark Wilbert returned to these sites to repeat the surveys.[2] The main changes in the forest during the 50-year interval were the maturation of the trees, the elimination of grazing, and limited logging and road building. The researchers found that the abundance ranks of different species (from least abundant to most abundant) of forest-interior species had changed remarkably little between the 1940s and 1990s. Although the relative abundance of Wood Thrush declined (as it has at many other sites in eastern North America), the abundance ranks of Brown Creeper, Winter

Wren, Black-throated Green Warbler, Blackburnian Warbler, and Scarlet Tanager increased. Other forest-interior species showed little change. Maturation of the forest should create better conditions for these species, so it isn't surprising that the populations of some species increased and that woodland birds in general showed no significant change in relative abundance. It is surprising, however, that the early successional shrub specialists also showed no significant change. This group included such common species as Chestnut-sided Warbler and Dark-eyed Junco. Small-scale logging and natural disturbances such as windstorms, ice storms, and fire created sufficient shrubland and young forest for these species. This forest is extensive enough that natural disturbances frequently create openings. Moreover, it is large enough, and far enough away from farming areas, that parasitic cowbirds are rare. The result is a diverse and relatively stable bird community with large populations of both early successional and forest-interior species. The breeding bird community has changed little during the past 50 years, suggesting that large blocks of forest like those in the Unicoi Mountains can sustain biological diversity over the long term.

Another study indicates that large blocks of forest can serve as sources of young woodland birds that may support populations over a larger region. George Farnsworth and Theodore Simons monitored the fate of 416 Wood Thrush nests in a variety of forest types in Great Smoky Mountains National Park in Tennessee.[3] This is in one of the most heavily forested regions in eastern North America. Cowbirds were rare, just as they were in the nearby Unicoi Mountains. They parasitized only seven of the 416 nests, and even these nests produced a surprisingly large number of thrush fledglings. Predation of eggs and nestlings was much more important, however. A variety of predators, from snakes to small rodents to black bears, caused the failure of 225 nests. Although the nest predation rate was higher and the nest success rate was lower than expected on the basis of studies of Wood Thrush nests in large forests in the Midwest and Pennsylvania, the Wood Thrush population in Great Smoky Mountains National Park apparently produces more than enough young to replace itself. Moreover, reproductive success remained at this moderate level throughout the five years of the study and was similar in various types of forest at different elevations, suggesting that Great Smoky Mountains National Park is a dependable source of Wood Thrush young. These young not only sustain the population in the national park but may also help bolster populations in the surrounding region.

The exchange of young birds among populations becomes especially critical when the habitat a species needs has been reduced to a few isolated, island-like patches. This is the situation in the northeastern United States for Grasshopper Sparrows and Upland Sandpipers, both of which need extensive areas of open grassland to nest. The distribution of isolated populations of these species and other species of grassland birds has been revealed by a massive survey of grasslands in New York and New England sponsored by the Massachusetts Audubon Society.[4] Numerous volunteers counted birds on 100-meter-diameter circular plots in grasslands throughout the region. More than 4,000 sites were surveyed between 1997 and 2000. The study revealed that grassland bird populations are widely scattered and that many populations are small, with only a few individuals. The regional populations of Grasshopper Sparrow and Upland Sandpiper are anchored by a few sites with large populations in coastal Maine, the Connecticut River Valley of Massachusetts and Connecticut, and (for Grasshopper Sparrow) western New York State. The many small populations in these regional metapopulations probably depend on immigrants from larger sites, so the large sites should have a high priority for conservation.

Grassland birds have been declining in most parts of North America, and the need to maintain and restore their habitats is increasingly recognized. Another group of early successional species, the shrubland specialists, have received much less attention even though most of these species also have been declining in many regions. One problem is that these species are often lumped with species that live in edge and residential habitats. Many of the shrubland species require open expanses of low woody vegetation, however, and they are not found in gardens, on golf courses, or along forest edges with a narrow band of shrubs. Consequently, they have declined in regions where most of the low woody vegetation has grown into forest or has been developed. Both land managers and researchers have now turned their attention to this group, a response that is a logical extension of the concern about grassland birds. In the summer of 2001, a landmark series of papers on early successional woody habitats was published as a special section in *Wildlife Society Bulletin*.[5] These papers provide us with a much clearer picture of the scope of the problem and the potential solutions.

In this special issue, Margaret Trani and her colleagues analyzed the distribution of early successional forest in the eastern United States.[6] They used data from the USDA Forest Service Forest Inventory and Analysis to assess changes in low, woody habitat (the "seedling-sapling" and "non-

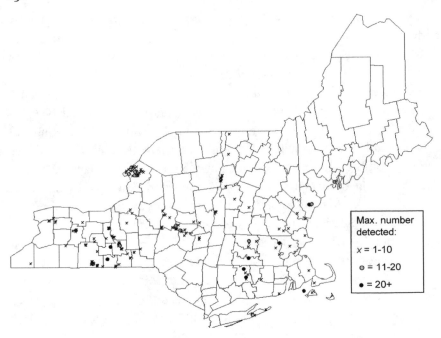

FIG. A.1. Distribution of populations of Grasshopper Sparrows in New York and New England based on surveys conducted by Massachusetts Audubon Society from 1997 to 2000. From Jones et al., 2001. Reprinted with permission of the Massachusetts Audubon Society.

stocked" categories used by foresters) since 1946. This habitat has declined to less than 20 percent of the forests in New England, the Middle Atlantic region, and interior regions of the South (such as Tennessee and Kentucky). Other regions, however, have much higher proportions of early successional habitat. Eastern Ohio has large areas of recently abandoned farmland while Maine, the Great Lakes states, and the southern coastal plain have large areas of regenerating young forest because of timber harvesting. Not surprisingly, most early successional bird species declined in southern New England and the Middle Atlantic states,[7] but many of the species associated with regenerating northern forests in Maine or the Great Lakes region increased.[8] For example, Spruce Grouse, Nashville Warbler, and Lincoln's Sparrow all showed overall increases in abundance on Breeding Bird Survey routes in eastern North America. In contrast, many early successional species have declined in coastal areas of the southeastern United States even though early successional (seedling-sapling) vegetation covers 35 percent

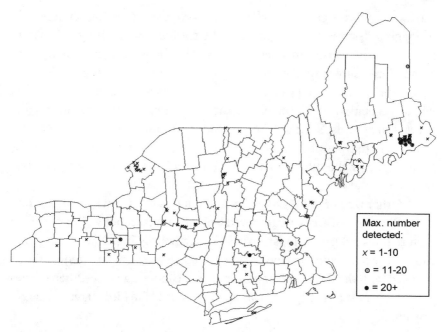

FIG. A.2. Distribution of populations of Upland Sandpipers in New York and New England based on surveys conducted by Massachusetts Audubon Society from 1997 to 2000. From Jones et al., 2001. Reprinted with permission of the Massachusetts Audubon Society.

of the forestland.[9] Much of the early successional vegetation in this region is in heavily managed pine plantations, however, where trees have low diversity and rapid growth rates, and these areas may not provide appropriate habitat for most species of shrubland birds.[10] This suggests that we need to pay close attention to the quality as well as the quantity of early successional vegetation.

In the same special section of the *Wildlife Society Bulletin*, Craig Lorimer discussed the different types of early successional habitat that result from various types of disturbances.[11] He made an especially important distinction between "successional habitat" and "young forest habitat." Successional habitat results from the colonization of bare or cleared ground by plants, so it develops on abandoned farms, abandoned and dried beaver ponds, and riversides where new soil has been deposited by floodwaters. "Pioneer species" of herbs, shrubs, and trees dominate these areas, often forming an impenetrable bramble or thicket of shrubs and vines. In contrast, young forest grows up when a fire, a windstorm, an insect outbreak, or logging destroys the tree canopy. The sprouts and seedlings of mature forest trees often grow up quickly, resulting in a dense layer of saplings with some of the surviving shrubs and herbs of the original forest. Although both habitats have been called shrubland and both are dominated by low, woody vegetation, the structure of the vegetation is different. Some early successional birds, such as Chestnut-sided and Blue-winged warblers, are found in both types of habitat, whereas other species tend to be more common in one or the other.[12] These differences have not been well studied, however.

Another type of early successional habitat is created when single trees or small groups of trees are removed from the forest canopy. This occurs in mature forests during windstorms and selective logging. Some forest-interior bird species, such as Hooded and Cerulean warblers, are frequently associated with the dense vegetation in these "canopy gaps."[13] Other species (including most of the species classified as "shrubland birds") require larger openings.

Even species that we typically associate with continuous mature forests may depend on early successional habitats in unexpected ways. For example, Bicknell's Thrush (*Catharus bicknelli*), a species usually described as a specialist on stunted but mature balsam fir–red spruce forests on mountaintops in the Northeast, is also found (albeit at lower densities) in recent clearcuts with young growth of white birch (*Betula papyrifera*), balsam fir, and pin cherry (*Prunus pensylvanica*) in New Brunswick.[14] Because of the pace of logging in New Brunswick, a period with few regenerating clearcuts is anticipated in the future, which may threaten the population of Bicknell's Thrush in the province.

Some birds that nest in forests and are generally considered forest specialists may use early successional habitats before or after the breeding season. For example, at the end of the breeding season Wood Thrushes move from their breeding areas in the forest to patches of early successional habitat. This pattern is seen in both fledglings and molting adults.[15] Angela Anders and her colleagues demonstrated this by strapping radio transmitters to the backs of nestlings and following their movements for up to 60 days, until the transmitter battery died.[16] The researchers were able to track 15 individuals from the time they fledged until after they left the family group and became independent. Initially the young birds stayed with their parents. Most family groups remained on the original nesting territory, but two families moved into other areas. After three weeks each young bird dispersed on its own. They dispersed an average of 1.2 miles (2 km), and as far as 2.9 miles (4.7 km), from the home territory before they settled into a small home range. They moved for several days before settling, passing through a variety of habitats, including upland forest, floodplain forest, old fields, and regenerating clearcuts. The majority settled in disturbed habitats with a dense shrub layer such as clearcuts with a 3–5-meter-high canopy or pine forest that had been burned periodically. The dense shrub layer in these habitats probably provided good protection from predators. Hawks took at least three of the 49 juveniles with radio transmitters before they dispersed from their home territories, so predators are clearly a danger to these young birds. The shrubby vegetation also provided a ready source of food, including blackberries (Rubus allegheniensis), which the young birds were often observed eating. Interestingly, while following the radio-tagged thrushes, the observers saw other forest bird species (including many juveniles) using the same patches of early successional habitat. These included Red-eyed Vireos and typical forest warblers such as Ovenbirds, Black-and-white Warblers, Worm-eating Warblers, and Kentucky Warblers.

A number of studies have shown that forest-nesting birds frequently use low, early successional habitat to rest and refuel after long migratory journeys in the fall and spring. On the Savannah River Site in South Carolina, for example, substantially higher densities of forest-nesting birds were captured in mist nets in forest openings than in the nearby bottomland forest.[17] The openings were all "group selection cuts" that had been created by harvesting small groups of trees. Mist nets sample low vegetation effectively but often miss birds that feed in the forest canopy, so this could partially account for the difference. Forest understory species such as Hooded Warbler were captured primarily in openings, however, even though they are susceptible to capture in the shrub layer of forests. Moreover, the density of migrants was much higher in large openings (120 ft or 40 m in diameter) than in smaller openings, suggesting that these birds were seeking out patches of low vegetation during migratory stopovers.

Additional evidence for the importance of early successional habitats to fall migrants comes from a long-term study of seven abandoned farm fields in New Jersey.[18] Migrants concentrated primarily in the fields that were covered with shrubs. Over the 18 years of the study, trees overtopped the dense shrubs in some fields. As sites became forested, fewer birds stopped during fall migration. During the same period, however, migrant abundance re-

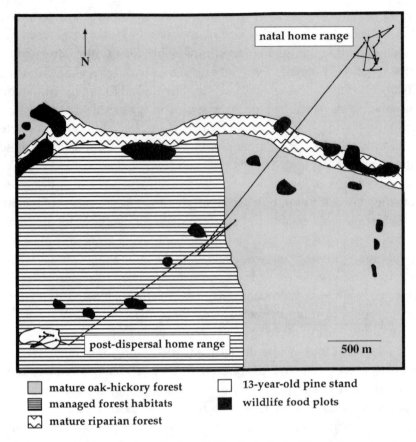

FIG. A.3. Movements of a juvenile Wood Thrush tracked with a radio transmitter after leaving the nest. Each point represents a daily observation; a straight line connects adjacent days. The "natal home range" is the area used by the family group before the young bird dispersed. The "post-dispersal home range" is the area used by the young bird after it left the family group and traveled independently for several days. From Anders et al., 1998. Reproduced with permission from *The Auk*.

mained high in fields that changed from open shrubland to dense shrubland. Also, migrants were consistently more abundant in shrubland than in young forest during the same autumn. Many of the birds using shrublands were feeding on berries such as those of panicled dogwood (*Cornus racemosa*), and migrants were most abundant in old fields when shrubs were in fruit. The high density of berries in shrublands may make them especially valuable to fall migrants, including some forest species that feed almost exclusively on insects during the summer but switch to fruit in autumn.

In response to these studies, we need to change how we think about the habitat requirements of forest-nesting neotropical migrants. It is well known that these species have distinctly different habitat requirements in the temperate zone and tropics, but their summer requirements have been pictured as relatively simple. Each pair needs an acre or two of mature forest for a breeding territory that contains food, cover, and nest site for parents and young. Many species thrive only in the interior of large forests, so protection of large blocks of continuous forest is the best way to sustain their breeding habitat. It is now clear, however, that many of these forest specialists use, and perhaps even need, patches of early successional habitat. Shrubby openings in the forest protect the young during the first, vulnerable weeks of independence and provide food and shelter for birds as they migrate south in the fall or north in the spring. Wildlife managers have long known that such species as American Woodcock (*Scolopax minor*), Northern Bobwhite, and black bear need a mosaic of successional stages to provide their needs at different seasons. It appears that we need to apply similar thinking to some of the most specialized forest-interior birds.

Preserving the biological diversity of any region requires a range of habitat types, including those created by natural disturbances. If there are no natural or artificial disturbances generating grassland, shrubland, and young forest, then many species will be in trouble. This will clearly affect grassland and shrubland specialists most directly, but it may also harm forest specialists that use early successional habitats at key points in their life cycles. Only immense parks or preserves can sustain a full range of early successional and mature forest habitats, so in most regions land managers will need to cooperate to ensure that these habitats are adequately represented across the regional landscape. This will be easier to achieve if different parks, nature preserves, and wildlife areas are managed with different goals. Research on the history and probable frequency of natural disturbances can help determine how much of each habitat type is needed. Instead of attempting to re-create conditions in 1600 or some other specific historical period, however, it is usually more practical to estimate the "historical range of variation" in the amount of each habitat type.[19] This approach recognizes that the proportions of grassland, shrubland, young forest, and old-growth forest have shifted constantly over the past few thousand years as the climate has changed and people have modified the land by hunting, burning, and farming. Staying within this historical range of variation should ensure that all native species will have enough habitat to survive over the long term.

NOTES

1. Dunwiddie, P., D. Foster, D. Leopold, and R. T. Leverett. 1996. Old-growth forests of southern New England, New York, and Pennsylvania. Pages 126–143 in M. B. Davis (editor), Eastern old-growth forests: Prospects for rediscovery and recovery. Island Press, Washington, D.C.

2. Haney, J. C., D. S. Lee, and M. Wilbert. 2001. A half-century comparison of breeding birds in the southern Appalachians. Condor 103: 268–277.

3. Farnsworth, G. L., and T. R. Simons. 1999. Factors affecting nesting success of Wood Thrushes in Great Smoky Mountains National Park. Auk 116: 1975–1082.

4. Jones, A. L., W. G. Shriver, and P. D. Vickery. 2001. Regional inventory of grassland birds in New England and New York, 1997–2000. Massachusetts Audubon Society Grassland Conservation Program, Lincoln, Mass.

5. Askins, R. A. 2001. Sustaining biological diversity in early successional communities: The challenge of managing unpopular habitats. Wildlife Society Bulletin 29: 407–412.

6. Trani, M. K., R. T. Brooks, T. L. Schmidt, V. A. Rudis, and C. M. Gabbard. 2001. Patterns and trends of early successional forests in the eastern United States. Wildlife Society Bulletin 29: 413–424.

7. Askins, R. A. 1993. Population trends in grassland, shrubland, and forest birds in eastern North America. Pages 1–34 in D. M. Power (editor), Current Ornithology, vol. 11. Plenum Press, New York.

8. Hunter, W. C., D. A. Buehler, R. A. Canterbury, J. L. Confer, and P. B. Hamel. 2001. Conservation of disturbance-dependent birds in eastern North America. Wildlife Society Bulletin 29: 440–455.

9. Krementz, D. G., and J. S. Christie. 2000. Clearcut size and scrub-successional bird assemblages. Auk 117: 913–924.

10. Dickson, J. G., F. R. Thompson III, R. N. Conner, and K. E. Franzreb. 1995. Silviculture in central and southeastern oak–pine forests. Pages 245–266 in T. E. Martin and D. M. Finch, Ecology and management of neotropical migratory birds. Oxford University Press, New York.

11. Lorimer, C. G. 2001. Historical and ecological roles of disturbance in eastern North American forests: 9000 years of change. Wildlife Society Bulletin 29: 425–439.

12. Askins, R. A. 2001 (see note 5, above).

13. Brawn, J. D., S. K. Robinson, and F. R. Thompson III. 2001. The role of disturbance in the ecology and conservation of birds. Annual Review of Ecology and Systematics 32: 251–276; Hunter et al., 2001. (see note 8, above).

14. Nixon, E. A., S. B. Holmes, and A. W. Diamond. 2001. Bicknell's Thrushes (*Catharus bicknelli*) in New Brunswick clear cuts: Their habitat associations and co-occurrence with Swainson's Thrushes (*Catharus ustulatus*). Wilson Bulletin 113: 33–40.

15. Hunter et al., 2001 (see note 8, above).

16. Anders, A. D., J. Faaborg, and F. R. Thompson III. 1998. Postfledgling dispersal, habitat use, and home-range size of juvenile Wood Thrushes. Auk 115: 349–358.

17. Kilgo, J. C., K. V. Miller, and W. P. Smith. 1999. Effects of group-selection timber harvest in bottomland hardwoods on fall migrant birds. Journal of Field Ornithology 70: 404–413.

18. Suthers, H. B., J. M. Bickal, and P. G. Rodewald. 2000. Use of successional habitat and fruit resources by songbirds during autumn migration in central New Jersey. Wilson Bulletin 112: 249–260.

19. Thompson, F. R., III, and R. M. DeGraaf. 2001. Conservation approaches for woody-early successional communities in the eastern USA. Wildlife Society Bulletin 29: 483–494.

Appendix 1
Scientific Names of Organisms Other Than Birds

Common names are alphabetized within taxonomic groups.

Common Name	Scientific Name
PLANTS	
Arizona sycamore	Platanus wrightii
Ashe juniper	Juniperus ashei
balsam fir	Abies balsamea
balsam poplar	Populus balsamifera
big-tooth maple	Acer grandidentatum
birch	Betula spp.
birdfoot violet	Viola pedata
black bayberry	Myrica heterophylla
blackberrry	Rubus spp.
blazing star	Liatris borealis
blueberry	Vaccinium spp.
blue grama	Bouteloua gracilis
blue spruce	Picea pungens
bluestem	Schizachyrium spp.
bristlecone pine	Pinus aristata
bushy rockrose	Helianthemum dumosum
buttonwood	Conocarpus erectus
cane or bamboo	Arundinaria gigantea
Caribbean pine	Pinus caribaea
catclaw acacia	Acacia greggii
clover	Trifolium spp.
coastal redwood	Sequoia sempervirens
cocklebur	Xanthium strumarium
cypress (baldcypress)	Taxodium distichum
desert mistletoe	Phoradendron californicum
desert willow	Chilopsis linearis
Douglas-fir	Pseudotsuga menziesii
eastern hemlock	Tsuga canadensis
Engelmann spruce	Picea engelmannii
fir	Abies spp.
Fremont cottonwood	Populus fremontii
Goodding willow	Salix gooddingii
jack pine	Pinus banksiana
Jeffrey pine	Pinus jeffreyi
juniper	Juniperus spp.

honey mesquite	*Prosopis glandulosa*
inkweed	*Suaeda torreyana*
jack-in-the-pulpit	*Arisaema* spp.
Lacey oak	*Quercus glaucoides*
larch	*Larix* spp.
little bluestem	*Schizachyrium scoparium*
limber pine	*Pinus flexilis*
loblolly pine	*Pinus taeda*
lodgepole pine	*Pinus contorta*
longleaf pine	*Pinus palustris*
low gallberry holly (inkberry)	*Ilex glabra*
marram grass	*Ammophila breviligulata*
marsh-elder	*Iva frutescens*
mesquite	*Prosopsis juliflora*
paper birch	*Betula papyrifera*
palo verde	*Cercidium floridum*
pine	*Pinus* spp.
pinyon pine	*Pinus edulis*
pond pine	*Pinus serotina*
ponderosa pine	*Pinus ponderosa*
quail bush	*Atriplex lentiformis*
quaking aspen	*Populus tremuloides*
red spruce	*Picea rubens*
redbay	*Persea borbonia*
rhododendron	*Rhododendron* spp.
sage	*Artemisia tridentata*
saltbush	*Atriplex polycarpa*
saltcedar	*Tamarix chinensis*
sandplain agalinis	*Agalinis acuta*
seepwillow	*Baccharis glutinosa*
sickle-leaved golden aster	*Chrysopsis falcata*
shortleaf pine	*Pinus echinata*
silverberry	*Elaegnus commutata*
Sitka spruce	*Picea sitchensis*
slash pine	*Pinus elliottii*
smooth sumac	*Rhus glabra*
southern magnolia	*Magnolia grandiflora*
spruce	*Picea* spp.
subapline fir	*Abies lasiocarpa*
sunflower	*Helianthus* spp.
sycamore	*Platanus occidentalis*
sweetbay magnolia	*Magnolia virginiana*
sweetfern	*Comptonia peregrina*
sweetgum	*Liquidambar styraciflua*
table mountain pine	*Pinus pungens*
titi	*Cyrilla racemiflora*
timothy	*Phleum pratense*
Virginia live oak	*Quercus virginiana*

water oak	Quercus nigra
water tupelo	Nyssa aquatica
western hemlock	Tsuga heterophylla
western snowberry	Symphoricarpos occidentalis
western redcedar	Thuja plicata
white fir	Abies concolor
white pine	Pinus strobus
white spruce	Picea glauca
whitebark pine	Pinus albicaulis
wild indigo	Baptisia tinctoria
wiregrass	Aristida stricta
wolfberry	Lycium spp.

FUNGI

| red-heart fungus | Phellinus pini |
| white pine blister rust | Cronartium ribicola |

FISH

| sand lance | Ammodytes hexapterus |
| surf smelt | Hypomesus pretiosus |

INSECTS

carpenter ant	Camponotus spp.
mountain pine beetle	Dendroctonus ponderosae
saddled prominent caterpillar	Heterocampa guttivitta
southern pine beetle	Dendroctonus frontalis
spruce budworm	Choristoneura fumiferana
whitespotted sawyer	Monochamus scutellatus
wood roach	Parcoblatta spp.

REPTILES

diamondback rattlesnake	Crotalus spp.
gray rat snake	Elaphe obsoleta spiloides
green turtle	Chelonia mydas
rat snake	Elaphe spp.

MAMMALS

African elephant	Loxodonta africana
beaver	Castor canadensis
bison	Bison bison
black bear	Ursus americanus
black-footed ferret	Mustela nigripes
blacktail prairie dog	Cynomys ludovicianus
caribou	Rangifer tarandus
dusky-footed woodrat	Neotoma fuscipes
eastern chipmunk	Tamias striatus
elk	Cervus elaphus
fisher	Martes pennanti
ground sloth	Megalonyx spp.
kangaroo rat	Dipodomys spp.

long-nosed peccary	*Mylohyus nasutus*
lynx	*Lynx canadensis*
marten	*Martes americana*
mastodont	*Mammut americanum*
moose	*Alces americana*
mountain lion	*Felis concolor*
mule deer	*Odocoileus hemionus*
northern flying squirrel	*Glaucomys sabrinus*
opossum	*Didelphis virginiana*
plains zebra	*Equus burchelli*
prairie dog	*Cynomys spp.*
raccoon	*Procyon lotor*
red squirrel	*Tamiasciurus hudsonicus*
snowshoe hare	*Lepus americanus*
southeastern pocket gopher	*Geomys pinetis*
striped skunk	*Mephitis mephitis*
thirteen-lined ground squirrel	*Citellus tridecemlineatus*
Tule elk	*Cervus elaphus nannodes*
wildebeest	*Connochaetes taurinus*
wolf	*Canis lupus*
wolverine	*Gulo luscus*

Appendix 2
Scientific Names of Birds in Phylogenetic Order

This list is based on the database in Bird Brain 4.0 (Ideaform, Inc.), which follows the A.O.U. Check-list of North American Birds (American Ornithologists' Union, 1983) and "Supplements" in *The Auk* (1985–97) for birds of North America (including Central America, Mexico, and the West Indies). For species not included on the AOU Checklist, this list follows Sibley and Monroe 1990.

Common Name	Scientific Name
Swallow-tailed Kite	*Elanoides forficatus*
Mississippi Kite	*Ictinia mississippiensis*
Red-shouldered Hawk	*Buteo lineatus*
Ferruginous Hawk	*Buteo regalis*
American Kestrel	*Falco sparverius*
Japanese Quail	*Coturnix japonica*
Spruce Grouse	*Falcipennis canadensis*
Black Grouse	*Tetrao tetrix*
Eurasian Capercaillie	*Tetrao urogallus*
Greater Prairie-chicken	*Tympanuchus cupido*
Sharp-tailed Grouse	*Tympanuchus phasianellus*
Northern Bobwhite	*Colinus virginianus*
Mountain Quail	*Oreortyx pictus*
Killdeer	*Charadrius vociferus*
Mountain Plover	*Charadrius montanus*
Upland Sandpiper	*Bartramia longicauda*
Marbled Murrelet	*Brachyramphus marmoratus*
Long-billed Murrelet	*Brachyramphus perdix*
Kittlitz's Murrelet	*Brachyramphus brevirostris*
Domestic Pigeon (Rock Dove)	*Columba livia*
Scaly-naped Pigeon	*Columba squamosa*
White-winged Dove	*Zenaida asiatica*
Eared Dove	*Zenaida auriculata*
Passenger Pigeon	*Ectopistes migratorius*
Bridled Quail-Dove	*Geotrygon mystacea*
Carolina Parakeet	*Conuropsis carolinensis*
Black-billed Cuckoo	*Coccyzus erythropthalmus*
Yellow-billed Cuckoo	*Coccyzus americanus*
Greater Roadrunner	*Geococcyx californianus*
Elf Owl	*Micrathene whitneyi*
Great Gray Owl	*Strix nebulosa*

Spotted Owl	*Strix occidentalis*
Flammulated Owl	*Otus flammeolus*
Burrowing Owl	*Athene cunicularia*
Boreal Owl	*Aegolius funereus*
Chuck-will's-widow	*Caprimulgus carolinensis*
Calliope Hummingbird	*Stellula calliope*
Rufous Hummingbird	*Selasphorus rufus*
Broad-tailed Hummingbird	*Selasphorus platycercus*
Black-chinned Hummingbird	*Archilochus alexandri*
Elegant Trogon	*Trogon elegans*
Red-headed Woodpecker	*Melanerpes erythrocephalus*
Acorn Woodpecker	*Melanerpes formicivorus*
Gila Woodpecker	*Melanerpes uropygialis*
Red-bellied Woodpecker	*Melanerpes carolinus*
Red-naped Sapsucker	*Sphyrapicus nuchalis*
Red-breasted Sapsucker	*Sphyrapicus ruber*
Williamson's Sapsucker	*Sphyrapicus thyroideus*
Ladder-backed Woodpecker	*Picoides scalaris*
Downy Woodpecker	*Picoides pubescens*
Great Spotted Woodpecker	*Dendrocopos major*
Hairy Woodpecker	*Picoides villosus*
Strickland's Woodpecker	*Picoides stricklandi*
Red-cockaded Woodpecker	*Picoides borealis*
Three-toed Woodpecker	*Picoides tridactylus*
Black-backed Woodpecker	*Picoides arcticus*
Gilded Flicker	*Colaptes chrysoides*
Northern Flicker	*Colaptes auratus*
Pileated Woodpecker	*Dryocopus pileatus*
Black Woodpecker	*Dryocopus martius*
Ivory-billed Woodpecker	*Campephilus principalis*
Olive-sided Flycatcher	*Contopus cooperi*
Western Wood-Pewee	*Contopus sordidulus*
Eastern Wood-Pewee	*Contopus virens*
Greater Antillean Pewee	*Contopus caribeus*
Yellow-bellied Flycatcher	*Empidonax flaviventris*
Alder Flycatcher	*Empidonax alnorum*
Willow Flycatcher	*Empidonax traillii*
Least Flycatcher	*Empidonax minimus*
Pacific-slope Flycatcher	*Empidonax difficilis*
Cordilleran Flycatcher	*Empidonax occidentalis*
Vermilion Flycatcher	*Pyrocephalus rubinus*
Dusky-capped Flycatcher	*Myiarchus tuberculifer*
Ash-throated Flycatcher	*Myiarchus cinerascens*
Great Crested Flycatcher	*Myiarchus crinitus*
Brown-crested Flycatcher	*Myiarchus tyrannulus*
Eastern Kingbird	*Tyrannus tyrannus*
Loggerhead Shrike	*Lanius ludovicianus*
White-eyed Vireo	*Vireo griseus*

Bell's Vireo	*Vireo bellii*
Black-capped Vireo	*Vireo atricapillus*
Gray Vireo	*Vireo vicinior*
Yellow-throated Vireo	*Vireo flavifrons*
Warbling Vireo	*Vireo gilvus*
Philadelphia Vireo	*Vireo philadelphicus*
Red-eyed Vireo	*Vireo olivaceus*
Siberian Jay	*Perisoreus infaustus*
Gray Jay	*Perisoreus canadensis*
Blue Jay	*Cyanocitta cristata*
Steller's Jay	*Cyanocitta stelleri*
Western Scrub-Jay	*Aphelocoma californica*
Pinyon Jay	*Gymnorhinus cyanocephalus*
Clark's Nutcracker	*Nucifraga columbiana*
Black-billed Magpie	*Pica pica*
American Crow	*Corvus brachyrhynchos*
Fish Crow	*Corvus ossifragus*
Common Raven	*Corvus corax*
Sky Lark	*Alauda arvensis*
Horned Lark	*Eremophila alpestris*
Purple Martin	*Progne subis*
Violet-green Swallow	*Tachycineta thalassina*
Carolina Chickadee	*Poecile carolinensis*
Black-capped Chickadee	*Poecile atricapillus*
Mountain Chickadee	*Poecile gambeli*
Boreal Chickadee	*Poecile hudsonicus*
Siberian Tit	*Poecile cinctus*
Willow Tit	*Parus montanus*
Coal Tit	*Parus ater*
Great Tit	*Parus major*
Plain Titmouse	*Parus inornatus*
Verdin	*Auriparus flaviceps*
Red-breasted Nuthatch	*Sitta canadensis*
Pygmy Nuthatch	*Sitta pygmaea*
Brown-headed Nuthatch	*Sitta pusilla*
Eurasian Treecreeper	*Certhia familiaris*
Brown Creeper	*Certhia americana*
Cactus Wren	*Campylorhynchus brunneicapillus*
Rock Wren	*Catherpes mexicanus*
Winter Wren	*Troglodytes troglodytes*
Sedge Wren	*Cistothorus platensis*
House Wren	*Troglodytes aedon*
Ruby-crowned Kinglet	*Regulus calendula*
Golden-crowned Kinglet	*Regulus satrapa*
Willow Warbler	*Phylloscopus trochilus*
Blue-gray Gnatcatcher	*Polioptila caerulea*
Black-tailed Gnatcatcher	*Polioptila melanura*
Pied Flycatcher	*Ficedula hypoleuca*

European Robin	Erithacus rubecula
Redstart	Phoenicurus phoenicurus
Western Bluebird	Sialia mexicana
Mountain Bluebird	Sialia currucoides
Veery	Catharus fuscescens
Swainson's Thrush	Catharus ustulatus
Wood Thrush	Hylocichla mustelina
Song Thrush	Turdus philomelos
American Robin	Turdus migratorius
Varied Thrush	Ixoreus naevius
Northern Mockingbird	Mimus polyglottus
Sage Thrasher	Oreoscoptes montanus
Brown Thrasher	Toxostoma rufum
Crissal Thrasher	Toxostoma crissale
Phainopepla	Phainopepla nitens
Olive Warbler	Peucedramus taeniatus
Bachman's Warbler	Vermivora bachmanii
Blue-winged Warbler	Vermivora pinus
Golden-winged Warbler	Vermivora chrysoptera
Tennessee Warbler	Vermivora peregrina
Orange-crowned Warbler	Vermivora celata
Nashville Warbler	Vermivora ruficapilla
Lucy's Warbler	Vermivora luciae
Northern Parula	Parula americana
Yellow Warbler	Dendroica petechia
Chestnut-sided Warbler	Dendroica pensylvanica
Magnolia Warbler	Dendroica magnolia
Cape May Warbler	Dendroica tigrina
Black-throated Blue Warbler	Dendroica caerulescens
Yellow-rumped Warbler	Dendroica coronata
Townsend's Warbler	Dendroica townsendi
Black-throated Green Warbler	Dendroica virens
Golden-cheeked Warbler	Dendroica chrysoparia
Blackburnian Warbler	Dendroica fusca
Yellow-throated Warbler	Dendorica dominica
Olive-capped Warbler	Dendroica pityophila
Pine Warbler	Dendroica pinus
Kirtland's Warbler	Dendroica kirtlandii
Prairie Warbler	Dendroica discolor
Bay-breasted Warbler	Dendroica castanea
Cerulean Warbler	Dendroica cerulea
Black-and-white Warbler	Mniotilta varia
American Redstart	Setophaga ruticilla
Prothonotary Warbler	Protonotaria citrea
Worm-eating Warbler	Helmitheros vermivorus
Swainson's Warbler	Limnothlypis swainsonii
Ovenbird	Seiurus aurocapillus
Northern Waterthrush	Seiurus noveboracensis

Kentucky Warbler	*Oporornis formosus*
Mourning Warbler	*Oporornis philadelphia*
MacGillvray's Warbler	*Oporornis tolmiei*
Common Yellowthroat	*Geothlypis trichas*
Hooded Warbler	*Wilsonia citrina*
Wilson's Warbler	*Wilsonia pusilla*
Canada Warbler	*Wilsonia canadensis*
Red-faced Warbler	*Cardellina rubrifrons*
Painted Redstart	*Myioborus pictus*
Yellow-breasted Chat	*Icteria virens*
Summer Tanager	*Piranga rubra*
Scarlet Tanager	*Piranga olivacea*
Western Tanager	*Piranga ludoviciana*
Eastern Towhee	*Pipilo erythrophthalmus*
Spotted Towhee	*Pipilo maculatus*
Bachman's Sparrow	*Aimophila aestivalis*
Chipping Sparrow	*Spizella passerina*
Clay-colored Sparrow	*Spizella pallida*
Field Sparrow	*Spizella pusilla*
Vesper Sparrow	*Pooecetes gramineus*
Lark Sparrow	*Chondestes grammacus*
Sage Sparrow	*Amphispiza belli*
Black-throated Sparrow	*Amphispiza bilineata*
Lark Bunting	*Calamospiza melanocorys*
Savannah Sparrow	*Passerculus sandwichensis*
Baird's Sparrow	*Ammodramus bairdii*
Grasshopper Sparrow	*Ammodramus savannarum*
Henslow's Sparrow	*Ammodramus henslowii*
Le Conte's Sparrow	*Ammodramus leconteii*
Song Sparrow	*Melospiza melodia*
Fox Sparrow	*Passerella iliaca*
Lincoln's Sparrow	*Melospiza lincolnii*
White-throated Sparrow	*Zonotrichia albicollis*
White-crowned Sparrow	*Zonotrichia leucophrys*
Dark-eyed Junco	*Junco hyemalis*
Yellow-eyed Junco	*Junco phaeonotus*
Chestnut-collared Longspur	*Calcarius ornatus*
Smith's Longspur	*Calcarius pictus*
Lapland Longspur	*Calcarius lapponicus*
McCown's Longspur	*Calcarius mccownii*
Rose-breasted Grosbeak	*Pheucticus ludovicianus*
Black-headed Grosbeak	*Pheucticus melanocephalus*
Blue Grosbeak	*Guiraca caerulea*
Lazuli Bunting	*Passerina amoena*
Indigo Bunting	*Passerina cyanea*
Dickcissel	*Spiza americana*
Painted Bunting	*Passerina ciris*
Bobolink	*Dolichonyx oryzivorus*

Eastern Meadowlark	Sturnella magna
Western Meadowlark	Sturnella neglecta
Brown-headed Cowbird	Molothrus ater
Pine Grosbeak	Pinicola enucleator
Purple Finch	Carpodacus purpureus
Cassin's Finch	Carpodacus cassinii
White-winged Crossbill	Loxia leucoptera
Red Crossbill	Loxia curvirostra
Parrot Crossbill	Loxia pytyopsittacus
Scottish Crossbill	Loxia scotica
Common Redpoll	Carduelis flammea
Pine Siskin	Carduelis pinus
American Goldfinch	Carduelis tristis
Evening Grosbeak	Coccothraustes vespertinus

Notes

CHAPTER 1. Grassland Birds of the East Coast: Pleistocene Parkland to Hay Meadow

Epigraph: Johnson, 1960. Reprinted by permission of the publishers and the Trustees of Amherst College from *The Poems of Emily Dickinson*, Thomas H. Johnson, ed., Cambridge, Mass.: The Belknap Press of Harvard University Press, Copyright © 1951, 1955, 1979, 1983 by the President and Fellows of Harvard College.

1. Audubon, 1831.
2. Mayfield, 1988a.
3. Griscom, 1949.
4. Forbush, 1925: 449.
5. Askins, 1993; Peterjohn and Sauer, 1994.
6. Vickery, 1992.
7. Vickery, 1992.
8. Herkert, 1991; Bollinger and Gavin, 1992.
9. Whitcomb, 1987.
10. Whitcomb, 1987.
11. Day, 1953.
12. Day, 1953.
13. Conrad, 1935.
14. Bull, 1974.
15. Harper, 1911.
16. Conrad, 1935; Cain et al., 1937.
17. Svenson, 1936; Stalter and Lamont, 1987.
18. Antenen et al., 1994.
19. Taylor, 1923; Whitney, 1994: 85.
20. Olmsted, 1937.
21. Vickery, 1994.
22. Day, 1953.
23. Whitney, 1994: 104.
24. Day, 1953: 331.
25. Patterson and Sassaman, 1988.
26. Crosby, 1972; Cronon, 1983; Denevan, 1992; Whitney, 1994.
27. Kulikoff, 1986.
28. Russell, 1980.
29. Hammett, 1992.
30. Whitney, 1994: 102–103.
31. Patterson and Sassaman, 1988.
32. Doolittle, 1992.
33. Whitney, 1994: 109.
34. Olmsted, 1937.
35. Whitney, 1994: 115.
36. Russell, 1983.
37. Patterson and Sassaman, 1988.
38. Winne, 1988.
39. Vickery, 1994.
40. Shaffer, 1992: 33–34, 45–46.
41. Shaffer, 1992: 56–57.
42. Shaffer, 1992: 51–63.
43. Iseminger, 1997.
44. Iseminger, 1997.
45. Iseminger, 1997.
46. Lewis, 1907, as cited by Shaffer, 1992: 60.
47. Crosby, 1986: 213–215.
48. Crosby, 1986: 213.
49. Smith, 1989, 1995.
50. Dunwiddie, 1989.
51. Remillard et al., 1987.
52. Naiman et al., 1988; Coles and Orme, 1983; Howard and Larson, 1985; Remillard et al., 1987.
53. Whitney, 1994: 304.
54. Naiman et al., 1994.
55. Coles and Orme, 1983.
56. Runkle, 1990.
57. Borman and Likens, 1979; Seischab and Orwig, 1991.
58. Hosmer, 1908: 85.
59. Webb, 1988.
60. Webb, 1988.
61. Webb, 1988.
62. Kurtén and Anderson, 1980.
63. Guilday et al., 1964.

64. Guilday, 1962.
65. Dublin et al., 1990.
66. Andersson and Appelquist, 1990; Puchkov, 1992.
67. Martin and Klein, 1984.
68. Andersson and Appelquist, 1990.
69. Forbush, 1927: 367–368; Thomas, 1951; Hurley and Franks, 1976.
70. Hurley and Franks, 1976.
71. Lanyon, 1956.
72. Brooks, 1938.
73. Lindholt, 1988.
74. Vaughan, 1977.
75. Feduccia, 1985.
76. Smith, 1968.
77. Smith, 1968.
78. Wheelwright and Rising, 1993.
79. Stobo and McLaren, 1975: 89, 18.
80. Stobo and McLaren, 1971.
81. Gross, 1932.
82. Forbush, 1927: 41; Gross, 1932; Vaughan, 1977; Johnsgard, 1983.
83. Bull, 1974.
84. Gleason and Cronquist, 1991.
85. Vickery, 1994.
86. Hart, 1968; Askins, 1993.
87. Bollinger and Gavin, 1992.
88. Veit and Petersen, 1993; Bevier, 1994; Melvin, 1994.
89. Frawley and Best, 1991; Dunn et al., 1993.
90. Crossman, 1989; Lent and Litwin, 1989; Melvin, 1994.
91. Melvin, 1994.
92. Askins, 1993.
93. Whitmore and Hall, 1978; Whitmore, 1981; Zimmerman, 1988.
94. Renken and Dinsmore, 1987; Zimmerman, 1988.
95. Herkert, 1994a; Vickery, 1994.
96. Herkert, 1994a.
97. Vickery, 1994.
98. Herkert, 1994b.
99. Ells, 1995.
100. Whitmore and Hall, 1978.
101. Bollinger et al., 1990.

CHAPTER 2. Another Quiet Decline: Birds of the Eastern Thickets

Epigraph: E. H. Forbush, 1929, Massachusetts Department of Food and Agriculture.

1. Wilson and Bonaparte, 1877: 91.
2. Andrle and Carroll, 1988; Zeranski and Baptist, 1990.
3. Johnston and Odum, 1956.
4. Conner and Adkisson, 1975.
5. Thompson and Nolan (1973) showed that the chat population in a field in Indiana was highly mobile and transient. They marked individual chats with numbered leg bands. A large number of the males caught during the breeding season were apparently moving through the area because they were not captured again. Moreover, few of the resident males and none of the breeding females returned to the same field during the next breeding season, indicating that the population is highly nomadic.
6. Thompson and Nolan, 1973; Thompson, 1977; Ford, 1992.
7. Anderson and Shugart, 1974.
8. Dennis, 1958; Thompson and Nolan, 1973.
9. Conner and Adkisson, 1975; Butcher et al., 1981.
10. Forbush, 1929.
11. Sauer et al., 1995.
12. Sauer et al., 1995.
13. Hagan et al., 1992; Hussell et al., 1992; Askins, 1993.
14. Hill and Hagan, 1991.
15. Hill and Hagan, 1991; DeGraaf and Miller, 1996.
16. Hagan, 1993.
17. Forbush, 1929.
18. Audubon, 1831.
19. Peterson and Peterson, 1985.
20. Forbush, 1929.
21. Audubon, 1834: 223.
22. Feduccia, 1985.
23. Brewer et al., 1991.
24. Hamel et al., 1982.

25. Richardson et al., 1981.

26. Christensen et al., 1981.

27. Richardson et al., 1981.

28. Oliveri, 1993.

29. White, 1998.

30. Stephens, 1956; Bormann and Likens, 1979.

31. Henry and Swan, 1974.

32. Bormann and Likens, 1979.

33. Runkle, 1990; DeGraaf and Miller, 1996.

34. Seischab and Orwig, 1991.

35. Runkle, 1982.

36. Runkle, 1990.

37. Clebsch and Busing, 1989.

38. Williams, 1989: 435.

39. Dwight, 1969.

40. Irland, 1982; Clark, 1984; De-Graaf and Miller, 1996.

41. Morse, 1989: 263.

42. Litvaitis, 1993.

43. Black, 1950; Brooks and Birch, 1988.

44. Confer, 1992; Askins, 1993.

45. Litvaitis, 1993.

46. Alerich and Drake, 1995.

47. Mayfield, 1996; Lee et al., 1997; Haney et al., 1998.

48. Mayfield, 1992.

49. Mayfield, 1960: 5.

50. Mayfield, 1960: 6.

51. Mayfield, 1992.

52. Mayfield, 1992.

53. Mayfield, 1960: 14.

54. Mayfield, 1960: 13; Davis, 1976; Botkin et al., 1991.

55. Botkin et al. 1991.

56. Mayfield, 1988b.

57. Mayfield, 1993.

58. Mayfield, 1960: 6.

59. Radabaugh, 1974.

60. Mayfield, 1993.

61. Mayfield, 1960; Walkinshaw, 1983: 159; Mayfield, 1992.

62. Walkinshaw, 1983: 154, 157−158, 181; Mayfield, 1992.

63. Walkinshaw, 1983: 150, 151.

64. Mayfield, 1992.

65. Mayfield, 1993.

66. Haney et al., 1998.

67. Webb et al., 1977; Crawford et al., 1981; DeGraaf, 1991; Yahner, 1993, 1997.

68. Conner and Adkisson, 1975.

69. Hagan et al., 1997.

70. Likens et al., 1978.

71. Rudnicky and Hunter, 1993b.

72. Robert Askins, unpublished data.

73. Germaine et al., 1997.

74. Neiring and Goodwin, 1974; Dreyer and Niering, 1986.

75. Askins, 1990, 1994.

76. Dennis, 1958; Confer and Knapp, 1981; Chasko and Gates, 1982; Bramble et al., 1992; Askins, 1994;

77. Chasko and Gates, 1982.

78. Yahner, 1995.

CHAPTER 3. The Great Plains: Birds of the Shifting Mosaic

Epigraph: O. E. Rölvaag, *Giants in the Earth*, copyright © 1927. Reprinted with permission from HarperCollins Publishers.

1. Some prairie ecologists argue that processes such and fire and grazing are so integral to the ecology of grasslands that they do not constitute disturbances. Instead, the prevention of fire or grazing is a disturbance that can disrupt these systems, causing them to change into woodland or some other habitat (Evans et al., 1989). This is a reasonable position, but I have chosen to use the term "disturbance" to refer to a sudden major change in vegetation structure both in habitats such as the tallgrass prairie where this occurs frequently, and in habitats such as the northern hardwood forest where it may occur rarely. I am not implying that natural disturbances have negative or destructive effects on an ecosystem.

2. Hart and Hart, 1997.
3. Reichman, 1987: 106.
4. Anderson, 1990; Zimmerman, 1993: 116; Knight et al., 1994.
5. Reichman, 1987: 107–114.
6. Hulbert, 1988.
7. Herkert, 1994b; Henslow's Sparrows show a similar response to burning in the Konza Prairie of Kansas (Zimmerman, 1988, 1993).
8. Pylypec, 1991; Zimmerman, 1993: 28, 34–35.
9. Johnsgard, 1983: 328, 336, 323–325.
10. Kirsch, 1974.
11. Swengel (1996), however, found that Grasshopper Sparrows and Henslow's Sparrows were both substantially more abundant in prairie preserves that were maintained by late-summer haying rather than by spring burning in tallgrass prairie in southwestern Missouri. This is consistent with the hypothesis that these species evolved in grasslands molded by grazing rather than by frequent fires.
12. Miller et al., 1994.
13. Whicker and Detling, 1988.
14. King, 1959.
15. Whicker and Detling, 1988.
16. Bjorndal, 1980; McNaughton, 1984.
17. McNaughton, 1984.
18. Whicker and Detling, 1988.
19. Miller et al., 1990; Sharps and Uresk, 1990; Plumpton and Lutz, 1993; Miller et al., 1994.
20. Agnew et al., 1986.
21. Bent, 1929.
22. Graul, 1975; Graul and Webster, 1976; Kricher, 1993: 140; Knopf and Miller, 1994.
23. Knowles et al. 1982; Olson, 1985; Knopf, 1996c.
24. Graul, 1975.
25. Knopf, 1996a.
26. Bent, 1929; Graul and Webster, 1976; Sauer et al., 1995.
27. Miller et al., 1990.
28. Miller et al., 1994.
29. Whicker and Detling, 1988.
30. Merriam, 1902.
31. Barnes, 1993; Cully, 1993; Clark et al., 1989; Apa et al., 1991; Miller et al., 1994.
32. Roemer and Forrest, 1996.
33. Miller et al., 1994.
34. O'Meilia et al., 1982; Miller et al., 1994.
35. Kantrud, 1981.
36. With, 1994.
37. Plumb and Dodd, 1993.
38. Sauer et al., 1995.
39. Fretwell, 1979.
40. Sauer et al., 1995.
41. Gross, 1968; Finck, 1984.
42. Zimmerman, 1993: 45–46; Bryan and Best, 1994.
43. Basili and Temple, 1994.
44. Gross, 1968.
45. Gross, 1968.
46. Fretwell, 1986.
47. Zimmerman, 1993; Basili and Temple, 1994, 1995.
48. Fretwell, 1986.
49. Zimmerman, 1983.
50. Grzybowski, 1982. The Lapland and Smith's longspurs are tundra nesters that spend only the winter on the prairie.
51. Zimmerman, 1993: 24–25.
52. Knopf, 1994; Knopf, 1996b.
53. Knopf, 1994.
54. Whitney, 1994: 87, 259.
55. Leopold, 1949: 189. From A Sand County Almanac by Aldo Leopold. Copyright © 1949, 1977 by Oxford University Press. Used by permission of Oxford University Press.
56. Whitney, 1994: 258.
57. Bollinger et al., 1990.
58. Knopf, 1994; Hart and Hart, 1997.
59. Knopf, 1994.
60. Reichman, 1987: 2.
61. Zimmerman, 1993: 33.

62. Zimmerman, 1993: 36.
63. Herkert, 1991; Herkert, 1994b.
64. Johnson and Schwartz, 1993.
65. Johnson and Schwartz, 1993.
66. Brown and MacDonald, 1995.
67. Wuerthner, 1994.
68. Fleischner, 1994.
69. Fleischner, 1995; Knopf, 1996a.
70. Howe, 1994.
71. Fleischner, 1997.

CHAPTER 4. Lost Birds of the Eastern Forest

Epigraph: William Faulkner, Go Down, Moses, Copyright © 1942 by William Faulkner. Pages 202–203. Reprinted by permission of Random House, Inc.

1. Cheatham and Cheatham (1988) argue that the "big woodpecker" in Go Down, Moses was an Ivory-billed Woodpecker, a species chosen by Faulkner to symbolize the soon-to-be-lost wildness of the bottomland forest. Faulkner may have been referring to the Pileated Woodpecker, however, which was called "Lord God" in some localities in the Southeast (Tanner, 1942: 21).
2. Cronon, 1983: 4.
3. Shepard, 1961: 157.
4. Simon and Wildavsky, 1993; Budiansky, 1993, 1994.
5. Cheatham and Cheatham, 1988.
6. Williams, 1989: 361–365.
7. Williams, 1989: 111.
8. Whitney, 1994: 153.
9. Williams, 1989: 113–115.
10. Williams, 1989: 61–63; Whitney, 1994: 132–133.
11. Whitney, 1994: 154.
12. Williams, 1989: 114–115.
13. Williams, 1989: 228–233, 279–282.
14. Budiansky, 1994.
15. Whitney, 1987; Zumeta, 1991.
16. Whitney and Davis, 1986; Foster, 1993.
17. Pimm and Askins, 1995.
18. Williams, 1989: 468.
19. Hart, 1968, 1980; Williams, 1989: 471–472.
20. DeGraaf and Rudis, 1986: 321; Askins, 1993.
21. Whitney, 1994: 168–171; Pimm and Askins, 1995.
22. Williams, 1989: 238.
23. Pimm and Askins, 1995.
24. This estimate is calculated from the species-area equation; $S = cA^z$, where S = number of species, A = habitat area, and c and z are constants that depend on the group of organisms under consideration, the habitat, and the geography of the landscape. If the original area of habitat, Ao, is reduced to a new, smaller area, An, then the number of species should decline from the original number, So, to a smaller number appropriate to the reduced area (Sn). The proportion of species that survives is derived from the following equation: $Sn/So = (An/Ao)^z$. The value of z is equal to approximately 0.25 for large, remnant patches of habitat, such as islands isolated by the rise in sea level following the last glacial period, so it is an appropriate value to use for the large pieces of forest left by clearing of the eastern forest (Pimm and Askins, 1995).
25. Pimm and Askins, 1995.
26. Stotz et al., 1996.
27. Collins, 1990: 146–147, 162–163.
28. Stotz et al., 1996.
29. Schorger, 1955: 225.
30. Schorger, 1955: 200–201; Greenway, 1958: 304; Blockstein and Tordoff, 1985.
31. Schorger, 1955: 91; Bucher, 1992.
32. Matthiessen, 1987: 150–152.
33. Schorger, 1955: 218–222.
34. Blockstein and Tordoff, 1985.
35. Schorger, 1955: 145.
36. Blockstein and Tordoff, 1985.
37. Matthiessen, 1987: 161–165.

38. Schorger, 1936.
39. Blockstein and Tordoff, 1985; Bucher, 1992.
40. Bucher, 1992.
41. Schorger, 1955: 61–62; Bucher, 1992.
42. Schorger, 1955: 16.
43. Bucher, 1992.
44. Bucher, 1992.
45. Blockstein and Tordoff, 1985.
46. Leopold, 1949: 111. From A Sand County Almanac by Aldo Leopold. Copyright © 1949, 1977 by Oxford University Press. Used by permission of Oxford University Press.
47. Mitsch and Gosselink, 1993: 49, 62, 456–457.
48. Lewis, 1907, as cited by Shaffer, 1992: 60.
49. Bartram, 1791: 324.
50. Nature Conservancy of Louisiana, 1992; Noss et al., 1995: 41.
51. Wilson and Bonaparte, 1877.
52. Bartram, 1791: 93–94.
53. Buscemi, 1978.
54. Snyder et al., 1987.
55. Greenway, 1958: 323.
56. McKinley, 1980.
57. Greenway, 1958: 323.
58. Tanner, 1942.
59. Bent, 1939: 8–9; Tanner, 1942: 40–41.
60. Tanner, 1942: 41.
61. Hamel, 1986, 1988.
62. Dingle, 1953a.
63. Hamel, 1986: 7–9.
64. Dingle, 1953a.
65. Hamel, 1986: 24–25.
66. Dingle, 1953a; Hamel, 1986: 25.
67. Hamel 1986: 18, 26.
68. Remsen, 1986.
69. Bartram, 1791: 200.
70. Platt and Brantley, 1992.
71. Hughes, 1951.
72. Hamel, 1986: 11–12.
73. Hamel, 1986: 19.
74. Askins and Ewert, 1991; Lynch, 1991.

75. Bent, 1953: 18; Eddleman et al., 1980.
76. Dingle, 1953b: 31; Thomas et al., 1996.
77. Amacher, 1998.
78. Kilgo et al., 1998.

CHAPTER 5. Deep-forest Birds and Hostile Edges

Epigraph: Conrad Richter, The Trees, Copyright © 1940. Reprinted by permission of Random House, Inc.

1. Williams, 1936.
2. Hickey, 1937.
3. Askins, 1987.
4. Verner, 1985.
5. Robbins, 1978; Enemar et al., 1987.
6. Butcher et al., 1981.
7. Askins et al., 1990.
8. Robbins, 1979.
9. Serrao, 1985, Askins et al., 1990.
10. Robbins, 1979; Johnston and Winings, 1987; Leck et al., 1988; Litwin and Smith, 1992.
11. Askins et al., 1990.
12. Askins et al., 1990.
13. Ambuel and Temple, 1982.
14. Baird, 1990.
15. Holmes et al., 1986; Holmes and Sherry, 1988.
16. Hall, 1984.
17. Hall, 1988.
18. Kendeigh and Fawver, 1981; Wilcove, 1988.
19. Askins et al., 1990.
20. Robbins et al., 1989a.
21. Whitcomb et al., 1981, Howe, 1984, Freemark and Merriam, 1986, Temple, 1986, Askins et al., 1987, Wenny et al., 1993. See Askins et al. (1990) for a review of these and other studies.
22. Rosenberg et al., 1995.
23. Askins et al., 1987.
24. Robbins et al., 1989a.

25. Askins et al., 1987.

26. Robinson et al., 1995.

27. Temple and Cary, 1988.

28. Gates and Gysel, 1978.

29. Chasko and Gates, 1982.

30. Paton, 1994.

31. Wilcove, 1985.

32. Martin, 1987; Willebrand and Marcström, 1988; Reitsma et al., 1990; Whelan et al., 1994.

33. Haskell, 1995; Reitsma et al., 1990.

34. Robinson, 1988, 1992; Brawn and Robinson, 1996.

35. Porneluzi, 1993.

36. Hoover et al., 1995.

37. Gale et al. (1997) describe an exception to this general pattern. In a section of western Connecticut that is 70 percent forested, there was no difference in nesting success of Worm-eating Warblers in a large forest and a set of nine small forests. Despite this, the density of Worm-eating Warblers was substantially higher in the large forest.

38. Brittingham and Temple, 1983.

39. Rich et al., 1994.

40. Robinson, 1992.

41. Porneluzi, 1993; Hoover et al., 1995.

42. Roth and Johnson, 1993.

43. Holmes et al., 1986.

44. Hanners and Patton, 1998.

45. Hahn and Hatfield, 1995.

46. Robinson and Wilcove, 1994.

47. Trine, 1998.

48. Hoover and Brittingham, 1993.

49. Robinson et al., 1995a; Donovan et al., 1997.

50. Burke and Nol, 1998; Robinson, 1998.

51. Robinson et al., 1995b.

52. Apparently not all small woodlots in agricultural areas are population sinks, however. A recent study in a heavily farmed region of southwestern Ontario showed that Wood Thrushes had high nest success in both small and large woodlots even though only 14 percent of the region is forested (Friesen et al., 1999). The implication is that the negative impact of forest fragmentation will vary from region to region depending on the abundance of nest predators and cowbirds.

53. Holmes et al., 1992.

54. Friesen et al., 1995.

55. Conner and Adkisson, 1975; Mauer et al., 1981.

56. Thompson et al., 1992.

57. Welsh and Healy, 1993.

58. DeGraaf, 1995.

59. Rudnicky and Hunter, 1993a.

60. DeGraaf, 1995; DeGraaf and Maier, 1996.

61. Hanski et al., 1996.

62. King et al., 1996.

63. Kurzejeski et al., 1993.

64. Ewert and Askins, 1991.

65. Terborgh, 1989: 76–77.

66. Collins, 1990: 102.

67. Rappole, 1995: 147.

68. Dirzo and Garcia, 1992.

69. Sader and Joyce, 1988.

70. Gradwohl and Greenberg, 1980; Rappole, 1995: 43–45.

71. Sherry and Holmes, 1996; Willis, 1966.

72. Rappole, 1995: 33–42.

73. Hutto, 1989, 1992; Waide, 1980; Lynch, 1989; Petit et al., 1992; Blake and Loiselle, 1992; Askins et al., 1992; Wunderle and Waide, 1993; Confer and Holmes, 1995; Gram and Faaborg, 1997.

74. Hutto et al., 1986.

75. Karr, 1981; Kricher, and Davis, 1992, Gram and Faaborg, 1997.

76. Robbins et al., 1992a.

77. Lynch, 1989; Confer and Holmes, 1995.

78. Lynch, 1989; Petit et al., 1992; Robbins et al. 1992a.

79. Robbins et al., 1989b; Askins et al., 1990.

80. Diamond, 1991.

81. Petit, et al., 1995.
82. Robbins et al., 1992b.
83. James et al., 1996.
84. Sherry and Holmes, 1995.
85. Rappole et al., 1989.
86. Sherry and Holmes, 1996.
87. Stotz et al., 1996.
88. Robbins et al., 1992b; Mayfield, 1996, Haney et al., 1998.
89. Hutto, 1988; Rappole and McDonald, 1994.
90. Sherry and Holmes, 1992; Marra et al., 1993; Holmes et al., 1995. Also, see Johnson and Geupel (1996) for similar results for a California population of Swainson's Thrush.
91. Robbins et al. 1989b.
92. Sherry and Holmes, 1995.
93. Holmes et al., 1996; Sherry and Holmes, 1996.
94. Moore et al., 1995; Ewert and Hamas, 1995.
95. McCann et al., 1993.
96. McCann et al., 1993.
97. Moore et al., 1990.
98. Weisbrod et al., 1993.
99. Ewert and Hamas, 1995.
100. Martin, 1980.
101. Askins, 1993.

CHAPTER 6. Industrial Forestry and the Prospects for Northern Birds

Epigraph: Thoreau, 1864: 89.

1. Mönkkönen and Welsh, 1994.
2. Haila and Järvinen, 1990.
3. Erskine, 1977: 9; Haila and Järvinen, 1990.
4. Haila and Järvinen, 1990.
5. Erskine, 1977: 12.
6. Davis, 1976.
7. Erskine, 1977: 12.
8. Davis, 1976.
9. Johnson, 1992: 6.
10. Johnson, 1992: 5.
11. Johnson, 1992: 35.
12. Johnson, 1992: 1.
13. Johnson, 1992: 35-36, 100.
14. Wright, 1974; Van Wagner, 1978.
15. Telfer, 1992.
16. Telfer, 1992.
17. Bergeron and Harvey, 1997.
18. Syrjänen et al., 1994; Bergeron and Harvey, 1997.
19. Lorimer, 1977.
20. Blackford, 1955; Appelbaum and Haney, 1981.
21. Villard and Beninger, 1993.
22. Erskine, 1977: 35, 41; Titterington et al., 1979; Appelbaum and Haney, 1981; Welsh, 1988.
23. Erskine, 1977: 18, 23; Titterington et al., 1979; Welsh, 1988; Telfer, 1992; Kirk et al., 1996.
24. Zumeta, 1991.
25. Erskine, 1977: 58.
26. Whitney, 1987.
27. Frelich, 1995.
28. Hansson, 1992.
29. Frelich, 1995.
30. Hagan et al., 1997.
31. Hakkila, 1995: 8-9,11.
32. Hakkila, 1995: 12-13.
33. Hakkila, 1995: 35-37.
34. Hakkila, 1995: 9.
35. Mönkkonen and Welsh, 1994.
36. Väisänen et al., 1986.
37. Helle and Järvinen, 1986; Väisänen et al., 1986; Haila and Järvinen, 1990.
38. Virkkala, 1987.
39. Helle, 1985; Väisänen et al., 1986.
40. Väisänen et al., 1986.
41. Haila and Järvinen, 1990.
42. Virkkala, 1987.
43. Virkkala, 1987.
44. Virkkala, 1987.
45. Rolstad and Wegge, 1989.
46. Johnsgard, 1983: 224.
47. Johnsgard, 1983: 235.
48. Rolstad and Wegge, 1989.
49. Schmiegelow et al., 1997.
50. Haapanen, 1965.
51. Virkkala et al., 1994.

52. Angelstam and Mikusiánski, 1994.

53. Haila, 1994; Syrjänen et al., 1994.

54. Angelstam, 1996.

55. Syrjänen et al., 1994.

56. Helle, 1985; Väisänen et al., 1986.

57. Benkman, 1992.

58. Benkman, 1993a.

59. Bent, 1968: 508; Benkman, 1992.

60. Benkman, 1992.

61. Benkman, 1992.

62. Hahn, 1995.

63. Newton, 1972: 231.

64. Benkman, 1993b.

65. Benkman, 1993b.

66. Benkman, 1993b.

67. Benkman, 1992.

68. Groth, 1988; Benkman, 1993a. In contrast to Red Crossbills in North America, those in Europe feed primarily on spruce seeds (Newton, 1972: 229). Parrot Crossbills are the specialists on pine seeds in Europe.

69. Groth, 1988; Groth, 1993a; DeBenedictis, 1995.

70. Adkisson, 1996.

71. Adkisson, 1996.

72. Adkisson, 1996.

73. Groth, 1988.

74. Groth, 1993a; DeBenedictis, 1995.

75. Benkman, 1993b.

76. Groth, 1988.

77. Bock and Lepthien, 1976.

78. Bock and Lepthien, 1976.

79. Bock and Lepthien, 1976.

80. Newton, 1972: 228.

81. Pimm, 1990; Benkman, 1993c.

82. DeBenedictis, 1995.

83. Pimm, 1990; Benkman, 1993c

84. Thompson and Curran, 1995.

85. Benkman, 1992; Fuller, 1995: 5.

86. Shelford, 1978: 128.

87. Morris et al., 1958.

88. Morris et al., 1958.

89. MacArthur, 1958.

90. Mönkkonen and Welsh, 1994.

91. Schmiegelow and Hannon, 1993.

92. Mönkkonen and Welsh, 1994.

93. Bergeron and Harvey, 1997.

CHAPTER 7. Birds of the Western Mountain Slopes

Epigraph: Austin, 1924: 36-37; Mary Hunter Austin, The Land of Journeys' Ending. 1924. Reprinted AMS Press, Inc.

1. Hejl et al., 1995.

2. Dobkin, 1994.

3. Peet, 1988.

4. Freemark et al., 1995.

5. Peet, 1988.

6. Kricher and Morrison, 1993: 140; Hejl et al., 1995.

7. Hejl et al., 1995.

8. Hejl et al., 1995.

9. Hejl et al., 1995.

10. Peet, 1988.

11. Kricher and Morrison, 1993: 183; Hejl et al., 1995.

12. Peet, 1988.

13. Peet, 1988.

14. Before European settlement, dense forests of old ponderosa pines grew in the northern Black Hills of South Dakota (Shineman and Baker, 1997). These forests were apparently subject to infrequent catastrophic fires that killed most of the trees. However, open, parklike woodlands of ponderosa pine were maintained by frequent ground fires in many other regions.

15. Peet, 1988.

16. Romme and Despain, 1989.

17. Hutto, 1995.

18. Franklin, 1988.

19. Gruell, 1983.

20. Hutto, 1995.

21. Dobkin, 1994: 14; Hejl, 1994; Hejl et al., 1995; Hutto, 1995.

22. Marshall, 1963; Brawn and Balda, 1988.

23. Hutto, 1995.

24. Hutto, 1995.
25. Meslow, 1978.
26. Meslow and Wight, 1975.
27. Hejl et al., 1995.
28. These patterns are generally confirmed by Hutto's (1995) comparison of clearcuts and mature forests in the northern Rocky Mountains.
29. Szaro and Balda, 1979.
30. Dellasala et al., 1996.
31. Bryant et al. (1993) found that the last three species were primarily concentrated in old-growth forests (more than 200 years old) along the coast of British Columbia, but that Golden-crowned Kinglets were equally common in mature forests of different ages.
32. Peterson, 1982; Hejl et al., 1995; Hutto, 1998.
33. Rosenberg and Raphael, 1986.
34. Lehmkuhl et al., 1991. Also see McGarigal and McCombs, 1995.
35. Keller and Anderson, 1992.
36. Schieck et al., 1995.
37. Hejl et al., 1995.
38. Wilcove, 1986.
39. Carey, 1985.
40. Guitiérrez et al., 1995.
41. Barrows and Barrows, 1978.
42. Guitiérrez et al., 1995.
43. Guitiérrez et al., 1995.
44. Lamberson et al., 1992.
45. Marshall, 1988.
46. Bent, 1919: 142.
47. The Asian subspecies of the Marbled Murrelet recently was reclassified as a separate species called the Long-billed Murrelet, *Brachyramphus perdix* (Konyukhov and Kitaysky, 1995; American Ornithologists' Union, 1998).
48. Binford et al., 1975.
49. Marshall, 1988.
50. Ford and Brown, 1995.
51. Day, 1995.
52. Burger, 1995; Grenier and Nelson, 1995; Kuletz et al., 1995; Miller and Ralph, 1995; Raphael et al., 1995.

53. Burger, 1995; Hanners, 1995; Miller and Ralph, 1995.
54. Marshall, 1988; Singer et al., 1991; Naslund, 1993.
55. Naslund et al., 1995.
56. Marshall, 1988; Naslund, 1993.
57. Marshall, 1988; Piatt and Ford, 1993; Ralph, 1994; Ralph et al., 1995.
58. Marshall, 1988.
59. Hamer, 1995.
60. Hobson, 1990; Ralph et al., 1995.
61. Carter et al., 1995; Fry, 1995.
62. Piatt et al., 1990; Piatt and Naslund, 1995.
63. Ralph, 1994; Ralph et al., 1995.
64. Ralph, 1988.
65. Ralph, 1988; Andersen and Bessinger, 1995.
66. Beissinger, 1995.
67. Piatt and Naslund, 1995.
68. Nelson et al., 1995.
69. Burger, 1995; Nelson and Hamer, 1995.
70. Singer et al., 1991.
71. John M. Marzluff, personal communication.
72. Steadman et al., 1994.
73. Marzluff and Balda, 1992: 28.
74. Lanner, 1996: 16.
75. Balda, 1987.
76. Balda, 1987.
77. Marzluff and Balda, 1992: 6.
78. Marzluff and Balda, 1992: 34.
79. Balda and Bateman, 1971; Ligon, 1978.
80. Marzluff and Balda, 1992: 35.
81. Lanner, 1996: 35.
82. Marzluff and Balda, 1992: 32.
83. Marzluff and Balda, 1992: 36, 38.
84. Ligon, 1978.
85. Marzluff and Balda, 1992: 157.
86. Marzluff and Balda, 1992: 132, 155.
87. Ligon, 1978.
88. Marzluff and Balda, 1992: 191.
89. Marzluff and Balda, 1992: 160.

90. Marzluff and Balda, 1992: 195.
91. Bent, 1946: 306.
92. Sauer et al., 1995.
93. Leopold, 1949: 138.
94. Tomback, 1980.
95. Tomback, 1978.
96. Lanner, 1996: 26.
97. Lanner, 1996: 43, 44; Tomback, 1982.
98. Tomback and Linhart, 1990.
99. Lanner and Vander Wall, 1980.
100. Tomback, 1980.
101. Lanner, 1996: 48.
102. Lanner, 1996: 52–54.
103. Tomback, 1980.
104. Lanner, 1996: 122–125.
105. Sauer et al., 1995.
106. Benkman, 1993a.
107. Peterjohn and Sauer, 1994.
108. Sauer et al., 1995.
109. Hutto, 1992.
110. Tomback, 1978; Lanner, 1996: 57–58.

CHAPTER 8. Declining Birds of Southwestern Floodplains

Epigraph: Cather, 1927: 24; reprinted by permission of Random House.

1. Knopf et al., 1988.
2. Saab et al., 1995.
3. Johnson et al., 1977; Saab et al., 1995.
4. Carothers and Johnson, 1975; Hubbard, 1977.
5. Ohmart, 1994.
6. Sauer et al., 1995.
7. Ohmart, 1994.
8. Rosenberg et al., 1991: 9.
9. Ohmart and Anderson, 1982: 435–436.
10. Rosenberg et al., 1991: 6.
11. Ohmart and Anderson, 1982: 437.
12. Ohmart and Anderson, 1982: 437.
13. Rosenberg et al., 1991: 33.
14. Walsberg, 1975.
15. Larson, 1996.
16. Anderson and Ohmart, 1978; Ohmart and Anderson, 1982
17. Ohmart and Anderson, 1982.
18. Anderson and Ohmart, 1978; Rosenberg et al., 1991: 279.
19. Walsberg, 1977.
20. Rand and Rand, 1943.
21. These sally flights may be territorial displays rather than foraging maneuvers (Kenneth Rosenberg, personal communication).
22. Walsberg, 1977.
23. Walsberg, 1977.
24. Walsberg, 1977.
25. Walsberg, 1977.
26. Sauer et al., 1995.
27. Rosenberg et al., 1991: 85.
28. Yong and Finch, 1997.
29. Terrill and Ohmart, 1984.
30. Rosenberg et al., 1991: 35.
31. Strong and Bock, 1990.
32. Dobkin and Wilcox, 1986.
33. Hunter et al., 1985.
34. Rosenberg et al., 1991: 15–17.
35. Brady et al., 1985.
36. Rosenberg et al., 1991: 21–22.
37. Rosenberg et al., 1991: 37–38.
38. Ohmart and Anderson, 1982: 454.
39. Hunter et al., 1985.
40. Ohmart and Anderson, 1982: 461.
41. Ohmart and Anderson, 1982: 455.
42. Carothers and Johnson, 1975.
43. Carothers and Johnson, 1975.
44. Gaines, 1977.
45. Ohmart and Anderson, 1982: 460.
46. Carothers and Johnson, 1975.
47. Fox, 1977.
48. Emslie, 1987.
49. Ohmart, 1994.
50. Carothers, 1977; Ohmart, 1994.
51. Dave, 1977; Ohmart and Anderson, 1982: 457–48.
52. Kennedy, 1977; Ohmart, 1994.
53. Ohmart and Anderson, 1982: 456.
54. Ohmart, 1994.
55. Ohmart, 1994.

56. Parker et al., 1985.
57. Saab et al., 1995.
58. Knopf et al., 1988.
59. Krueper, 1993.
60. Ohmart, 1994.
61. Saab et al., 1995.
62. Knopf et al., 1988.
63. Ames, 1977.
64. Saab et al., 1995.
65. Rosenberg et al., 1991: 58–63.
66. Anderson and Ohmart, 1985.
67. Anderson and Ohmart, 1985.
68. Swenson and Mullins, 1985.
69. Rosenberg et al., 1991: 60–62.
70. Rosenberg et al., 1991: 62.
71. Nabhan, 1985.
72. Anderson and Ohmart, 1985.

CHAPTER 9. Red-cockaded Woodpeckers and the Longleaf Pine Woodland

Epigraph: Bartram, 1791: 51.

1. Phillips, 1994.
2. Frost, 1993.
3. Hudson, 1997: 146–147.
4. Peet and Allard, 1993.
5. Frost, 1993.
6. Frost, 1993; Noss et al., 1995.
7. See appendix E in Noss et al., 1995.
8. Walker, 1993.
9. Frost, 1993; Jackson, 1994.
10. Christensen, 1988; Wilson et al., 1995.
11. Frost, 1993.
12. Means and Grow, 1985; Rebertus et al., 1993.
13. Means and Grow, 1985.
14. Frost, 1993.
15. Means and Grow, 1985.
16. Frost, 1993.
17. Means and Grow, 1985.
18. Frost, 1993.
19. Robbins and Myers, 1992.
20. Streng et al., 1993.
21. Streng et al., 1993.
22. Means, 1996.
23. Streng et al., 1993.
24. Means and Grow, 1985.
25. Rebertus et al, 1993.
26. Robbins and Myers, 1992.
27. Engstrom et al., 1984.
28. Wilson et al., 1995.
29. Dunning, 1993.
30. LeGrand and Schneider, 1992.
31. Hardin et al., 1982; LeGrand and Schneider, 1992; Dunning, 1993.
32. Dunning and Watts, 1990.
33. Dunning and Watts, 1991.
34. Haggerty, 1988.
35. LeGrand and Schneider, 1992.
36. Dunning, 1993.
37. Shackelford and Conner, 1997.
38. Jackson, 1995.
39. Hanula and Franzreb, 1995.
40. Walters, 1991.
41. Jackson, 1974; McFarlane, 1992: 89.
42. Rudolph et al., 1990.
43. Jackson, 1974.
44. Rudolph and Conner, 1991.
45. Jackson, 1986; Conner and Rudolph, 1995b.
46. Conner et al., 1998.
47. Hooper, 1988; Rudolph and Conner, 1991.
48. Conner and Locke, 1982; Hooper et al., 1991b.
49. Conner and Rudolph, 1995b.
50. Hooper et al., 1990.
51. Hooper, 1988.
52. Conner and Rudolph, 1995b.
53. Walters, 1991.
54. Costa and Walker, 1995.
55. Costa and Walker, 1995.
56. Costa and Walker, 1995.
57. Engstrom and Evans, 1990.
58. McFarlane, 1992: 248.
59. Hooper et al., 1990; Walters, 1991; Watson et al., 1995.
60. Conner et al., 1991.
61. Conner and Rudolph, 1995a.
62. Rudolph and Conner, 1995.
63. Conner et al., 1991.
64. Todd Engstrom, personal communication.

65. Jackson, 1986.
66. Conner and Rudolph, 1995a.
67. Taylor and Hooper, 1991.
68. Taylor and Hooper, 1991.
69. Allen, 1991.
70. Walters, 1991.
71. Conner and Rudolph, 1995a.
72. Hooper et al., 1990; Walters, 1991.
73. Watson et al., 1995.
74. Jackson, 1986.
75. Seagle et al., 1987.
76. Jackson, 1988.
77. Lennartz, 1988; Hooper et al., 1991a.
78. Means, 1996.
79. Simberloff, 1993.
80. Conner et al. 1997.
81. Conner et al., 1997.
82. Walters, 1991; McFarlan, 1995.
83. Hooper et al., 1982.
84. Reed et al., 1988.
85. Jackson, 1994.
86. Conner and Rudolph, 1991; Hanula and Franzreb, 1995.
87. Allen et al. 1993.
88. Hess and Costa, 1995.
89. Rudolph et al., 1992.
90. Costa and Kennedy, 1994.
91. Pulliam et al., 1992.

CHAPTER 10. Landscape Ecology: The Key to Bird Conservation

Epigraph: Tanner, 1942: 89.

1. Tanner, 1942: xi.
2. Tanner, 1942: 94–96.
3. Buscemi, 1978.
4. Williams, 1989: 279–280.
5. Chapman, 1932.
6. Andrén, 1994.
7. Hall, 1988; Rabenold et al., 1998.
8. World Resources Institute, 1994.
9. Diamond et al., 1987.
10. Keiter, 1991.
11. Stevens, 1992.
12. Kricher and Morrison, 1993: 112.
13. Sexton, 1992.
14. Sexton, 1992.
15. Kricher and Morrison, 1993: 113.
16. Kricher and Morrison, 1993: 113.
17. Grzybowski et al., 1994.
18. Stevens, 1992.
19. Walker, 1993.

References

Adkisson, C. 1996. Red Crossbill. In A. Poole, P. Stettenheim, and F. Gill (editors), The birds of North America, no. 256. Academy of Natural Sciences, Philadelphia, and American Ornithologists' Union, Washington, D.C.

Agnew, W., D. W. Uresk, and R. M. Hansen. 1986. Flora and fauna associated with prairie dog colonies and adjacent ungrazed mixed-grass prairie in western South Dakota. Journal of Range Management 39: 135–139.

Alerich, C. L., and D. A. Drake. 1995. Forest statistics for New York: 1980 and 1993. Resource Bulletin NE-132, Northeastern Experiment Station, USDA Forest Service, Radnor, Pa., 249 pp.

Allen, D. H. 1991. An insert technique for constructing artificial Red-cockaded Woodpecker cavities. USDA Forest Service General Technical Report SE-73, Asheville, N.C.

Allen, D. H., K. E. Franzreb, and R. F. Escano. 1993. Efficacy of translocation strategies for Red-cockaded Woodpeckers. Wildlife Society Bulletin 21: 155–159.

Amacher, G. S., J. Sullivan, L. Shabman, L. Zepp, and D. Grebner. 1998. Reforestation of flooded farmland. Journal of Forestry 96: 10–17.

Ambuel, B., and S. A. Temple. 1982. Songbird populations in southern Wisconsin forests: 1954 and 1979. Journal of Field Ornithology 53: 149–158.

Andrén, H. 1994. Effects of habitat fragmentation on birds and mammals in landscapes with different proportions of suitable habitat: A review. Oikos 71: 355–366.

American Ornithologists' Union. 1983. The AOU check-list of North American birds. 6th ed. American Ornithologists' Union, Lawrence, Kan., 877 pp.

American Ornithologists' Union. 1998. The AOU check-list of North American birds. 7th ed. American Ornithologists' Union, Lawrence, Kan., 829 pp.

Ames, C. R. 1977. Wildlife conflicts in riparian management: Grazing. Pages 49–51 in R. R. Johnson and D. A. Jones (technical coordinators), Importance, preservation and management of riparian habitat: A symposium. USDA Forest Service General Technical Report RM-43, Fort Collins, Colo.

Andersen, H. L., and S. R. Beissinger. 1995. Preliminary observations on juvenile:adult ratios of Marbled Murrelets in Auke Bay, Southeast Alaska. Northwestern Naturalist 76: 79–81.

Anderson, R. C. 1990. The historic role of fire in the North American grassland. Pages 8–18 in S. L. Collins and L. L. Wallace (editors), Fire in the North American tallgrass prairies. University of Oklahoma Press, Norman.

Anderson, B. W., and R. D. Ohmart. 1978. Phainopepla utilization of honey mesquite forests in the Colorado River Valley. Condor 80: 334–338.

Anderson, B. W., and R. D. Ohmart. 1985. Riparian revegetation as a mitigating process in stream and river restoration. In J. A. Gore (editor), The restoration of rivers and streams: Theories and experience. Butterworth, Boston.

Anderson, S. H., and H. H. Shugart, Jr. 1974. Habitat selection of breeding birds in an east Tennessee deciduous forest. Ecology 55: 828–837.

Andersson, L., and T. Appelquist. 1990. The influence of the Pleistocene megafauna on the nemoral and the boreonemoral ecosystem: A hypothesis with implications for nature conservation strategy. Svensk Botanisk Tidskrift 84: 355–368.

Andrle, R. F., and J. H. Carroll. 1988. The atlas of breeding birds in New York State. Cornell University Press, Ithaca, N.Y., 551 pp.

Angelstam, P. 1996. The ghost of forest past—Natural disturbance regimes as a basis for reconstruction of biologically diverse forests in Europe. Pages 287–337 in R. M. De-Graaf and R. I. Miller (editors), Conservation of faunal diversity in forested landscapes. Chapman and Hall, New York.

Angelstam, P., and G. Mikusiánski. 1994. Woodpecker assemblages in natural and managed boreal and hemiboreal forest—A review. Annales Zoologici Fennici 31: 157–172.

Antenen, S., M. Jordan, K. Motivans, J. B. Washa, and R. Zaremba. 1994. Hempstead Plains Fire Management Plan, Nassau County, Long Island, N.Y. Report to the Long Island Chapter of the Nature Conservancy.

Apa, A. D., D. W. Uresk, and R. L. Linder. 1991. Impacts of black-tailed prairie dog rodenticides on nontarget passerines. Great Basin Naturalist 51: 301–309.

Appelbaum, S., and A. Haney. 1981. Bird populations before and after wildfire in a Great Lakes pine forest. Condor 83: 347–354.

Askins, R. A. 1987. Territories: A key to understanding bird behavior. American Birds 41: 35–40.

Askins, R. A. 1990. Birds of the Connecticut College Arboretum: Population changes over forty years. Connecticut College Arboretum Bulletin 31, 43 pp.

Askins, R. A. 1993. Population trends in grassland, shrubland, and forest birds in eastern North America. Pages 1–34 in D. M. Power (editor), Current Ornithology, vol. 11. Plenum, New York.

Askins, R. A. 1994. Open corridors in a heavily forested landscape: Impact on shrubland and forest-interior birds. Wildlife Society Bulletin 22: 339–347.

Askins, R. A., and D. N. Ewert. 1991. Impact of Hurricane Hugo on bird populations on St. John, U.S. Virgin Islands. Biotropica 23: 481–487.

Askins, R. A, D. N. Ewert, and R. Norton. 1992. Abundance of wintering migrants in fragmented and continuous forests in the U.S. Virgin Islands. Pages 197–206 in J. W. Hagan and D. W. Johnston (editors), Ecology and Conservation of Neotropical Migrant Landbirds. Smithsonian Institution Press, Washington, D.C.

Askins, R. A., J. F. Lynch, and R. Greenberg. 1990. Population declines in migratory birds in eastern North America. Pages 1–57 in D. M. Power (editor), Current Ornithology, vol. 7. Plenum, New York.

Askins, R. A., M. J. Philbrick, and D. S. Sugeno. 1987. Relationship between the regional abundance of forest and the composition of forest bird communities. Biological Conservation 39: 129–152.

Audubon, J. J. 1831. Ornithological biography, or an account of the birds of the United States of America, vol. 1. J. Dobson, Philadelphia, 512 pp.

Audubon, J. J. 1834. Ornithological biography, or an account of the birds of the United States of America, vol. 2. A. & C. Black, Edinburgh, 588 pp.

Austin, M. 1924 [1969]. The land of journeys' ending. AMS, New York, 459 pp.

Baird, T. H. 1990. Changes in breeding bird populations between 1930 and 1985 in the Quaker Run Valley of Allegany State Park, New York. New York State Museum Bulletin, no. 477, 41 pp.

Balda, R. P. 1987. Avian impacts on pinyon-juniper woodlands. Pages 525–533 in R. L. Everett (compiler), Proceedings of the Pinyon—Juniper Conference, Reno, Nevada, Jan. 13–16, 1986. USDA Forest Service General Technical Report INT-215, Ogden, Utah.

Balda, R. P., and G. C. Bateman. 1971. Flocking and annual cycle of the Piñon Jay, *Gymnorhinus cyanocephalus*. Condor 73: 287–302.

Barnes, A. M. 1993. A review of plague and its relevance to prairie dog populations and the black-footed ferret. U.S. Fish and Wildlife Service Biological Report 0 (13): 28–37.

Barrows, C., and K. Barrows. 1978. Roost characteristics and behavioral thermoregulation in the Spotted Owl. Western Birds 9: 1–8.

Bartram, W. 1791. Travels through North and South Carolina, Georgia, East and West Florida. James and Johnson, Philadelphia, 425 pp.

Basili, G. D., and S. A. Temple. 1994. Central plains of America and llanos of Venezuela: Where are Dickcissel populations limited? Abstracts of the 1994 North American Ornithological Conference, Missoula, Mont.

Basili, G. D., and S. A. Temple. 1995. A perilous migration. Natural History 104 (9): 40–47.

Beissinger, S. R. 1995. Population trends of the Marbled Murrelet projected from demographic analyses. Pages 385–393 in Ralph, C. J., G. L. Hunt, Jr., M. G. Raphael, and J. F. Piatt (technical coordinators), Ecology and conservation of the Marbled Murrelet. USDA Forest Service General Technical Report PSW-GTR-152, Albany, Calif.

Benkman, C. W. 1992. White-winged Crossbill. In A. Poole, P. Stettenheim, and F. Gill (editors), The birds of North America, no. 27. Academy of Natural Sciences, Philadelphia, and American Ornithologists' Union, Washington, D.C.

Benkman, C. W. 1993a. Adaptation to single resources and the evolution of crossbill (*Loxia*) diversity. Ecological Monographs 63: 305–325.

Benkman, C. W. 1993b. Logging, conifers, and the conservation of crossbills. Conservation Biology 7: 473–479.

Benkman, C. W. 1993c. The evolution, ecology, and decline of the Red Crossbill of Newfoundland. American Birds 47: 225–229.

Bent, A. C. 1919. Life histories of North American diving birds. Smithsonian Institution, United States National Museum Bulletin 107, 239 pp.

Bent, A. C. 1929. Life histories of North American shore birds. Smithsonian Institution, United States National Museum Bulletin 146, 412 pp.

Bent, A. C. 1939. Life histories of North American woodpeckers. Smithsonian Institution, United States National Museum Bulletin 174, 334 pp.

Bent, A. C. 1946. Life histories of North American jays, crows and titmice. Smithsonian Institution, United States National Museum Bulletin 191, 495 pp.

Bent, A. C. 1953. Life histories of North American wood warblers. Smithsonian Institution, United States National Museum Bulletin 203, 734 pp.

Bent, A. C. 1968. Life histories of North American cardinals, grosbeaks, buntings, towhees, finches, sparrows, and allies. Part 1. Smithsonian Institution, United States National Museum Bulletin 237, 602 pp.

Bergeron, Y., and B. Harvey. 1997. Basing silviculture on natural ecosystem dynamics: An approach applied to the southern boreal mixedwood forest of Quebec. Forest Ecology and Management 92: 235–242.

Bevier, L. R. (editor). 1994. The atlas of breeding birds of Connecticut. State Geological and Natural History Survey of Connecticut Bulletin 113.

Binford, L. C., B. G. Elliott, and S. W. Singer. 1975. Discovery of a nest and downy chick of the Marbled Murrelet. Wilson Bulletin 87: 303–319.

Bjorndal, K. Q. 1980. Nutrition and grazing behavior of the green turtle *Chelonia mydas*. Marine Biology 56: 147–154.

Black, J. D. 1950. The rural economy of New England. Harvard University Press, Cambridge, 796 pp.

Blackford, J. L. 1955. Woodpecker concentration in burned forest. Condor 57: 28–30.

Blake, J. G., and B. A. Loiselle. 1992. Habitat use by Neotropical migrants at La Selva Biological Station and Braulio Carrillo National Park, Costa Rica. Pages 257–272 in J. M. Hagan III and D. W. Johnston (editors), Ecology and conservation of neotropical migrant landbirds. Smithsonian Institution Press, Washington, D.C.

Blockstein, D. E., and H. B. Tordoff. 1985. A contemporary look at the extinction of the Passenger Pigeon. American Birds 39: 845–851.

Bock, C. E., and L. W. Lepthien. 1976. Synchronous eruptions of boreal seed-eating birds. American Naturalist 110: 559–571.

Bollinger, E. K., and T. A. Gavin. 1992. Eastern Bobolink populations: Ecology and conservation in an agricultural landscape. Pages 497–506 in J. M. Hagan III and D. W. Johnston (editors), Ecology and conservation of neotropical migrant landbirds. Smithsonian Institution Press, Washington, D.C.

Bollinger, E. K., P. B. Bollinger, and T. A. Gavin. 1990. Effects of hay-cropping on eastern populations of the Bobolink. Wildlife Society Bulletin 18: 142–150.

Borman, F. H., and G. E. Likens. 1979. Catastrophic disturbance and the steady state in northern hardwood forests. American Scientist 67: 660–669.

Botkin, D. B., D. A. Woodby, and R. A. Nisbet. 1991. Kirtland's Warbler habitats: A possible early indicator of climatic warming. Biological Conservation 56: 63–78.

Brady, W., D. R. Patton, and J. Paxson. 1985. The development of southwestern riparian gallery forests. Pages 39–43 in R. R. Johnson, C. D. Ziebell, D. R. Patton, P. F. Ffolliott, and R. H. Hamre (technical coordinators), Riparian ecosystems and their management: Reconciling conflicting uses. USDA Forest Service General Technical Report RM-120, Fort Collins, Colo.

Bramble, W. C., R. H. Yahner, and W. R. Byrnes. 1992. Breeding-bird population changes following right-of-way maintenance treatments. Journal of Arboriculture 18: 23–32.

Brawn, J. D., and R. P. Balda. 1988. The influence of silvicultural activity on ponderosa pine forest bird communities in the southwestern United States. Bird Conservation 3: 3–21.

Brawn, J. D., and S. K. Robinson. 1996. Source-sink population dynamics may complicate the interpretation of long-term census data. Ecology 77: 3–12.

Brewer, R., G. A. McPeek, and R. J. Adams, Jr. 1991. The atlas of breeding birds of Michigan. Michigan State University Press, East Lansing, 594 pp.

Brittingham, M. C., and S. A. Temple. 1983. Have cowbirds caused forest songbirds to decline? BioScience 33: 31–35.

Brooks, M. 1938. The eastern Lark Sparrow in the upper Ohio Valley. Cardinal 4: 181–200.

Brooks, R. T., and T. W. Birch. 1988. Changes in New England forests and forest owners: Implications for wildlife habitat resources and management. Transactions of the Fifty-third North American Wildlife and Natural Resources Conference, pp. 78–87.

Brown, J. H., and W. McDonald. 1995. Livestock grazing and conservation on southwestern rangelands. Conservation Biology 9: 1644–1647.

Bryan, G. G., and L. B. Best. 1994. Avian nest density and success in grassed waterways in Iowa rowcrop fields. Wildlife Society Bulletin 22: 583–592.

Bucher, E. H. 1992. The causes of the extinction of the Passenger Pigeon. Pages 1–36 in D. M. Power (editor), Current Ornithology, vol. 9. Plenum, New York.

Budiansky, S. 1993. The doomsday myths. U.S. News and World Report 115 (23): 81–83 (December 13).

Budiansky, S. 1994. Extinction of miscalculation? Nature 370: 105.

Bull, J., 1974. Birds of New York state. Doubleday / Natural History Press, Garden City, N.Y., 655 pp.

Burger, A. E. 1995. Inland habitat associations of Marbled Murrelets in British Columbia. Pages 151–161 in C. J. Ralph, G. L. Hunt, Jr., M. G. Raphael, and J. F. Piatt (technical coordinators), Ecology and conservation of the Marbled Murrelet. USDA Forest Service General Technical Report PSW-GTR-152, Albany, Calif.

Burke, D. M., and E. Nol. 1998. Influence of food abundance, nest-site habitat, and forest fragmentation on breeding Ovenbirds. Auk 115: 96–104.

Buscemi, D. 1978. The last American parakeet. Natural History 87: 10–12.

Butcher, G. S., W. A. Niering, W. J. Barry, and R. H. Goodwin. 1981. Equilibrium biogeography and the size of nature preserves: An avian case study. Oecologia 49: 29–37.

Byrant, A. A., J. P. L. Savard, and R. T. McLaughlin. 1993. Avian communities in old-growth and managed forests of western Vancouver Island, British Columbia. Technical Report Series no. 167, Canadian Wildlife Service, Pacific and Yukon Region, British Columbia, 115 pp.

Cain, S. A., M. Nelson, and W. McLean. 1937. Andropogonetum Hempsteadi: A Long Island grassland vegetation type. American Midland Naturalist 18: 334–350.

Carey, A. B. 1985. A summary of the scientific basis for Spotted Owl management. Pages 100–114 in R. J. Gutiérrez and A. B. Carey, Ecology and management of the Spotted Owl in the Pacific Northwest. General Technical Report PNW-185, Pacific Northwest Forest Experiment Station, U.S. Department of Agriculture, Portland, Ore.

Carothers, S. W. 1977. Importance, preservation and management of riparian habitats: An overview. Pages 2–4 in R. R. Johnson and D. A. Jones, Importance, preservation and management of riparian habitat: A symposium. USDA Forest Service General Technical Report RM-43, Fort Collins, Colo.

Carothers, S. W., and R. R. Johnson. 1975. Water management practices and their effects on nongame birds in range habitats. Pages 210–222 in D. R. Smith (technical coordinator), Proceedings of the symposium on management of forest and range habitats for nongame birds. USDA Forest Service General Technical Report WO-1, Washington, D.C.

Carter, H. R., M. L. C. McAllister, and M. E. Isleib. 1995. Mortality of Marbled Murrelets in gill nests in North America. Pages 271–283 in C. J. Ralph, G. L. Hunt, Jr., M. G. Raphael, and J. F. Piatt (technical coordinators), Ecology and conservation of the Marbled Murrelet. USDA Forest Service General Technical Report PSW-GTR-152, Albany, Calif..

Cather, W. 1927. Death comes for the archbishop. Knopf, New York, 299 pp.

Chapman, F. M. 1932. Notes on the plumage of North American birds; ninety-fourth paper. Bird Lore 34: 328–330.

Chasko, G. G., and J. E. Gates. 1982. Avian habitat suitability along a transmission-line corridor in an oak-hickory forest region. Wildlife Monographs 82, 41 pp.

Cheatham, G., and J. Cheatham. 1988. The Ivory Billed Woodpecker in Faulkner's Go Down, Moses. ANQ 1: 61–63.

Christensen, N. L. 1988. Vegetation of the southeastern coastal plain. Pages 317–363 in M. G. Barbour and W. D. Billings, North American Terrestrial Vegetation. Cambridge University Press, New York, 434 pp.

Christensen, N. L., R. B. Burchell, A. Liggett, and E. L. Simms. 1981. The structure and development of pocosin vegetation. Pages 43–61 in C. J. Richardson (editor), Pocosin wetlands: An integrated analysis of coastal plain freshwater bogs in North Carolina. Hutchinson Ross, Stroudsburg, Pa.

Clark, T. D. 1984. The greening of the South: The recovery of land and forest. University Press of Kentucky, Lexington, 168 pp.

Clark, T. W., D. Hinckley, and T. Rich. 1989. The prairie dog ecosystem: Managing for biological diversity. Montana BLM Wildlife Technical Bulletin 2: 13–23.

Clebsch, E. C., and R. T. Busing. 1989. Secondary succession, gap dynamics, and community structure in a southern Appalachian cove forest. Ecology 70: 728–735.

Coles, J. M., and B. J. Orme. 1983. Homo sapiens or Castor fiber? Antiquity 57: 95–102.

Collins, M. 1990. The last rain forests. Oxford University Press, New York, 200 pp.

Confer, J. L. 1992. Golden-winged Warbler. In A. Poole, P. Stettenheim, and F. Gill (editors), The birds of North America, no. 20. Academy of Natural Sciences, Philadelphia, and American Ornithologists' Union, Washington, D.C.

Confer, J. L., and R. T. Holmes. 1995. Neotropical migrants in undisturbed and human-altered forests in Jamaica. Wilson Bulletin 107: 577–589.

Confer, J. L., and K. Knapp. 1981. Golden-winged Warblers and Blue-winged Warblers: The relative success of a habitat specialist and a generalist. Auk 98: 108–114.

Conner, R. N., and C. S. Adkisson. 1975. Effects of clearcutting on the diversity of breeding birds. Journal of Forestry 73: 781–785.

Conner, R. N., and B. A. Locke. 1982. Fungi and Red-cockaded Woodpecker cavity trees. Wilson Bulletin 94: 64–70.

Conner, R. N., and D. C. Rudolph. 1991. Forest habitat loss, fragmentation, and Red-cockaded Woodpecker populations. Wilson Bulletin 103: 446–457.

Conner, R. N., and D. C. Rudolph. 1995a. Losses of Red-cockaded Woodpecker cavity trees to southern pine beetles. Wilson Bulletin 107: 81–92.

Conner, R. N., and D. C. Rudolph. 1995b. Excavation dynamics and use patterns of Red-cockaded Woodpecker cavities: Relationships with cooperative breeding. Pages 343–352 in D. L. Kulhavy, R. G. Hooper, and R. Costa (editors), Red-cockaded Woodpecker: Recovery, ecology and management. Center for Applied Studies in Forestry, College of Forestry, Stephen F. Austin State University, Nacogdoches, Texas.

Conner, R. N., D. C. Rudolph, D. L. Kulhavy, and A. E. Snow. 1991. Causes of mortality of Red-cockaded Woodpecker cavity trees. Journal of Wildlife Management 55: 531–537.

Conner, R. N., D. C. Rudolph, R. R. Schaeffer, and D. Saenz. 1997. Long-distance dispersal of Red-cockaded Woodpeckers in Texas. Wilson Bulletin 109: 157–160.

Conner, R. N., D. Saenz, D. C. Rudolph, W. G. Ross, and D. L. Kulhavy. 1998. Red-cockaded Woodpecker nest-cavity selection: Relationships with cavity age and resin production. Auk 115: 447–454.

Conrad, H. S. 1935. The plant associations of central Long Island. A study in descriptive plant sociology. American Midland Naturalist 16: 433–516.

Costa, R. 1994. Red-cockaded Woodpecker translocations, 1989–1994: State-of-our-knowledge. Pages 74–81 in American Zoo and Aquarium Association Annual Conference Proceedings, Zoo Atlanta, Atlanta.

Costa, R., and J. L. Walker. 1995. Red-cockaded Woodpecker. Pages 86–89 in E. T. LaRoe, G. S. Farris, C. E. Puckett, et al. (editors). Our living resources: A report to the nation on the distribution, abundance, and health of U.S. plants, animals, and ecosystems. U.S. National Biological Service, Washington, D.C.

Crawford, H. S., R. G. Hooper, and R. W. Titterington. 1981. Songbird population response to silvicultural practices in central Appalachian hardwoods. Journal of Wildlife Management 45: 680–692.

Cronon, W. 1983. Changes in the land. Indians, colonists, and the ecology of New England. Hill and Wang, New York, 241 pp.

Crosby, A. W. Jr. 1972. The Columbian exchange: Biological and cultural consequences of 1492. Greenwood Press, Westport, Conn.

Crosby, A. W. Jr. 1986. Ecological imperialism. The biological expansion of Europe, 900–1900. Cambridge University Press, New York, 368 pp.

Crossman, T. I. 1989. Habitat use of Grasshopper and Savannah sparrows at Bradley International Airport and management recommendations. Unpublished master's thesis, University of Connecticut.

Cully, J. F. Jr. 1993. Plague, prairie dogs, and black-footed ferrets. U.S. Fish and Wildlife Service Biological Report 0 (13): 38–49.

Dave, G. A. 1977. Management alternatives for the riparian habitat in the Southwest. Pages 59–61 in R. R. Johnson and D. A. Jones, Importance, preservation and management of riparian habitat: A symposium. USDA Forest Service General Technical Report RM-43, Fort Collins, Colo.

Davis, M. B. 1976. Pleistocene biogeography of temperate deciduous forests. Geoscience and Man 13: 13–26.

Day, G. M. 1953. The Indian as an ecological factor in the northeastern forest. Ecology 34: 329–346.

Day, R. H. 1995. New information on Kittlitz's Murrelet nests. Condor 97: 271–273.

DeBenedictis, P. A. 1995. Red Crossbills, one through eight. Birding 27: 494–501.

DeGraaf, R. M. 1991. Breeding bird assemblages in managed northern hardwood forests in New England. Pages 155–171 in J. E. Rodiek and E. G. Bolen. Wildlife and habitats in managed landscapes. Island Press, Washington, D.C.

DeGraaf, R. M. 1995. Nest predation rates in managed and reserved extensive northern hardwoods forest. Forest Ecology and Management 79: 227–234.

DeGraaf, R. M., and T. J. Maier. 1996. Effect of egg size on predation by white-footed mice. Wilson Bulletin 108: 535–539.

DeGraaf, R. M., and R. I. Miller. 1996. The importance of disturbance and land-use history in New England: Implications for forested landscapes and wildlife conservation. Pages 3–35 in R. M. DeGraaf and R. I. Miller (editors), Conservation of faunal diversity in forested landscapes. Chapman and Hall, New York.

DeGraaf, R. M., and D. D. Rudis. 1986. New England wildlife: Habitat, natural history, and distribution. General Technical Report NE-108, Northeastern Forest Experiment Station, U.S. Department of Agriculture, Broomall, Pa.

Dellasala, D. A., J. C. Hagar, K. A. Engel, W. C. McComb, R. L. Fairbanks, and E. G. Campbell. 1996. Effects of silvicultural modifications of temperate rainforest on breeding and wintering bird communities, Prince of Wales Island, southeast Alaska. Condor 98: 706–721.

Denevan, W. M. 1992. The pristine myth: The landscape of the Americas in 1492. Annals of the Association of American Geographers 82: 369–385.

Dennis, J. V. 1958. Some aspects of the breeding ecology of the Yellow-breasted Chat (Icteria virens). Bird-Banding 29: 169–183.

Diamond, A. W. 1991. Assessment of the risks from tropical deforestation to Canadian songbirds. Transactions of the North American Wildlife and Natural Resources Conference 56: 177–194.

Diamond, J. M., K. D. Bishop, and S. van Balen. 1987. Bird survival in an isolated Javan woodland: Island or mirror? Conservation Biology 1: 132–142.

Dingle, E. 1953a. Bachman's Warbler. Pages 67–74 in A. C. Bent (editor), Life histories of North American warblers. Part 2. Smithsonian Institution, United States National Museum Bulletin 203.

Dingle, E. 1953b. Prothonotary Warbler. Pages 17–30 in A. C. Bent (editor), Life histories of North American warblers. Part 2. Smithsonian Institution, United States National Museum Bulletin 203.

Dirzo, R., and M. C. Garcia. 1992. Rates of deforestation in Los Tuxtlas, a neotropical area in southeast Mexico. Conservation Biology 6: 84–90.

Dobkin, D. S., and B. A. Wilcox. 1986. Analysis of natural forest fragments: Riparian birds of the Toiyabe Mountains, Nevada. Pages 293–299 in J. Verner, M. L. Morrison, and C. J. Ralph (editors), Wildlife 2000: Modeling habitat relationships of terrestrial vertebrates. University of Wisconsin Press, Madison.

Donovan, T. M., P. W. Jones, E. M. Annand, and F. R. Thompson III. 1997. Variation in local-scale effects: Mechanisms and landscape context. Ecology 78: 2064–2075.

Doolittle, W. E. 1992. Agriculture in North America on the eve of contact: A reassessment. Annals of the Association of American Geographers 82: 386–401.

Dreyer, G. D., and W. A. Niering. 1986. Evaluation of two herbicide techniques on electric transmission rights-of-way: Development of relatively stable shrublands. Environmental Management 10: 113–118.

Dublin, H. T., A. R. E. Sinclair, and J. McGlade. 1990. Elephants and fire as causes of multiple stable states in the Serengeti-Mara woodlands. Journal of Animal Ecology 59: 1147–1164.

Dunn, C. P., F. Stearns, G. R. Guntenspergen, and D. M. Sharpe. 1993. Ecological benefits of the Conservation Reserve Program. Conservation Biology 7: 132–139.

Dunning, J. B. Jr. 1993. Bachman's Sparrow. In A. Poole and F. Gill (editors), The birds of North America, no. 38. Academy of Natural Sciences, Philadelphia, and American Ornithologists' Union, Washington, D.C.

Dunning, J. B. Jr., and B. D. Watts. 1990. Regional differences in habitat occupancy by Bachman's Sparrow. Auk 107: 463–472.

Dunning, J. B. Jr., and B. D. Watts. 1991. Habitat occupancy by Bachman's Sparrow in the Francis Marion National Forest before and after Hurricane Hugo. Auk 108: 723–725.

Dunwiddie, P. W. 1989. Forest and heath: The shaping of vegetation on Nantucket Island. Journal of Forest History 33: 126–133.

Dwight, T. 1969. Travels in New England and New York, 4 vols. Belknap Press of Harvard University Press, Cambridge.

Eddleman, W. R., K. E. Evans, and W. H. Elder. 1980. Habitat characteristics and management of Swainson's Warblers in southern Illinois. Wildlife Society Bulletin 8: 228–233.

Ells, S. F. 1995. Breeding Henslow's Sparrows in Lincoln, Massachusetts, 1994. Bird Observer 23: 113–115.

Emslie, S. D. 1987. Age and diet of fossil California Condors in Grand Canyon, Arizona. Science 237: 768–770.

Enemar, A., B. Sjostrand, and S. Svensson. 1978. The effect of observer variability on bird census results obtained by a territory mapping technique. Ornis Scandinavia 9: 31–39.

Engstrom, R. T., and G. W. Evans. 1990. Hurricane damage to Red-cockaded Woodpecker (Picoides borealis) cavity trees. Auk 107: 608–610.

Engstrom, R. T., R. L. Crawford, and W. W. Baker. 1984. Breeding bird populations in relation to changing forest structure following fire exclusion: A 15-year study. Wilson Bulletin 96: 437–450.

Erskine, A. J. 1977. Birds in boreal Canada. Canadian Wildlife Service Report Series, no. 41, Ottawa, Canada.

Evans, E. W., J. M. Biggs, E. J. Finck, D. J. Gibson, S. W. James, D. W. Kaufman, and T. R. Seastedt. 1989. Is fire a disturbance in grasslands? Pages 159–161 in T. B. Bragg and J. Stubbendieck (editors), Proceedings of the Eleventh North American Prairie Conference. University of Nebraska Printing, Lincoln.

Ewert, D. N., and R. A. Askins. 1991. Flocking behavior of migratory warblers in winter in the Virgin Islands. Condor 93: 864–868.

Ewert, D. N., and M. J. Hamas. 1995. Ecology of migratory landbirds during migration in the Midwest. Pages 200–208 in F. R. Thompson (editor), Management of midwestern landscapes for the conservation of neotropical migratory birds. U.S. Forest Service General Technical Report NC–187, North Central Forest Experiment Station, St. Paul, Minn.

Faulkner, W. 1942. Go down, Moses, and other stories. Random House, New York, 383 pp.

Feduccia, A. 1985. Catesby's birds of colonial America. University of North Carolina Press, Chapel Hill, 176 pp.

Finck, E. J. 1984. Male Dickcissel behavior in primary and secondary habitats. Wilson Bulletin 96: 672–680.

Fleischner, T. L. 1994. Ecological costs of livestock grazing in western North America. Conservation Biology 8: 629–644.

Fleischner, T. L. 1995. Grazing and advocacy. Conservation Biology 9: 233–234.

Fleischner, T. L. 1997. Review of "Beyond the Rangeland Conflict: Toward a West that Works" by Dan Daggert. Journal of Wildlife Management 61: 582–584.

Forbush, E. H. 1925. Birds of Massachusetts and other New England states. Part 1. Massachusetts Department of Agriculture, 481 pp.

Forbush, E. H. 1927. Birds of Massachusetts and other New England states. Part 2. Massachusetts Department of Agriculture, 461 pp.

Forbush, E. H. 1929. Birds of Massachusetts and other New England states. Part 3. Massachusetts Department of Agriculture, 466 pp.

Ford, C., and M. Brown. 1995. Unusual Marbled Murrelet nest. Wilson Bulletin 107: 178–179.

Ford, T. B. 1992. Management of thickets for Yellow-breasted Chat. Connecticut Warbler 12: 93–100.

Foster, D. R. 1993. Land-use history and forest transformations in central New England. Pages 91–110 in M. J. McDonell and S. T. A. Pickett (editors), Land-use history and forest transformations in central New England. Springer-Verlag, New York.

Fox, K. 1977. Importance of riparian ecosystems: Economic considerations. Pages 19–22 in R. R. Johnson and D. A. Jones (technical coordinators), Importance, preservation and management of riparian habitat: A symposium. USDA Forest Service General Technical Report RM-43, Fort Collins, Colo.

Frawley, B. J., and L. B. Best. 1991. Effects of mowing on breeding bird abundance and species composition in alfalfa fields. Wildlife Society Bulletin 19: 135–142.

Freemark, K. E., and H. G. Merriam. 1986. Importance of area and habitat heterogeneity to bird assemblages in temperate forest fragments. Biological Conservation 36: 115–141.

Freemark, K. E., J. B. Dunning, S. J. Hejl, and J. R. Probst. 1995. A landscape ecology perspective for research, conservation and management. Pages 381–421 in T. E. Martin and D. M. Finch (editors), Ecology and management of neotropical migratory birds. Oxford University Press, New York.

Frelich, L. E. 1995. Old forest in the Lake States today and before European settlement. Natural Areas Journal 15: 157–167.

Fretwell, S. 1986. Distribution and abundance of the Dickcissel. Pages 211–242 in R. F. Johnston (editor), Current Ornithology, vol 4. Plenum, New York.

Fretwell, S. D. 1979. Dickcissel extinction predicted before the year 2000. Bird Watch 7: 1–3.

Friesen, L., M. D. Cadman, and R. J. McKay. 1999. Nesting success of neotropical migrant songbirds in a highly fragmented landscape. Conservation Biology 13: 338–346.

Friesen, L. E., P. F. J. Eagles, and R. J. MacKay. 1995. Effects of residential development on forest-dwelling neotropical migrant songbirds. Conservation Biology 9: 1408–1414.

Frost, C. C. 1993. Four centuries of changing landscape patterns in the longleaf pine ecosystem. Pages 17–43 in S. M. Hermann (editor), The longleaf pine ecosystem: Ecology, restoration, and management. Proceedings of the Eighteenth Tall Timbers Fire Ecology Conference. Tall Timbers Research, Inc., Tallahassee, Fla.

Fry, D. M. 1995. Pollution and fishing threats to Marbled Murrelets. Pages 257–260 in C. J. Ralph, G. L. Hunt, Jr., M. G. Raphael, and J. F. Piatt (technical coordinators), Ecology and conservation of the Marbled Murrelet. USDA Forest Service General Technical Report PSW-GTR-152, Albany, Calif.

Fuller, R. J. 1995. Bird life of woodland and forest. Cambridge University Press, New York.

Gaines, D. A. 1977. The valley riparian forests of California: Their importance to bird populations. Pages 107–110 in R. R. Johnson and D. A. Jones (technical coordinators), Importance, preservation and management of riparian habitat: A symposium. USDA Forest Service General Technical Report RM-43, Fort Collins, Colo.

Gale, G. A., L. A. Hanners, and S. R. Patton. 1997. Reproductive success of Worm-eating Warblers in a forested landscape. Conservation Biology 11: 246–250.

Gates, J. E., and L. W. Gysel. 1978. Avian nest dispersion and fledging success in field-forest ecotones. Ecology 59: 871–883.

Germaine, S. S., S. H. Vessey, and D. E. Capen. 1997. Effects of small forest openings on the breeding bird community in a Vermont hardwood forest. Condor 99: 708–718.

Gleason, H. A., and A. Cronquist. 1991. Manual of vascular plants of northeastern United States and adjacent Canada. 2nd ed. New York Botanical Garden, New York.

Gradwohl, J., and R. Greenberg. 1980. The formation of antwren flocks on Barro Colorado Island, Panamá. Auk 97: 385–395.

Gram, W. K., and J. Faaborg. 1997. The distribution of neotropical migrant birds wintering in the El Cielo Biosphere Reserve, Tamaulipas, Mexico. Condor 99: 658–670.

Graul, W. D. 1975. Breeding biology of the Mountain Plover. Wilson Bulletin 87: 6–31.

Graul, W. D., and L. E. Webster. 1976. Breeding status of the Mountain Plover. Condor 78: 265–267.

Greenway, J. C. 1958. Extinct and vanishing birds of the world. American Committee for International Wild Life Protection, Special Publication no. 13, New York, 518 pp.

Grenier, J. J., and S. K. Nelson. 1995. Marbled Murrelet habitat associations in Oregon. Pages 191–204 in C. J. Ralph, C. J., G. L. Hunt, Jr., M. G. Raphael, and J. F. Piatt (technical coordinators), Ecology and conservation of the Marbled Murrelet. USDA Forest Service General Technical Report PSW-GTR-152, Albany, Calif.

Griscom, L. 1949. The birds of Concord. Harvard University Press, Cambridge, Mass., 340 pp.

Gross, A. O. 1968. Dickcissel. Pages 158–191 in O. L. Austin, Jr. (editor), Life histories of North American grosbeaks, buntings, towhees, finches, sparrows, and allies. Part 1. U.S. National Museum Bulletin 237.

Gross, A. O. 1932. Heath hen. Pages 264–280 in A. C. Bent (editor), Life histories of North American gallinaceous birds. U.S. National Museum Bulletin 162.

Groth, J. G. 1988. Resolution of cryptic species in Appalachian red crossbills. Condor 90: 745–760.

Groth, J. G. 1993. Evolutionary differentiation in morphology, vocalizations, and allozymes among nomadic sibling species in the North American Red Crossbill (Loxia curvirostra) complex. University of California Publications in Zoology 127: 1–143.

Gruell, G. E. 1983. Fire and vegetation trends in the northern Rockies: Interpretations from 1971–1982 photographs. USDA Forest Service General Technical Report INT-158, Ogden, Utah.

Grzybowski, J. A. 1982. Population structure of grassland bird communities during winter. Condor 84: 137–151.

Grzybowski, J. A., D. J. Tazik, and G. D. Schnell. 1994. Regional analysis of Black-capped Vireo breeding habitats. Condor 96: 512–544.

Guilday, J. E. 1962. The Pleistocene local fauna of the Natural Chimneys, Augusta County, Virginia. Annals of Carnegie Museum 36: 87–122.

Guilday, J. E., P. S. Martin, and A. D. McGrady. 1964. New Paris no. 4: A Pleistocene cave deposit on Bedford County, Pennsylvania. Bulletin of the National Speleological Society 26: 121–194.

Guitiérrez, R. J., A. B. Franklin, and W. S. Lahaye. 1995. Spotted Owl (Strix occidentalis), in A. Poole and F. Gill (editors), The birds of North America, no. 179. Academy of Natural Sciences, Philadelphia, and American Ornithologists' Union, Washington, D.C.

Haapanen, A. 1965. Bird fauna of the Finnish forests in relation to forest succession, Part I. Annales Zoologici Fennici 2: 153–195.

Hagan, J. M. III. 1993. Decline of the Rufous-sided Towhee in the eastern United States. Auk 110: 863–874.

Hagan, J. M. III, T. L. Lloyd-Evans, J. L. Atwood, and D. S. Wood. 1992. Long-term changes in migratory landbirds in the northeastern United States: Evidence from migration capture data. Pages 115–130 in J. M. Hagan III and D. W. Johnston (editors),

Ecology and conservation of neotropical migrant landbirds. Smithsonian Institution Press, Washington, D.C.

Hagan, J. M. III, P. S. McKinley, A. L. Meehan, and S. L. Grove. 1997. Diversity and abundance of landbirds in a northeastern industrial forest landscape. Journal of Wildlife Management 61: 718–735.

Haggerty, T. M. 1988. Aspects of the breeding biology and productivity of Bachman's Sparrow in central Arkansas. Wilson Bulletin 100: 247–255.

Hahn, D. C., and J. S. Hatfield. 1995. Parasitism at the landscape scale: Cowbirds prefer forests. Conservation Biology 9: 1415–1424.

Hahn, T. P. 1995. Integration of photoperiodic and food cues to time changes in reproductive physiology by an opportunistic breeder, the Red Crossbill, *Loxia curvirostra* (Aves: Carduelinae). Journal of Experimental Zoology 272: 213–226.

Haila, L. 1994. Vertebrate distributions relative to clear-cut edges in a boreal forest landscape. Landscape Ecology 9: 105–115.

Haila, Y., and O. Järvinen. 1990. Northern conifer forests and their bird species assemblages. Pages 61–85 in A. Keast (editor), Biogeography and ecology of forest bird communities. SPB Academic Publishing, The Hague, Netherlands.

Hakkila, P. 1995. Procurement of timber for the Finnish forest industries. Research Paper 557, Finnish Forest Research Institute, Vantaa, 73 pp.

Hall, G. A. 1984. Population decline of neotropical migrants in an Appalachian forest. American Birds 38: 14–18.

Hall, G. A. 1988. Birds of the southern Appalachian subalpine forest. Pages 101–117 in J. Jackson (editor), Bird Conservation 3, University of Wisconsin Press, Madison.

Hamel, P. B. 1986. Bachman's Warbler. A species in peril. Smithsonian Institution Press, Washington, D.C., 40 pp.

Hamel, P. B. 1988. Bachman's Warbler. Pages 625–635 in W. J. Chandler (editor), Audubon Wildlife Report, 1988–1989. Academic Press, New York.

Hamel, P. B., H. E. LeGrand, Jr., M. R. Lennartz, and S. A. Gauthreaux, Jr. 1982. Bird habitat relationships on southeastern forest lands. USDA Forest Service General Technical Report SE-22, Asheville, N.C, 417 pp.

Hamer, T. E. 1995. Inland habitat associations of Marbled Murrelets in western Washington. Pages 163–175 in C. J. Ralph, G. L. Hunt, Jr., M. G. Raphael, and J. F. Piatt (technical coordinators), Ecology and conservation of the Marbled Murrelet. USDA Forest Service General Technical Report PSW-GTR-152, Albany, Calif.

Hammett, J. E. 1992. The shapes of adaptation: Historical ecology of anthropogenic landscapes in the southeastern United States. Landscape Ecology 7: 121–135.

Haney, J. C., D. S. Lee, and M. Walsh-McGehee. 1998. A quantitative analysis of winter distribution and habitats of Kirtland's Warblers in the Bahamas. Condor 100: 201–217.

Hanners, L. A., and S. R. Patton. 1998. Worm-eating Warbler. In A. Poole, P. Stettenheim, and F. Gill (editors), The birds of North America, no. 367. Academy of Natural Sciences, Philadelphia, and American Ornithologists' Union, Washington, D.C.

Hansson, L. 1992. Landscape ecology of boreal forests. Trends in Ecology and Evolution 7: 299–302.

Hanski, I., T. Fenske, and G. J. Niemi. 1996. Lack of edge effect in nesting success of breeding birds in managed forest landscapes. Auk 113: 578–585.

Hanula, J. L., and K. E. Franzreb. 1995. Arthropod prey of nestling Red-cockaded Woodpeckers in the upper coastal plain of South Carolina. Wilson Bulletin 107: 485–495.

Hardin, K. I., T. S. Baskett, and K. E. Evans. 1982. Habitat of Bachman's Sparrow breeding on Missouri glades. Wilson Bulletin 94: 208–212.

Harper, R. M. 1911. The Hempstead Plains: A natural prairie on Long Island. Bulletin of the American Geographic Society 43: 351–360.

Hart, J. F. 1968. Loss and abandonment of cleared land in the eastern United States. Annals of the Association of American Geographers 58: 417–440.

Hart, J. F. 1980. Land use change in a Piedmont county. Annals of the Association of American Geographers 70: 492–527.

Hart, R. H., and J. A. Hart. 1997. Rangelands of the Great Plains before Eurpoean settlement. Rangelands 19: 4–11.

Haskell, D. G. 1995. A reevaluation of the effects of forest fragmentation on rates of bird-nest predation. Conservation Biology 9: 1316–1318.

Hejl, S. J. 1994. Human-induced changes in bird populations in coniferous forests in western North America during the past 100 years. Studies in Avian Biology 15: 232–246.

Hejl, S. J., R. L. Hutto, C. R. Preston, and D. M. Finch. 1995. Effects of silvicultural treatments in the Rocky Mountains. Pages 220–244 in T. E. Martin and D. M. Finch (editors), Ecology and management of neotropical migratory birds. Oxford University Press, New York.

Helle, P. 1985. Effects of forest fragmentation on bird densities in northern boreal forests. Ornis Fennica 62: 35–41.

Helle, P., and O. Järvinen. 1986. Population trends of North Finnish land birds in relation to their habitat selection and changes in forest structure. Oikos 46: 107–115.

Henry, J. D., and J. M. A. Swan. 1974. Reconstructing forest history from live and dead plant material—An approach to the study of forest succession in southwest New Hampshire. Ecology 55: 772–783.

Herkert, J. R. 1991. Prairie birds of Illinois: Population response to two centuries of habitat change. Illinois Natural History Survey Bulletin 34: 393–399.

Herkert, J. R. 1994a. The effect of habitat fragmentation on midwestern grassland bird communities. Ecological Applications 4: 461–471.

Herkert, J. R. 1994b. Breeding bird communities of midwestern prairie fragments: The effects of prescribed burning and habitat-area. Natural Areas Journal 14: 128–135.

Hess, C. A., and R. Costa. 1994. Augmentation from the Apalachicola National Forest: The development of a new management technique. Pages 385–388 in D. L. Kulhavy, R. G. Hooper, and R. Costa (editors), Red-cockaded Woodpecker: Recovery, ecology and management. Center for Applied Studies in Forestry, College of Forestry, Stephen F. Austin State University, Nacogdoches, Texas.

Hickey, J. J. 1937. *Bird-Lore's* first breeding-bird census. Bird Lore 39: 373–374.

Hilfiker, E. L. 1991. Beavers: Water, wildlife, and history. Windswept Press, Interlaken, N.Y., 198 pp.

Hill, N. P., and J. M. Hagan III. 1991. Population trends of some northeastern North American landbirds: A half-century of data. Wilson Bulletin 103: 165–182.

Hobson, K. A. 1990. Stable isotope analysis of Marbled Murrelets: Evidence for freshwater feeding and determination of trophic level. Condor 92: 897–903.

Holmes, R. T., and T. W. Sherry. 1988. Assessing population trends of New Hampshire forest birds: Local vs. regional patterns. Auk 105: 756–768.

Holmes, R. T., P. P. Marra, and T. W. Sherry. 1995. Habitat-specific demography of breeding Black-throated Blue Warblers (*Dendroica caerulescens*): Implications for population dynamics. Journal of Animal Ecology 65: 183–195.

Holmes, R. T., T. W. Sherry, and F. W. Sturges. 1986. Bird community dynamics in a temperate deciduous forest: Long-term trends at Hubbard Brook. Ecological Monographs 56: 201–220.

Holmes, R. T., T. W. Sherry, P. P. Marra, and K. E. Petit. 1992. Multiple brooding and productivity of a neotropical migrant, the Black-throated Blue Warbler (*Dendroica caerulescens*), in an unfragmented temperate forest. Auk 109: 321–333.

Hooper, R. G. 1988. Longleaf pines used for cavities by Red-cockaded Woodpeckers. Journal of Wildlife Management 52: 392–398.

Hooper, R. G., D. L. Krusac, and D. L. Carlson. 1991a. An increase in a population of Red-cockaded Woodpeckers. Wildlife Society Bulletin 19: 277–286.

Hooper, R. G., M. R. Lennartz, and H. D. Muse. 1991b. Heart rot and cavity tree selection by Red-cockaded Woodpeckers. Journal of Wildlife Management 55: 323–327.

Hooper, R. G., L. J. Niles, R. F. Harlow, and G. W. Wood. 1982. Home ranges of Red-cockaded Woodpeckers in coastal South Carolina. Auk 99: 675–682.

Hooper, R. G., J. C. Watson, and R. E. F. Escano. 1990. Hurricane Hugo's initial effects on Red-cockaded Woodpeckers in the Francis Marion National Forest. Transactions of the North American Wildlife and Natural Resources Conference 55: 220–224.

Hoover, J. P., and M. C. Brittingham. 1993. Regional variation in cowbird parasitism of Wood Thrushes. Wilson Bulletin 105: 228–238.

Hoover, J. P., M. C. Brittingham, and L. J. Goodrich. 1995. Effects of forest patch size on nesting success of Wood Thrushes. Auk 112: 146–155.

Hosmer, J. K. 1908. Winthrop's journal: "History of New England, 1630–1649." Charles Scribner's and Sons, New York.

Howard, R. J., and J. S. Larson. 1985. A stream habitat classification system for beaver. Journal of Wildlife Management 49: 19–25.

Howe, H. F. 1994. Managing species diversity in tallgrass prairie: Assumptions and implications. Conservation Biology 8: 691–704.

Howe, R. W. 1984. Local dynamics of bird assemblages in small forest habitat islands in Australia and North America. Ecology 65: 1585–1601.

Hubbard, J. P. 1977. Importance of riparian ecosystems: Biotic considerations. Pages 14–18 in R. R. Johnson and D. A. Jones (technical coordinators), Importance, preservation and management of riparian habitat: A symposium. USDA Forest Service General Technical Report RM-43, Fort Collins, Colo.

Hudson, C. 1997. Knights of Spain, warriors of the sun. University of Georgia Press, Athens, 561 pp.

Hughes, R. H. 1951. Observations of cane (*Arundinaria*) flowers, seed and seedlings in the North Carolina coastal plain. Bulletin of the Torrey Botanical Club 78: 113–121.

Hulbert, L. C. 1988. Causes of fire effects in tallgrass prairie. Ecology 69: 46–58.

Hunter, W. C., B. W. Anderson, and R. D. Ohmart. 1985. Summer avian community composition of *Tamarix* habitats in three southwestern desert riparian systems. Pages 128–134 in R. R. Johnson, C. D. Ziebell, D. R. Patton, P. F. Ffolliott, and R. H. Hamre (technical coordinators), Riparian ecosystems and their management: Reconciling conflicting uses. USDA Forest Service General Technical Report RM-120, Fort Collins, Colo.

Hurley, R. J., and E. C. Franks. 1976. Changes in the breeding ranges of two grassland birds. Auk 93: 108–115.

Hussell, D. J. T., M. H. Mather, and P. H. Sinclair. 1992. Trends in numbers of tropical- and temperate-wintering migrant landbirds in migration at Long Point, Ontario, 1961–1988. Pages 101–114 in J. M. Hagan III and D. W. Johnston (editors), Ecology and conservation of neotropical migrant landbirds. Smithsonian Institution Press, Washington, D.C.

Hutto, R. L. 1988. Is tropical deforestation responsible for the reported declines in neotropical migrant populations? American Birds 42: 375–379.

Hutto, R. L. 1989. The effect of habitat alteration on migratory land birds in a West Mexican tropical forest: A conservation perspective. Conservation Biology 3: 138–148.

Hutto, R. L. 1992. Habitat distributions of migratory landbird species in western Mexico. Pages 221–239 in J. M. Hagan III and D. W. Johnston (editors), Ecology and conservation of neotropical migrant landbirds. Smithsonian Institution Press, Washington, D.C.

Hutto, R. L. 1995. Composition of bird communities following stand-replacement fires in northern Rocky Mountain (U.S.A.) conifer forests. Conservation Biology 9: 1041–1058.

Hutto, R. L. 1998. Using landbirds as an indicator species group. Pages 75–92 in J. M. Marzluff and R. Sallabanks (editors), Avian conservation: Research and Management. Island Press, Covelo, Calif.

Hutto, R. L., S. M. Pletschet, and P. Hendricks. 1986. A fixed-radius point count method for nonbreeding and breeding season use. Auk 103: 593–602.

Irland, L. C. 1982. Wildlands and woodlots. University Press of New England, Hanover, N.H., 217 pp.

Iseminger, W. R. 1997. Culture and environment in the American Bottom: The rise and fall of Cahokia Mounds. Pages 38–57 in A. Hurley (editor), Common fields: An environmental history of St. Louis. Missouri Historical Society Press, St. Louis.

Jackson, J. A. 1974. Gray rat snakes versus Red-cockaded Woodpeckers: Predator-prey adaptations. Auk 91: 342–347.

Jackson, J. A. 1986. Biopolitics, management of federal lands, and the conservation of the Red-cockaded Woodpecker. American Birds 40: 1162–1168.

Jackson, J. A. 1988. The southeastern pine forest ecosystem and its birds: Past, present, and future. Pages 119–159 in J. Jackson (editor), Bird Conservation 3. University of Wisconsin Press, Madison.

Jackson, J. A. 1994. Red-cockaded Woodpecker. In A. Poole and F. Gill (editors), The birds of North America, no. 85. Academy of Natural Sciences, Philadelphia, and American Ornithologists' Union, Washington, D.C.

Jackson, J. A. 1995. The Red-cockaded Woodpecker: Two hundred years of knowledge, twenty years under the Endangered Species Act. Pages 42–48 in D. L. Kulhavy, R. G. Hooper, and R. Costa (editors), Red-cockaded Woodpecker: Recovery, ecology and management. Center for Applied Studies in Forestry, College of Forestry, Stephen F. Austin State University, Nacogdoches, Texas.

James, F. C., C. F. McCulloch, and D. A. Wiedenfeld. 1996. New approaches to the analysis of population trends in land birds. Ecology 77: 13–27.

Johnsgard, P. A. 1983. The grouse of the world. University of Nebraska Press, Lincoln, 413 pp.

Johnson, D. H., and M. D. Schwartz. 1993. The Conservation Reserve Program and grassland birds. Conservation Biology 7: 934–937.

Johnson, E. A. 1992. Fire and vegetation dynamics. Studies from the North American boreal forest. Cambridge University Press, Cambridge, 129 pp.

Johnson, M. D., and G. R. Geupel. 1996. The importance of productivity to the dynamics of a Swainson's Thrush population. Condor 98: 133–141.

Johnson, T. H. (editor). 1960. The complete poems of Emily Dickinson. Little, Brown, Boston, 770 pp.

Johnston, D. W., and E. P. Odum. 1956. Breeding bird populations in relation to plant succession on the Piedmont of Georgia. Ecology 37: 50–62.

Johnston, D. W., and D. I. Winings. 1987. Natural history of Plummers Island, Maryland. XXVII: The decline of forest birds on Plummers Island, Maryland and vicinity. Proceedings of the Biological Society of Washington 100: 762–768.

Kantrud, H. A. 1981. Grazing intensity effects on the breeding avifauna of North Dakota native grasslands. Canadian Field Naturalist 95: 404–417.

Karr, J. R. 1981. Surveying birds with mist nets. Pages 62–67 in C. J. Ralph and M. J. Scott (editors), Estimating numbers of terrestrial landbirds. Studies in Avian Biology, no. 6.

Keiter, R. B. 1991. An introduction to the ecosystem management debate. Pages 3–18 in R. B. Keiter and M. S. Boyce (editors), The Greater Yellowstone Ecosystem: Redefining America's wilderness heritage. Yale University Press, New Haven.

Keller, M. E., and S. H. Anderson. 1992. Avian use of habitat configurations created by forest cutting in southeastern Wyoming. Condor 94: 55–65.

Kendeigh, S. C., and B. J. Fawver. 1981. Breeding bird populations in the Great Smoky Mountains, Tennessee and North Carolina. Wilson Bulletin 93: 218–242.

Kennedy, C. E. 1977. Wildlife conflicts in riparian management: Water. Pages 52–58 in R. R. Johnson and D. A. Jones (technical coordinators), Importance, preservation and management of riparian habitat: A symposium. USDA Forest Service General Technical Report RM-43, Fort Collins, Colo.

Kilgo, J. C., R. A. Sargent, B. R. Chapman, and K. V. Miller. 1998. Effect of stand width and adjacent habitat on breeding bird communities in bottomland hardwoods. Journal of Wildlife Management 62: 72–83.

King, D. J., C. R. Griffin, and R. M. DeGraaf. 1996. Effects of clearcutting on habitat use and reproductive success of the Ovenbird in forested landscapes. Conservation Biology 10: 1380–1386.

King, J. A. 1959. The social behavior of prairie dogs. Scientific American 201 (4): 128–140.

Kirk, D. A., A. W. Diamond, K. A. Hobson, and A. R. Smith. 1996. Breeding bird communities of the western and northern Canadian boreal forest: Relationship to forest type. Canadian Journal of Zoology 74: 1749–1770.

Kirsch, L. M. 1974. Habitat management considerations for prairie chickens. Wildlife Society Bulletin 2: 124–129.

Knight, C. L., J. M. Briggs, and M. D. Nellis. 1994. Expansion of gallery forest on Konza Prairie Research Natural Area, Kansas, USA. Landscape Ecology 9: 117–125.

Knopf, F. L. 1988. Conservation of riparian ecosystems in the United States. Wilson Bulletin 100: 272–284.

Knopf, F. L. 1994. Avian assemblages on altered grasslands. Pages 247–257 in J. R. Jehl, Jr., and N. K. Johnson (editors), A century of avifaunal change in western North America. Studies in Avian Biology, no. 15, Cooper Ornithological Society.

Knopf, F. L. 1996a. Perspectives on grazing nongame bird habitats. Pages 51–58 in P. R. Krausman (editor), Rangeland wildlife. Society for Range Management, Denver.

Knopf, F. L. 1996b. Prairie legacies—Birds. Pages 135–148 in F. B. Samson and F. L. Knopf (editors), Prairie conservation: Preserving North America's most endangered ecosystem. Island Press, Covelo, Calif.

Knopf, F. L. 1996c. Mountain Plover. In A. Poole and F. Gill (editors), The birds of North America, no. 211. Academy of Natural Sciences, Philadelphia, and American Ornithologists' Union, Washington, D.C.

Knopf, F. L., and B. J. Miller. 1994. Charadrius montanus—Montane, grassland, or bare-ground plover? Auk 111: 504–506.

Knopf, F. L., J. A. Sedgwick, and R. W. Cannon. 1988. Guild structure of a riparian avifauna relative to seasonal cattle grazing. Journal of Wildlife Management 52: 280–290.

Knowles, C. J., C. J. Stoner, and S. P. Gieb. 1982. Selective use of black-tailed prairie dog towns by Mountain Plovers. Condor 84: 71–74.

Konyukhov, N. B., and A. S. Kitaysky. 1995. The Asian race of the Marbled Murrelet. Pages 23–29 in C. J. Ralph, G. L. Hunt, Jr., M. G. Raphael, and J. F. Piatt (technical coordinators), Ecology and conservation of the Marbled Murrelet. USDA Forest Service General Technical Report PSW-GTR-152, Albany, Calif.

Kricher, J. C., and W. E. Davis, Jr. 1992. Patterns of avian species richness in disturbed and undisturbed habitats in Belize. Pages 240–246 in J. M. Hagan III and D. W. Johnston (editors), Ecology and conservation of neotropical migrant landbirds. Smithsonian Institution Press, Washington, D.C.

Kricher, J. C., and G. Morrison. 1993. A field guide to the ecology of western forests. Houghton Mifflin, New York, 554 pp.

Krueper, D. J. 1993. Effects of land use practices on western riparian ecosystems. Pages 321–330 in D. M. Finch and P. W. Stangel (editors), Status and management of neotropical migratory birds. USDA Forest Service General Technical Report RM-229, Rocky Mountain Forest and Range Experiment Station, Fort Collins, Colo.

Kuletz, K. J., D. K. Marks, N. L. Naslund, N. J. Goodson, and M. B. Cody. 1995. Inland habitat suitability for the Marbled Murrelet in southcentral Alaska. Pages 141–149 in C. J. Ralph, G. L. Hunt, Jr., M. G. Raphael, and J. F. Piatt (technical coordinators), Ecology and conservation of the Marbled Murrelet. USDA Forest Service General Technical Report PSW-GTR-152, Albany, Calif.

Kulikoff, A. 1986. Tobacco and slaves: The development of southern cultures in the Chesapeake, 1680–1800. University of North Carolina Press, Chapel Hill.

Kurtén, B., and E. Anderson. 1980. Pleistocene mammals of North America. Columbia University Press, New York, 442 pp.

Kurzejeski, E. W., R. L. Clawson, R. B. Renken, S. L. Sheriff, and L. D. Vangilder. 1993. Experimental evaluation of forest management: The Missouri Ozark Ecosystem Project. Transactions of the Fifty-eighth North American Wildlife and Natural Resources Conference, pp. 599–609.

Lamberson, R. H., R. McKelvey, B. R. Noon, and C. Voss. 1992. A dynamic analysis of Northern Spotted Owl viability in a fragmented forest landscape. Conservation Biology 6: 505–512.

Lanner, R. M. 1981. The piñon pine. University of Nevada Press, Reno, 208 pp.

Lanner, R. M. 1996. Made for each other: A symbiosis of birds and pines. Oxford University Press, New York, 160 pp.

Lanner, R. M., and S. B. Vander Wall. 1980. Dispersal of limber pine seed by Clark's Nutcracker. Journal of Forestry 78: 637–639.

Lanyon, W. E. 1956. Ecological aspects of the sympatric distribution of meadowlarks in the north-central states. Ecology 37: 98–108.

Larson, D. L. 1996. Seed dispersal by specialist versus generalist foragers: The plant's perspective. Oikos 76: 113–120.

Leck, C. F., B. G. Murray, Jr., and J. Swineboard. 1988. Long-term changes in the breeding bird populations of a New Jersey forest. Biological Conservation 46: 145–157.

Lee, D. S., M. Walsh-McGehee, and J. C. Haney. 1997. A history, biology, and re-evaluation of the Kirtland's Warbler habitat in the Bahamas. Bahamas Journal of Science 4: 19–29.

Lehmkuhl, J. F., L. E. Ruggiero, and P. A. Hall. 1991. Landscape patterns of forest fragmentation and wildlife richness and abundance in the southern Washington Cascade range. Pages 425–442 in L. F. Ruggierro, K. B. Aubry, A. B. Carey, and M. H. Huff (technical coordinators), Wildlife and vegetation of unmanaged Douglas-fir forests. USDA Forest Service General Technical Report PNW-GTR-285, Pacific Northwest Forest and Range Experiment Station, Portland, Ore.

LeGrand, H. E. Jr., and K. J. Schneider. 1992. Bachman's Sparrow, *Aimophila aetivalis*. Pages 299–313 in K. J. Schneider and D. M. Pence (editors), Migratory nongame birds of management concern in the Northeast. U.S. Department of the Interior, Fish and Wildlife Service, Newton's Corner, Mass.

Lennartz, M. R. 1988. The Red-cockaded Woodpecker: Old-growth species in a second-growth landscape. Natural Areas Journal 8: 160–165.

Lent, R. A., and T. S. Litwin. 1989. Bird-habitat relationships as a guide to ecologically-based management at Floyd Bennett Field, Gateway National Recreation Area. Part 1: Baseline study. Seatuck Research Program, Cornell Laboratory of Ornithology, Islip, N.Y.

Leopold, A. 1949. A Sand County almanac and sketches here and there. Oxford University Press, New York, 226 pp.

Lewis, T. H. 1907. The narrative of the expedition of Hernando De Soto by the gentleman of Elvas. Charles Scribner's Sons, New York.

Ligon, J. D. 1978. Reproductive interdependence of Pinyon Jays and pinyon pines. Ecological Monographs 48: 111–126.

Likens, G. E., F. H. Bormann, R. S. Pierce, and W. A. Reiners. 1978. Recovery of a deforested ecosystem. Science 199: 429–496.

Lindholt, P. J. 1988. John Josselyn, colonial traveler: A critical edition of two voyages to New-England. University Press of New England, Hanover, N.H.

Litvaitis, J. A. 1993. Response of early successional vertebrates to historic changes in land use. Conservation Biology 7: 866–873.

Litwin, T. S., and C. R. Smith. 1992. Factors influencing the decline of neotropical migrants in a northeastern forest fragment: Isolation, fragmentation or mosaic effects? Pages 483–496 in J. M. Hagan III and D. W. Johnston (editors), Ecology and conservation of neotropical migrant landbirds. Smithsonian Institution Press, Washington, D.C.

Lorimer, C. G. 1977. The presettlement forest and natural disturbance cycle of northeastern Maine. Ecology 58: 139–148.

Lynch, J. F. 1987. Responses of breeding bird communities to forest fragmentation. Pages 123–140 in D. A. Saunders, G. W. Arnold, A. A. Burbidge, and A. J. M. Hopkins (editors), Nature conservation: The role of remnants of native vegetation. Surrey Beatty & Sons, Chipping Norton, Australia.

Lynch, J. F. 1989. Distribution of overwintering nearctic migrants in the Yucatan Peninsula. I: General patterns of occurrence. Condor 91: 515–544.

Lynch, J. F. 1991. Effects of Hurricane Gilbert on birds in a dry tropical forest in the Yucatan Peninsula. Biotropica 23: 488–496.

MacArthur, R. H. 1958. Population ecology of some warblers of northeastern coniferous forests. Ecology 39: 599–619.

Marra, P. P., T. W. Sherry, and R. T. Holmes. 1993. Territorial exclusion by a long-distance migrant in Jamaica: A removal experiment with American Redstarts (Setophaga ruticilla). Auk 110: 565–572.

Marshall, D. B. 1988. Status of the Marbled Murrelet in North America, with special emphasis on populations in California, Oregon, and Washington. U.S. Fish and Wildlife Service, Biological Report 88 (30), 19 pp.

Martin, P. S., and R. G. Klein (editors). 1984. Quaternary extinctions: A prehistoric revolution. University of Arizona Press, Tucson, 892 pp.

Martin, T. E. 1980. Diversity and abundance of spring migratory birds using habitat islands in the Great Plains. Condor 82: 430–439.

Martin, T. E. 1987. Artificial nest experiments: Effects of nest appearance and type of predator. Condor 89: 925–928.

Marzluff, J. M., and R. P. Balda. 1992. The Pinyon Jay: Behavioral ecology of a colonial and cooperative corvid. Poyser, London, 317 pp.

Matthiessen, P. 1987. Wildlife in America. Elisabeth Sifton Books, Viking, New York, 332 pp.

Mauer, B. A., L. B. McArthur, and R. C. Whitmore. 1981. Effects of logging on guild structure of a forest bird community in West Virginia. American Birds 35: 11–13.

Mayfield, H. F. 1960. The Kirtland's Warbler. Cranbrook Institute of Science, Bloomfield Hills, Mich., 242 pp.

Mayfield, H. F. 1988a. Changes in bird life at the western end of Lake Erie. Part 1. American Birds 42: 393–398.

Mayfield, H. F. 1988b. Where were Kirtland's Warblers during the last Ice Age? Wilson Bulletin 100: 659–660.

Mayfield, H. F. 1992. Kirtland's Warbler. In A. Poole, P. Stettenheim, and F. Gill (editors), The birds of North America, no. 19. Academy of Natural Sciences, Philadelphia, and American Ornithologists' Union, Washington, D.C.

Mayfield, H. F. 1993. Kirtland's Warblers benefit from large forest tracts. Wilson Bulletin 105: 351–353.

Mayfield, H. F. 1996. Kirtland's Warblers in winter. Birding 28: 34–39.

McCann, J. M., S. E. Mabey, L. J. Niles, C. Bartlett, and P. Kerlinger. 1993. A regional study of coastal migratory stopover habitat for neotropical migrant songbirds: Land management implications. Transactions of the Fifty-eighth North American Wildlife and Natural Resources Conference, pp. 398–407.

McFarlane, R. W. 1992. A stillness in the pines: The ecology of the Red-cockaded Woodpecker. W. W. Norton, New York, 270 pp.

McFarlane, R. W. 1995. The relationship between body size, trophic position and foraging territory among woodpeckers. Pages 303–308 in D. L. Kulhavy, R. G. Hooper, and R. Costa (editors), Red-cockaded Woodpecker: Recovery, ecology and management. Center for Applied Studies in Forestry, College of Forestry, Stephen F. Austin State University, Nacogdoches, Texas.

McGarigal, K., and W. C. McComb. 1995. Relationship between landscape structure and breeding birds in the Oregon Coast Range. Ecological Monographs 65: 235–260.

McKinley, D. 1980. The balance of decimating factors and recruitment in extinction of the Carolina Parakeet. Indiana Audubon Quarterly 58: 8–18, 51–61, 103–114.

McNaughton, S. J. 1984. Grazing lawns: Animal in herds, plant form, and coevolution. American Naturalist 124: 863–886.

Means, D. B. 1996. Longleaf pine forest, going, going, . . . Pages 210–229 in M. B. Davis (editor), Eastern old-growth forests: Prospects for rediscovery and recovery. Island Press, Washington, D.C.

Means, D. B., and G. Grow. 1985. The endangered longleaf pine community. ENFO Report, Environmental Information Center of the Florida Conservation Foundation, Winter Park, Fla., 10 pp.

Melvin, S. 1994. Military bases provide habitat for rare grassland birds. Massachusetts Division of Fisheries and Wildlife, Natural Heritage News 4: 3.

Merriam, C. H. 1902. The prairie dog of the Great Plains. Yearbook of the Department of Agriculture 1901: 257–270.

Meslow, E. C. 1978. The relationship of birds to habitat structure—Plant communities and successional stages. Pages 12–18 in R. M. DeGraaf (technical coordinator), Proceedings of the Workshop on Nongame Bird Habitat Management in the Coniferous Forests of the Western United States. USDA Forest Service General Technical Report PNW-64, Pacific Northwest Forest and Range Experiment Station, Portland, Ore.

Meslow, E. C., and H. M. Wight. 1975. Avifauna and succession in Douglas-fir forests in the Pacific Northwest. Pages 266–271 in D. R. Smith (technical coordinator), Proceedings of the Symposium on the Management of Forest and Range Habitats for Nongame Birds. USDA Forest Service General Technical Report WO-1, Washington, D.C.

Miller, B., C. Wemmer, D. Biggins, and R. Reading. 1990. A proposal to conserve black-footed ferrets and the prairie dog ecosystem. Environmental Management 14: 763–769

Miller, B., G. Ceballos, and R. Reading. 1994. The prairie dog and biotic diversity. Conservation Biology 8: 677–681.

Miller, S. L., and C. J. Ralph. 1995. Relationship of Marbled Murrelets with habitat characteristics at inland sites in California. Pages 205–215 in C. J. Ralph, G. L. Hunt, Jr., M. G. Raphael, and J. F. Piatt (technical coordinators), Ecology and conservation of the Marbled Murrelet. USDA Forest Service General Technical Report PSW-GTR-152, Albany, Calif.

Mitsch, W. J., and J. G. Gosselink. 1993. Wetlands. 2nd ed. Van Nostrand Reinhold, New York, 722 pp.

Mönkkönen, M., and D. A. Welsh. 1994. A biogeographical hypothesis on the effects of human-caused landscape changes on the forest bird communities of Europe and North America. Annales Zoologici Fennici 31: 61–70.

Moore, F. R., P. Kerlinger, and T. R. Simons. 1990. Stopover on a Gulf Coast barrier island by spring trans-Gulf migrants. Wilson Bulletin 102: 487–500.

Moore, F. R., S. A. Gauthreaux, Jr., P. Kerlinger, and T. R. Simons. 1995. Habitat requirements during migration: Important link in conservation. Pages 121–144 in T. E. Martin and D. M. Finch (editors), Ecology and management of neotropical migratory birds. Oxford University Press, New York.

Morris, R. F., W. F. Cheshire, C. A. Miller, and D. G. Mott. 1958. The numerical response of avian and mammalian predators during a gradation of the spruce budworm. Ecology 39: 487–494.

Morse, D. H. 1989. American warblers. An ecological and behavioral perspective. Harvard University Press, Cambridge, 406 pp.

Nabhan, G. P. 1985. Riparian vegetation and indigenous southwestern agriculture: Control of erosion, pests, and microclimate. Pages 232–236 in R. R. Johnson, C. D.

Ziebell, D. R. Patton, P. F. Ffolliott, and R. H. Hamre (technical coordinators), Riparian ecosystems and their management: Reconciling conflicting uses. USDA Forest Service General Technical Report RM-120, Fort Collins, Colo.

Naiman, R. J., J. M. Melillo, and J. E. Hobbie. 1988. Ecosystem alteration of boreal forest streams by beaver (*Castor canadensis*). Ecology 67: 1254–1269.

Naiman, R. J., G. Pinay, C. A. Johnston, and J. Pastor. 1994. Beaver influences on the long-term biogeochemical characteristics of boreal forest drainage networks. Ecology 75: 905–921.

Naslund, N. L. 1993. Why do Marbled Murrelets attend old-growth forest nesting areas year-round? Auk 110: 594–602.

Naslund, N. L., K. J. Kuletz, M. B. Cody, and D. K. Marks. 1995. Tree and habitat characteristics and reproductive success at Marbled Murrelet tree nests in Alaska. Northwestern Naturalist 76: 12–25.

Nature Conservancy of Louisiana. 1992. The forested wetlands of the Mississippi River: An ecosystem in crisis. Nature Conservancy, Baton Rouge, La., 24 pp.

Nelson, S. K., and T. E. Hamer. 1995. Nest success and the effects of predation on Marbled Murrelets. Pages 89–97 in C. J. Ralph, G. L. Hunt, Jr., M. G. Raphael, and J. F. Piatt (technical coordinators), Ecology and conservation of the Marbled Murrelet. USDA Forest Service General Technical Report PSW-GTR-152, Albany, Calif.

Newton, I. 1972. Finches. Collins, London, 288 pp.

Niering, W. A., and R. H. Goodwin. 1974. Creation of relatively stable shrublands with herbicides: Arresting "succession" on rights-of-way and pastureland. Ecology 55: 784–795.

Noss, R. F., E. T. LaRoe III, and J. M. Scott. 1995. Endangered ecosystems of the United States: A preliminary assessment of loss and degradation. Biological Report 28, National Biological Service, Washington, D.C.

Ohmart, R. D. 1994. The effects of human-induced changes on the avifauna of western riparian habitats. Studies in Avian Biology 15: 273–285.

Ohmart, R. D., and B. W. Anderson. 1982. North American desert riparian ecosystems. Pages 433–466 in G. L. Bender (editor), Reference handbook on the deserts of North America. Greenwood, Westport, Conn.

Ohmart, R. D., W. O. Deason, and C. Burke. 1977. A riparian case history: The Colorado River. Pages 35–46 in R. R. Johnson and D. A. Jones (technical coordinators), Importance, preservation and management of riparian habitat: A symposium. USDA Forest Service General Technical Report RM-43, Fort Collins, Colo.

Oliveri, S. F. 1993. Bird responses to habitat changes in Baxter State Park, Maine. Maine Naturalist 1: 145–154.

O'Meilia, M. E., F. L. Knopf, and J. C. Lewis. 1982. Some consequences of competition between prairie dogs and beef cattle. Journal of Range Management 35: 580–585.

Olmsted, C. E. 1937. Vegetation of certain sand plains of Connecticut. Botanical Gazette 99: 209–300.

Olson, S. L. 1985. Mountain Plover food items on and adjacent to a prairie dog town. Prairie Naturalist 17: 83–90.

Parker, M., F. J. Wood, Jr., B. H. Smith, and R. G. Elder. 1985. Erosional downcutting in lower order riparian ecosystems: Have historical changes been caused by removal of beaver? Pages 35–38 in R. R. Johnson, C. D. Ziebell, D. R. Patton, P. F. Ffolliott, and R. H. Hamre (technical coordinators), Riparian ecosystems and their management: Reconciling conflicting uses. USDA Forest Service General Technical Report RM820, Fort Collins, Colo.

Paton, P. W. 1994. The effect of edge on avian nest success: How strong is the evidence? Conservation Biology 8: 17–26.

Patterson, W. A. III, and K. E. Sassaman. 1988. Indian fires in the prehistory of New England. Pages 107–135 in G. P. Nichols (editor), Holocene human ecology in northeastern North America. Plenum, New York.

Peet, R. K. 1988. Forests of the Rocky Mountains. Pages 63–101 in M. G. Barbour and W. D. Billings (editors), North American Terrestrial Vegetation. Cambridge University Press, New York, 434 pp.

Peet, R. K., and D. J. Allard. 1993. Longleaf pine vegetation of the southern Atlantic and eastern Gulf Coast regions: A preliminary classification. Pages 45–81 in S. M. Hermann (editor), The longleaf pine ecosystem: Ecology, restoration, and management. Proceedings of the Eighteenth Tall Timbers Fire Ecology Conference. Tall Timbers Research, Inc., Tallahassee, Fla.

Peterjohn, B., and J. R. Sauer. 1994. The North American Breeding Bird Survey. Birding 26: 386–398.

Peterson, R. T., and V. M. Peterson. 1985. Commentaries on Audubon's Birds of America. Abbeville, New York, 239 pp.

Peterson, S. R. 1982. A preliminary survey of forest bird communities in northern Idaho. Northwest Science 56: 287–298.

Petit, D. R., J. F. Lynch, R. L. Hutto, J. G. Blake, and R. B. Waide. 1995. Habitat use and conservation in the neotropics. Pages 145–197 in T. E. Martin and D. M. Finch (editors), Ecology and management of neotropical migratory birds. Oxford University Press, New York.

Petit, D. R., L. J. Petit, and K. G. Smith. 1992. Habitat associations of migratory birds overwintering in Belize, Central America. Pages 246–256 in J. M. Hagan III and D. W. Johnston (editors), Ecology and conservation of neotropical migrant landbirds. Smithsonian Institution Press, Washington, D.C.

Phillips, J. D. 1994. Forgotten hardwood forests of the coastal plain. Geographical Review 84: 162–171.

Piatt, J. F., and R. G. Ford. 1993. Distribution and abundance of Marbled Murrelets in Alaska. Condor 95: 662–669.

Piatt, J. F., and N. L. Naslund. 1995. Abundance, distribution, and population status of Marbled Murrelets in Alaska. Pages 285–294 in C. J. Ralph, G. L. Hunt, Jr., M. G. Raphael, and J. F. Piatt (technical coordinators), Ecology and conservation of the Marbled Murrelet. USDA Forest Service General Technical Report PSW-GTR-152, Albany, Calif.

Piatt, J. F., C. J. Lensink, W. Butler, M. Kendziorek, and D. R. Nysewander. 1990. Immediate impact of the "Exxon Valdez" oil spill on marine birds. Auk 107: 387–397.

Pimm, S. L. 1990. The decline of the Newfoundland crossbill. Trends in Ecology and Evolution 5: 350–351.

Pimm, S. L., and R. A. Askins. 1995. Forest losses predict bird extinctions in eastern North America. Proceedings of the National Academy of Sciences USA 92: 9343–9347.

Platt, S. G., and C. G. Brantley. 1992. The management and restoration of switchcane (Louisiana). Restoration and Management Notes 10: 84–85.

Plumb, G. E., and J. L. Dodd. 1993. Foraging ecology of bison and cattle on a mixed prairie: Implications for natural area management. Ecological Applications 3: 631–643.

Plumpton, D. L., and R. S. Lutz. 1993. Nesting habitat use by Burrowing Owls in Colorado. Journal of Raptor Research 27: 175–179.

Porneluzi, P., J. C. Bednarz, L. J. Goodrich, N. Zawada, and J. Hoover. 1993. Reproductive performance of territorial Ovenbirds occupying forest fragments and a contiguous forest in Pennsylvania. Conservation Biology 7: 618–622.

Puchkov, P. V. 1992. Uncompensated Wuermian extinctions. Part 2. Transformation of the environment by giant herbivores. Vestnik Zoologii o(1): 58–66.

Pulliam, H. R., J. B. Dunning, Jr., and J. Liu. 1992. Population dynamics in complex landscapes: A case study. Ecological Applications 2: 165–177.

Pylypec, B. 1991. Impact of fire on bird populations in a fescue prairie. Canadian Field-Naturalist 105: 346–349.

Rabenold, K. N., P. T. Fauth, B. W. Goodner, J. A. Sadowski, and P. G. Parker. 1998. Response of avian communities to disturbance by an exotic insect in spruce-fir forests of the southern Appalachians. Conservation Biology 12: 177–188.

Radabaugh, B. E. 1974. Kirtland's Warbler and its Bahama wintering ground. Wilson Bulletin 86: 374–383.

Ralph, C. J. 1994. Evidence of changes in populations of the Marbled Murrelet in the Pacific Northwest. Studies in Avian Biology 15: 286–292.

Ralph, C. J., G. L. Hunt, Jr., M. G. Raphael, and J. F. Piatt. 1995. Ecology and conservation of the Marbled Murrelet in North America: An overview. Pages 3–22 in C. J. Ralph, G. L. Hunt, Jr., M. G. Raphael, and J. F. Piatt (technical coordinators), Ecology and conservation of the Marbled Murrelet. USDA Forest Service General Technical Report PSW-GTR-152, Albany, Calif.

Rand, A. L., and R. M. Rand. 1943. Breeding notes on the Phainopepla. Auk 60: 333–341.

Raphael, M. G., J. A. Young, and B. M. Galleher. 1995. A landscape-level analysis of Marbled Murrelet habitat in western Washington. Pages 177–189 in C. J. Ralph, G. L. Hunt, Jr., M. G. Raphael, and J. F. Piatt (technical coordinators), Ecology and conservation of the Marbled Murrelet. USDA Forest Service General Technical Report PSW-GTR-152, Albany, Calif.

Rappole, J. H. 1995. The ecology of migrant birds: A neotropical perspective. Smithsonian Institution Press, Washington, D.C., 269 pp.

Rappole, J. H., and M. V. McDonald. 1994. Cause and effect in population declines of migratory birds. Auk 111: 652–660.

Rappole, J. H., M. A. Ramos, and K. Winker. 1989. Wintering wood thrush movements and mortality in southern Veracruz. Auk 106: 402–410.

Rebertus, A. J., G. B. Williamson, and W. J. Platt. 1993. Impact of temporal variation in fire regime on savanna oaks and pines. Pages 215–225 in S. M. Hermann (editor), The longleaf pine ecosystem: Ecology, restoration, and management. Proceedings of the Eighteenth Tall Timbers Fire Ecology Conference. Tall Timbers Research, Inc., Tallahassee, Fla.

Reed, J. M., P. D. Doerr, and J. R. Walters. 1988. Minimum viable population size of the Red-cockaded Woodpecker. Journal of Wildlife Management 52: 385–391.

Reichman, O. J. 1987. Konza prairie. A tallgrass natural history. University Press of Kansas, Lawrence, 226 pp.

Reitsma, L. R., R. T. Holmes, and T. W. Sherry. 1990. Effects of red squirrels, *Tamiasciurus hudsonicus*, and eastern chipmunks, *Tamias striatus*, on nest predation in a northern hardwood forest: An artificial nest experiment. Oikos 57: 375–380.

Remillard, M. M., G. K. Gruendling, and D. J. Bogucki. 1987. Disturbance by beaver (*Castor canadensis* Kuhl) and increased landscape heterogeneity. Pages 103–122 in M. G. Turner (editor), Landscape heterogeneity and disturbance. Springer-Verlag, New York.

Remsen, J. V. Jr. 1986. Was Bachman's Warbler a bamboo specialist? Auk 103: 216–219.

Renken, R. B and J. J. Dinsmore. 1987. Nongame bird communities on managed grasslands in North Dakota. Canadian Field Naturalist 101: 551–557.

Rich, A. C., D. S. Dobkin, and L. J. Niles. 1994. Defining forest fragmentation by corridor width: The influence of narrow forest-dividing corridors on forest-nesting birds in southern New Jersey. Conservation Biology 8: 1109–1121.

Richardson, C. J., R. Evans, and D. Carr. 1981. Pocosins: An ecosystem in transition. Pages 3–19 in C. J. Richardson (editor), Pocosin wetlands: An integrated analysis of coastal plain freshwater bogs in North Carolina. Hutchinson Ross, Stroudsburg, Pa.

Richter, C. 1940. The trees. Knopf, New York, 154 pp.

Robbins, C. S. 1978. Census techniques for forest birds. Pages 142–163 in R. M. DeGraaf (editor), Proceedings of the Workshop on Management of Southern Forests for Nongame Birds. USDA Forest Service General Technical Report SE-14, Asheville, N.C.

Robbins, C. S. 1979. Effect of forest fragmentation on bird populations. Pages 198–212 in R. M. DeGraaf and K. E. Evans (editors), Management of North-Central and Northeastern Forests for Nongame Birds, Workshop Proceedings. U.S. Forest Service General Technical Report NC-51, North Central Forest Experiment Station, St. Paul, Minn.

Robbins, C. S., D. K. Dawson, and B. A. Dowell. 1989a. Habitat area requirements of breeding forest birds of the Middle Atlantic states. Wildlife Monographs 103, 34 pp.

Robbins, C. S., J. R. Sauer, R. Greenberg, and S. Droege. 1989b. Population declines in North American birds that migrate to the neotropics. Proceedings of the National Academy of Sciences 86: 7658–7662.

Robbins, C. S., B. A. Dowell, D. K. Dawson, J. A. Colón, R. Estrada, A. Sutton, R. Sutton, and D. Weyer. 1992a. Comparison of neotropical migrant landbird populations wintering in tropical forest, isolated forest fragments, and agricultural habitats. Pages 207–220 in J. M. Hagan III and D. W. Johnston (editors), Ecology and conservation of neotropical migrant landbirds. Smithsonian Institution Press, Washington, D.C.

Robbins, C. S., J. W. Fitzpatrick, and P. B. Hamel. 1992b. A warbler in trouble: *Dendroica cerulea*. Pages 549–562 in J. M. Hagan III and D. W. Johnston (editors), Ecology and conservation of neotropical migrant landbirds. Smithsonian Institution Press, Washington, D.C.

Robbins, L. E., and R. L. Myers. 1992. Seasonal effects of prescribed burning in Florida: A review. Miscellaneous Publication no. 8, Tall Timbers Research, Tallahassee, Fla., 96 pp.

Robinson, S. K. 1988. Reappraisal of the costs and benefits of habitat heterogeneity for nongame wildlife. Transactions of the Fifty-third North American Wildlife and Natural Resources Conference, pp. 145–155.

Robinson, S. K. 1992. Population dynamics of breeding neotropical migrants in a fragmented Illinois landscape. Pages 408–418 in J. M. Hagan III and D. W. Johnston (editors), Ecology and conservation of neotropical migrant landbirds. Smithsonian Institution Press, Washington, D.C.

Robinson, S. K. 1998. Another threat posed by forest fragmentation: Reduced food supply. Auk 115: 1–3.

Robinson, S. K., and D. S. Wilcove. 1994. Forest fragmentation in the temperate zone and its effects on migratory songbirds. Bird Conservation International 4: 233–249.

Robinson, S. K., S. I. Rothstein, M. C. Brittingham, L. J. Petit, and J. A. Grzybowski. 1995a. Ecology and behavior of cowbirds and their impact on host populations. Pages 428–460 in T. E. Martin and D. M. Finch (editors), Ecology and management of neotropical migratory birds. Oxford University Press, New York.

Robinson, S. K., F. R. Thompson III, T. M. Donovan, D. R. Whitehead, and J. Faaborg. 1995b. Regional forest fragmentation and the nesting success of migratory birds. Science 267: 1987–1990.

Roemer, D. M., and S. C. Forrest. 1996. Prairie dog poisoning in northern Great Plains: An analysis of programs and policies. Environmental Management 20: 349–359.

Rolstad, J., and Wegge, P. 1989. Capercaillie *Tetrao urogallus* populations and modern forestry—A case for landscape ecological studies. Finnish Game Research 46: 43–52.

Rölvaag, O. E. 1927. Giants in the earth. Harper and Row, New York, 472 pp.

Romme, W. H., and D. G. Despain. 1989. The Yellowstone fires. Scientific American 261 (5): 36–46.

Rosenberg, K. V., and M. G. Raphael. 1986. Effects of forest fragmentation on vertebrates in Douglas-fir forests. Pages 263–272 in J. Verner, M. L. Morrison, and C. J. Ralph (editors), Wildlife 2000: Modeling habitat relationships of terrestrial vertebrates. University of Wisconsin Press, Madison.

Rosenberg, K. V., A. A. Dhondt, D. L. Tessaglia, J. D. Lowe, P. Senesac, and S. K. Gregory. 1995. A tale of four tanagers. Project Tanager confirms pilot study results. Birdscope 9 (2): 4–5.

Rosenberg, K. V., R. D. Ohmart, W. C. Hunter, and B. W. Anderson. 1991. Birds of the lower Colorado River Valley. University of Arizona Press, Tucson, 416 pp.

Roth, R. R., and R. K. Johnson. 1993. Long-term dynamics of a Wood Thrush population breeding in a forest fragment. Auk 110: 37–48.

Rudnicky, T. C., and M. L. Hunter, Jr. 1993a. Avian nest predation in clearcuts, forests, and edges in a forest-dominated landscape. Journal of Wildlife Management 57: 358–364.

Rudnicky, T. C., and M. L. Hunter, Jr. 1993b. Reversing the fragmentation perspective: Effects of clearcut size on bird species richness in Maine. Ecological Applications 3: 357–366.

Rudolph, D. C., and R. N. Conner. 1991. Cavity tree selection by Red-cockaded Woodpeckers in relation to tree age. Wilson Bulletin 103: 458–467.

Rudolph, D. C., and R. N. Conner. 1995. The impact of southern pine beetle induced mortality on Red-cockaded Woodpecker cavity trees. Pages 208–213 in D. L. Kulhavy, R. G. Hooper, and R. Costa (editors), Red-cockaded Woodpecker: Recovery, ecology and management. Center for Applied Studies in Forestry, College of Forestry, Stephen F. Austin State University, Nacogdoches, Texas.

Rudolph, D. C., R. N. Conner, D. K. Carrie, and R. R. Schaefer. 1992. Experimental reintroduction of Red-cockaded Woodpeckers. Auk 109: 914–916.

Rudolph, D. C., H. Kyle, and R. N. Conner. 1990. Red-cockaded Woodpeckers vs. rat snakes: The effectiveness of the resin barrier. Wilson Bulletin 102: 14–22.

Runkle, J. R. 1982. Patterns of disturbance in some old-growth mesic forests in eastern North America. Ecology 63: 1533–1546.

Runkle, J. R. 1990. Gap dynamics in an Ohio *Acer-Fagus* forest and speculations on the geography of disturbance. Canadian Journal of Forest Research 20: 632–641.

Russell, E. W. B. 1983. Indian-set fires in the forests of the northeastern United States. Ecology 64: 78–88.

Russell, H. S. 1980. Indian New England before the Mayflower. University Press of New England, Hanover, N.H., 284 pp.

Saab, V. A., C. E. Bock, T. D. Rich, and D. S. Dobkin. 1995. Livestock grazing effects in western North America. Pages 311–353 in T. E. Martin and D. M. Finch (editors), Ecology and management of neotropical migratory birds. Oxford University Press, New York.

Sader, S. A., and A. T. Joyce. 1988. Deforestation rates and trends in Costa Rica, 1940–1983. Biotropica 20: 11–19.

Sauer, J. R., S. Schwartz, B. G. Peterjohn, and J. E. Hines. 1995. North American breeding bird home page. Version 95.1. Unpublished computer program, Patuxent Environmental Science Center, Laurel, Md.

Schieck, J., K. Lertzman, B. Nyberg and R. Page. 1995. Effects of patch size on birds in old-growth montane forests. Conservation Biology 9: 1072–1084.

Schmiegelow, F. K. A., and S. J. Hannon. 1993. Adaptive management, adaptive science and the effects of forest fragmentation on boreal birds in northern Alberta. Transactions of the North American Wildlife and Natural Resources Conference 58: 584–598.

Schmiegelow, F. K. A., C. S. Machtans, and S. J. Hannon. 1997. Are boreal birds resilient to forest fragmentation? An experimental study of short-term community responses. Ecology 78: 1914–1932.

Schorger, A. W. 1936. The great Wisconsin nesting of 1871. Proceedings of the Linnaean Society 48: 1–26.

Schorger, A. W. 1955. The passenger pigeon: Its natural history and extinction. University of Wisconsin Press, Madison, 424 pp.

Seagle, S. W., R. A. Lancia, D. A. Adams, M. R. Lennartz, and H. A. Devine. 1987. Integrating timber and Red-cockaded Woodpecker habitat management. Transactions of the North American Wildlife and Natural Resources Conference 52: 41–52.

Seischab, F. K., and D. Orwig. 1991. Catastrophic disturbances in the presettlement forests of western New York. Bulletin of the Torrey Botanical Club 114: 330–335.

Serrao, J. 1985. Decline of forest songbirds. Records of New Jersey Birds 11: 5–9.

Sexton, C. 1992. The Golden-cheeked Warbler. Birding 24: 373–376.

Shackelford, C. E., and R. N. Conner. 1997. Woodpecker abundance and habitat use in three forest types in eastern Texas. Wilson Bulletin 109: 614–629.

Shaffer, L. N. 1992. Native Americans before 1492. The Moundbuilding centers of the eastern woodlands. M. E. Sharpe, Armonk, N.Y., 124 pp.

Sharps, J. C., and D. W. Uresk. 1990. Ecological review of black-tailed prairie dogs and associated species in western South Dakota. Great Basin Naturalist 5: 339–345.

Shelford, V. E. 1978. The ecology of North America. University of Illinois Press, Urbana, 610 pp.

Shepard, O. (editor). 1961. The heart of Thoreau's journals. Dover Publications, New York, 228 pp.

Sherry, T. W., and R. T. Holmes. 1992. Population fluctuations in a long-distance migrant: Demographic evidence for the importance of breeding season events in the American Redstart. Pages 431–442 in J. M. Hagan III and D. W. Johnston (editors), Ecology and conservation of neotropical migrant landbirds. Smithsonian Institution Press, Washington, D.C.

Sherry, T. W., and R. T. Holmes. 1995. Summer versus winter limitation of populations: What are the issues and what is the evidence? Pages 85–120 in T. E. Martin and D. M. Finch (editors), Ecology and management of neotropical migratory birds. Oxford University Press, New York.

Sherry, T. W., and R. T. Holmes. 1996. Winter habitat quality, population limitation, and conservation of neotropical-nearctic migrant birds. Ecology 77: 36–48.

Shinneman, D. J., and W. L. Baker. 1997. Nonequilibrium dynamics between catastrophic disturbances and old-growth forests in ponderosa pine landscapes of the Black Hills. Conservation Biology 11: 1276–1288.

Sibley, C. G., and B. L. Monroe, Jr. 1990. Distribution and taxonomy of birds of the world. Yale University Press, New Haven, 1,111 pp.

Simberloff, D. 1993. Species-area and fragmentation effects on old-growth forests: Prospects for longleaf pine communities. Pages 1–13 in S. M. Hermann (editor), The longleaf pine ecosystem: Ecology, restoration, and management. Proceedings of the Eighteenth Tall Timbers Fire Ecology Conference. Tall Timbers Research, Inc., Tallahassee, Fla.

Simon, J. L., and A. Wildavsky. 1993. Facts, not species, are imperiled. New York Times, May 13.

Singer, S. W., N. L. Naslund, S. A. Singer, and C. J. Ralph. 1991. Discovery and observations of two tree nests of the Marbled Murrelet. Condor 93: 330–339.

Smith, B. D. 1989. Origins of agriculture in eastern North America. Science 246: 1566–1571.

Smith, B. D. 1995. The origins of agriculture in the Americas. Evolutionary Anthropology 3: 174–184.

Smith, W. P. 1968. Eastern Henslow's Sparrow. Pages 776–778 in O. L. Austin, Jr. (editor), Life histories of North American cardinals, grosbeaks, buntings, towhees, finches, sparrows, and allies. Part Two. Dover Publications, New York.

Snyder, N. F. R., J. W. Wiley, and C. B. Kepler. 1987. The parrots of Luquillo: Natural history and conservation of the Puerto Rican Parrot. Western Foundation of Vertebrate Zoology, Los Angeles.

Stalter, R., and E. E. Lamont. 1987. Vegetation of the Hempstead Plains, Mitchell Field, Long Island, New York. Bulletin of the Torrey Botanical Club 114: 330–335.

Steadman, D. W., J. Arroyo-Cabrales, E. Johnson, and A. F. Guzman. 1994. New information on the late Pleistocene birds from San Josecito Cave, Nuevo León, Mexico. Condor 96: 577–589.

Stephens, E. P. 1956. The uprooting of trees: A forest process. Soil Science Society of America Proceedings 20: 113–116.

Stobo, W. T., and I. A. McLaren. 1971. Late winter distribution of the Ipswich Sparrow. American Birds 25: 941–944.

Stobo, W. T., and I. A. McLaren. 1975. The Ipswich Sparrow. Nova Scotian Institute of Science, Halifax, Nova Scotia, Canada.

Stotz, D. F., J. W. Fitzpatrick, T. A. Parker III, and D. K. Moskovits. 1996. Neotropical birds: Ecology and conservation. University of Chicago Press, Chicago, 478 pp.

Streng, D. R., J. S. Glitzenstein, and W. J. Platt. 1993. Evaluating effects of season of burn on longleaf pine forests: A critical literature review and some results from an ongoing long-term study. Pages 227–263 in S. M. Hermann (editor), The longleaf pine ecosystem: Ecology, restoration, and management. Proceedings of the Eighteenth Tall Timbers Fire Ecology Conference. Tall Timbers Research, Inc., Tallahassee, Fla.

Strong, T. R., and C. E. Bock. 1990. Bird species distribution patterns in riparian habitats in southeastern Arizona. Condor 92: 866–885.

Svenson, H. K. 1936. The early vegetation of Long Island. Brooklyn Botanic Garden Journal 25: 207–227.

Swengel, S. R. 1996. Management responses of three species of declining sparrows in tallgrass prairie. Bird Conservation International 6: 241–253.

Swenson, E. A., and C. L. Mullins. 1985. Revegetating riparian trees in southwestern floodplains. Pages 135–138 in R. R. Johnson, C. D. Ziebell, D. R. Patton, P. F. Ffolliott, and R. H. Hamre (technical coordinators), Riparian ecosystems and their management: Reconciling conflicting uses. USDA Forest Service General Technical Report RM-120, Fort Collins, Colo.

Syrjänen, K., R. Kalliola, A. Puolasmaa, and J. Mattsson. 1994. Landscape structure and forest dynamics in subcontinental Russian European taiga. Annales Zoologici Fennici 31: 19–34.

Szaro, R. C., and R. P. Balda. 1979. Effects of harvesting ponderosa pine on nongame bird populations. Research Paper RM-212, Rocky Mountain Forest and Range Experiment Station, USDA Forest Service, Fort Collins, Colo., 8 pp.

Tanner, J. T. 1942. The Ivory-billed Woodpecker. National Audubon Society, New York, 111 pp.

Taylor, N. 1923. The vegetation of Long Island, Part 1. The vegetation of Montauk: A study of grassland and forest. Brooklyn Botanic Garden Memoirs, vol. 2. Brooklyn, N.Y.

Taylor, W. E., and R. G. Hooper. 1991. A modification of Copeyon's drilling technique for making artificial Red-cockaded Woodpecker cavities. USDA Forest Service General Technical Report SE-72, Asheville, N.C., 31 pp.

Telfer, E. S. 1992. Wildfire and the historical habitats of the boreal forest avifauna. Pages 27–37 in D. H. Kuhnke (editor), Birds in the boreal forest. Northern Forestry Centre, Northwest Region, Forestry Canada.

Temple, S. A. 1986. Predicting impacts of habitat fragmentation on forest birds: A comparison of two models. Pages 301–304 in J. Verner, M. L. Morrison, and C. J. Ralph (editors), Wildlife 2000: Modeling habitat relationships of terrestrial vertebrates. University of Wisconsin Press, Madison.

Temple, S. A., and J. R. Cary. 1988. Modeling Dynamics of habitat-interior bird populations in fragmented landscapes. Conservation Biology 2: 340–347.

Terborgh, J. 1989. Where have all the birds gone? Princeton University Press, Princeton, N.J., 207 pp.

Terrill, S. B., and R. D. Ohmart. 1984. Facultative extension of fall migration by Yellow-rumped Warblers. Auk 101: 427–438.

Thomas, B. G., E. P. Wiggers, and R. L. Clawson. 1996. Habitat selection and breeding status of Swainson's Warblers in southern Missouri. Journal of Wildlife Management 60: 611–616.

Thomas, E. S. 1951. Distribution of Ohio animals. Ohio Journal of Science 51: 153–167.

Thompson. C. F. 1977. Experimental removal and replacement of territorial male Yellow-breasted Chats. Auk 94: 107–113.

Thompson, C. F., and V. Nolan, Jr. 1973. Population biology of the Yellow-breasted Chat (Icteria virens L.) in southern Indiana. Ecological Monographs 43: 145–171.

Thompson, F. R. III, W. D. Dijak, T. G. Kulowiec, and D. A. Hamilton. 1992. Breeding bird populations in Missouri Ozark forests with and without clearcutting. Journal of Wildlife Management 56: 23–29.

Thompson, I. D., and W. J. Curran. 1995. Habitat suitability of marten of second-growth balsam forests in Newfoundland. Canadian Journal of Zoology 73: 2059–2064.

Thoreau, H. D. 1864 [reprinted 1906]. The Maine woods. Houghton Mifflin, New York. 364 pp.

Titterington, R. W., H. S. Crawford, and B. N. Burgason. 1979. Songbird responses to commercial clear-cutting in Maine spruce-fir forests. Journal of Wildlife Management 43: 602–609.

Tomback, D. F. 1978. Foraging strategies of Clark's Nutcracker. Living Bird, 16th Annual (1977): 123–161.

Tomback, D. F. 1980. How nutcrackers find their seed stores. Condor 82: 10–19.

Tomback, D. F. 1982. Dispersal of whitebark pine seeds by Clark's Nutcracker: A mutualism hypothesis. Journal of Animal Ecology 51: 451–467.

Tomback, D. F., and Y. B. Linhart. 1980. The evolution of bird-dispersed pines. Evolutionary Ecology 4: 185–219.

Trine, C. L. 1998. Wood Thrush population sinks and implications for the scale of regional conservation strategies. Conservation Biology 12: 576–585.

Turner, O. 1849. Pioneer history of the Holland Purchase of western New York. Jewett, Thomas, Buffalo, N.Y., 666 pp.

Väisänen, R. A., O. Järvinen, and P. Rauhala. 1986. How are extensive, human-caused habitat alterations expressed on the scale of local bird populations in boreal forests? Ornis Scandinavica 17: 282–292.

Van Wagner, C. E. 1978. Age-class distribution and the forest fire cycle. Canadian Journal of Forest Research 8: 220–227.

Vaughan, A. T. 1977. New England's prospect (by William Wood). University of Massachusetts Press, Amherst, 132 pp.

Veit, R. R., and W. R. Petersen. 1993. Birds of Massachusetts. Massachusetts Audubon Society, Lincoln, 514 pp.

Verner, J. 1985. Assessment of counting techniques. Pages 247–302 in R. F. Johnston (editor), Current Ornithology, vol. 2. Plenum, New York.

Vickery, P. D. 1992. A regional analysis of endangered, threatened, and special concern birds in the northeastern United States. Transactions of the Northeast Section of the Wildlife Society 48: 1–10.

Vickery, P. D. 1994. Effects of habitat area on the distribution of grassland birds in Maine. Conservation Biology 8: 1087–1097.

Villard, P., and C. W. Beninger. 1993. Foraging behavior of male Black-backed and Hairy woodpeckers in a forest burn. Journal of Field Ornithology 64: 71–76.

Virkkala, R. 1987. Effects of forest management on birds breeding in northern Finland. Annales Zoologici Fennici 24: 281–294.

Virkkala, R., A. Rajasärkkä, R. A. Väisänen, R. Vickholm, and E. Virolainen. 1994. Conservation value of nature reserves: Do hole-nesting birds prefer protected forests in southern Finland? Annales Zoologici Fennici 31: 173–186.

Waide, R. B. 1980. Resource partitioning between migrant and resident birds: The use of irregular resources. Pages 337–352 in A. Keast and E. S. Morton, Migrant birds in the neotropics: Ecology, behavior, distribution and conservation. Smithsonian Institution Press, Washington, D.C.

Walker, J. 1993. Rare vascular plant taxa associated with the longleaf pine ecosystems: Patterns in taxonomy and ecology. Pages 105–125 in S. M. Hermann (editor), The longleaf pine ecosystem: Ecology, restoration, and management. Proceedings of the Eighteenth Tall Timbers Fire Ecology Conference. Tall Timbers Research, Inc., Tallahassee, Fla.

Walkinshaw, L. H. 1983. The Kirtland's Warbler: The natural history of an endangered species. Cranbrook Institute of Science, Bloomfield Hills, Mich., 207 pp.

Walsberg, G. E. 1975. Digestive adaptations of Phainopepla nitens associated with the eating of mistletoe berries. Condor 77: 169–174.

Walsberg, G. E. 1977. Ecology and energetics of contrasting social systems in Phainope-

pla nitens (Aves: Ptilogonatidae). University of California Publications in Zoology 108: 1–63.

Walters, J. R. 1991. Application of ecological principles to the management of endangered species: The case of the Red-cockaded Woodpecker. Annual Review of Ecology Systematics 22: 505–523.

Watson, J. C., R. G. Hooper, D. L. Carlson, W. E. Taylor, and T. E. Milling. 1995. Restoration of the Red-cockaded Woodpecker population on the Francis Marion National Forest: Three years post Hugo. Pages 172–182 in D. L. Kulhavy, R. G. Hooper, and R. Costa (editors), Red-cockaded Woodpecker: Recovery, ecology and management. Center for Applied Studies in Forestry, College of Forestry, Stephen F. Austin State University, Nacogdoches, Texas.

Webb, T. III. 1988. Eastern North America. Pages 385–414 in B. Huntley and T. Webb III (editors), Vegetation history. Kluwer Academic Publishers, Boston.

Webb, W. L., D. F. Behrend, and B. Saisorn. 1977. Effect of logging on songbird populations in a northern hardwood forest. Wildlife Monographs 55, 35 pp.

Weisbrod, A. R., C. J. Burnett, J. G. Turner, and D. W. Warner. 1993. Migrating birds at a stopover site in the Saint Croix River Valley. Wilson Bulletin 105: 265–284.

Welsh, C. J. E., and W. M. Healy. 1993. Effect of even-aged timber management on bird species diversity and composition in northern hardwoods of New Hampshire. Wildlife Society Bulletin 21: 143–154.

Welsh, D. A. 1988. Meeting the habitat needs of non-game forest wildlife. Forestry Chronicle 64: 262–266.

Wenny, D. G., R. L. Clawson, J. Faaborg, and S. L. Sheriff. 1993. Population density, habitat selection and minimum area requirements of three forest-interior warblers in central Missouri. Condor 95: 968–979.

Wheelwright, N. T., and J. D. Rising. 1993. Savannah Sparrow (*Passerculus sandwichensis*). In A. Poole and F. Gill (editors), The birds of North America, no. 45. Academy of Natural Science, Philadelphia; American Ornithologists' Union, Washington, D.C.

Whelan, C. J., M. L. Dilger, D. Robson, N. Hallyn, and S. Dilger. 1994. Effects of olfactory cues on artificial-nest experiments. Auk 111: 945–952.

Whicker, A. D., and J. K. Detling. 1988. Ecological consequences of prairie dog disturbances. BioScience 38: 778–785.

Whitcomb, R. F. 1987. North American forests and grassland: Biotic conservation. Pages 163–176 in D. A. Saunders, G. W. Arnold, A. A. Burbidge, and A. J. M. Hopkins (editors), Nature conservation: The role of remnants of native vegetation. Surrey Beatty & Sons, Chipping Norton, Australia.

Whitcomb, R. F., C. S. Robbins, J. F. Lynch, B. L. Whitcomb, M. K. Klimkiewicz, and D. Bystrak. 1981. Effects of forest fragmentation on avifauna of the eastern deciduous forest. Pages 125–205 in R. L. Burgess and D. M. Sharpe (editors), Forest island dynamics in man-dominated landscapes. Springer-Verlag, New York.

White, R. H. 1998. Changes in the avian communities along successional gradients caused by tornadoes in hardwood forests. Master's thesis, Empire State College.

Whitmore, R. C. 1981. Structural characteristics of Grasshopper Sparrow habitat. Journal of Wildlife Management 45: 811–814.

Whitmore, R. C., and G. A. Hall. 1978. The response of passerine species to a new resource: Reclaimed surface mines in West Virginia. American Birds 32: 6–9.

Whitney, G. G. 1987. An ecological history of the Great Lakes forest of Michigan. Journal of Ecology 75: 667–684.

Whitney, G. G. 1994. From coastal wilderness to fruited plain. Cambridge University Press, New York, 451 pp.

Whitney, G. G., and W. C. Davis. 1986. From primitive woods to cultivated woodlots: Thoreau and the forest history of Concord, Massachusetts. Journal of Forest History 30: 70–81.

Wilcove, D. S. 1985. Nest predation in forest tracts and the decline of migratory songbirds. Ecology 66: 1211–1214.

Wilcove, D. S. 1986. Owls and old-growth. Trends in Ecology and Evolution 1: 113–114.

Wilcove, D. S. 1988. Changes in the avifauna of the Great Smoky Mountains: 1947–1983. Wilson Bulletin 100: 256–271.

Willebrand, T., and V. Marcström. 1988. On the danger of using dummy nests to study predation. Auk 105: 378–379.

Williams, A. B. 1936. The composition and dynamics of a beech-maple climax community. Ecological Monographs 6: 317–408.

Williams, M. 1989. Americans and their forests: A historical geography. Cambridge University Press, New York.

Willis, E. O. 1966. The role of migrant birds at swarms of army ants. Living Bird 5: 187–231.

Wilson, A., and C. L. Bonaparte. 1877. American ornithology or the natural history of the birds of the United States. Vol. 1. J. W. Boulton, New York. 408 pp.

Wilson, C. W., R. E. Masters, and G. A Bukenhofer. 1995. Breeding bird response to pine-grassland community restoration for Red-cockaded Woodpeckers. Journal of Wildlife Management 59: 56–67.

Winne, J. C. 1988. History of vegetation and fire on the Pineo Ridge blueberry barrens in Washington County, Maine. Unpublished master's thesis, University of Maine.

Winthrop, John. 1642. [1908]. History of New England. Edited by J. F. Hosmer. Barnes and Noble, New York.

With, K. A. 1994. McCown's Longspur (Calcarius mccownii). In A. Poole, P. Stettenheim, and F. Gill (editors), The birds of North America, no. 96. Academy of Natural Sciences, Philadelphia, and American Ornithologists' Union, Washington, D.C.

World Resources Institute. 1994. World resources, 1994–95. Oxford University Press, New York.

Wright, H. E. Jr. 1974. Landscape development, forest fires, and wilderness management. Science 186: 487–495.

Wuerthner, G. 1994. Subdivisions versus agriculture. Conservation Biology 8: 905–908.

Wunderle, J. M. Jr., and R. B. Waide. 1993. Distribution of overwintering nearctic migrants in the Bahamas and Greater Antilles. Condor 95: 904–933.

Yahner, R. H. 1993. Effects of long-term forest clear-cutting on wintering and breeding birds. Wilson Bulletin 105: 239–255.

Yahner, R. H. 1995. Forest-dividing corridors and neotropical migrant birds. Conservation Biology 9: 476–477.

Yahner, R. H. 1997. Long-term dynamics of bird communities in a managed forested landscape. Wilson Bulletin 109: 595–613.

Yong, W., and D. M. Finch. 1997. Migration of the Willow Flycatcher along the middle Rio Grande. Wilson Bulletin 109: 253–268.

Zeranski, J. D., and T. R. Baptist. 1990. Connecticut birds. University Press of New England, Hanover, N.H., 328 pp.

Zimmerman, J. L. 1983. Cowbird parasitism of Dickcissels in different habitats and at different nest densities. Wilson Bulletin 95: 7–22.

Zimmerman, J. L. 1988. Breeding season habitat selection by the Henslow's Sparrow (Ammodramus henslowii) in Kansas. Wilson Bulletin 100: 17–24.

Zimmerman, J. L. 1993. The birds of Konza. The avian ecology of the tallgrass prairie. University Press of Kansas, Lawrence, 186 pp.

Zumeta, D. C. 1991. Protecting biological diversity: A major challenge for Minnesota's forestry in the 1990s. Journal of the Minnesota Academy of Science 56: 24–33.

Index

For scientific names of animals and plants, see appendixes. Birds are listed by type of bird, such as Warblers.

Acacia, catclaw, 192
Acorns, 87–88, 144, 233
Adirondack Mountains, 12
Adkisson, Curtis, 146
Africa, 60
Agalinis, 20
Agriculture: clearing of floodplains for, 196, 197; clearing of forests for, 76–78, 88; decline in acreage of cleared farmland, 39–41; Dickcissels in agricultural habitats, 66–67; fallow farmlands, 22, 25, 28–29, 39, 66; grassland birds' use of farmland, 21–22; irrigation for, 196, 197; in Latin America, 124; of Moundbuilders, 9–12, 89; prairie birds' use of farmland, 66–67, 71, 72; regrowth of forests in abandoned farmlands, 79, 80, 128; slash-and-burn agriculture, 8, 39. See also specific crops
Airfields, 22, 23, 25
Alabama River, 90
Alberta, Canada, forests, 134–35, 141
Alcids (auks), 168
Alder, 127
Algonquin Provincial Park, Ontario, 12
Allegany State Park, N.Y., 34, 35, 105–6
Allegheny Mountains, 7
Alpine meadows, 156, 158
Amazon Basin, 76, 78, 84
Amazonia, 125
Ames, Charles, 202
Ammodramus henslowii susurrans, 19
Anders, Angela, 252
Anderson, Bertin, 202
Andes Mountains, 124, 125, 128, 235
Angelina National Forest, Tex., 222
Antbirds, 122
Ants, 225; army ants, 122; carpenter ants, 93, 217
Antwrens, 122
Apalachicola National Forest, Fla., 210, 226
Appalachian Mountains, 37, 38, 82, 96, 99, 104, 107, 117, 146, 147
Aquifer, 198

Archeological record, 9–12, 90
Area sensitivity, 23, 49, 140–42, 165–67, 203, 238, 241
Argentina, 86, 235
"Asbestos forests," 37
Ash, 57, 93
Aspens, 79, 131–32, 134–37, 142, 152, 195, 199; quaking aspen, 120, 141, 157
Asters, 20
Atchafalaya Basin, 90
Au Sable River, 42, 44, 45
Audubon, John James, 2, 18, 32, 33
Audubon Sanctuaries, 230
Austin, Mary, 154

Badlands National Park, S.D., 61–62
Bahamas, 42, 45, 47–48, 125
Baird, Timothy, 105–6
Balsam woolly adelgid, 234
Bamboo, 94–95
Baraboo Hills, Wis., 113
Barataria Bay estuary, 90
Barley fields, 66
Barton Creek Habitat Preserve, Tex., 242
Bartram, John, 93
Bartram, William, 90–91, 95, 209
Baxter State Park, Maine, 34
Bayberry, 34
BBS. See Breeding Bird Survey (BBS)
Beans, 10
Bears, black, 75, 119, 246, 248, 255
Beaver meadows, 12–13, 14, 21, 33, 37
Beavers and beaver dams, 12–15, 21, 39, 52, 53, 75, 199, 203, 234, 236, 237, 244
Beech forests, 88, 100, 106
Beechnuts, 88, 144, 233
Beetles, 142; bark beetles, 225; engraver beetles, 93; southern pine beetles, 220–22, 225; wood-boring beetles, 93, 135, 217
Beissinger, Steven, 170
Belize, 122
Benkman, Craig, 181
Big Basin Redwoods State Park, Calif., 168, 169

Biological diversity, 155, 238–39, 240, 242, 244–45. *See also* Conservation

Birch, 79, 106, 120, 131, 132, 135, 138, 139, 142, 149, 195

Birds: scientific names of, 251–56. *See also* *specific types of birds*

Bison, 12, 55, 61–66, 69, 72, 85, 198, 234, 236, 237, 240

Black Hills, 278n14

Black Swamp, Ohio, 76

Blackberry, 28, 94, 253

Blazing star, 20, 21

Blowdowns, 34–38, 159

Blue grama, 62

Blueberry barrens, 7, 9, 19, 21, 43, 141

Bluebirds: Mountain Bluebirds, 160, 163, 190; Western Bluebirds, 156–57, 160, 190

Bluestem grass, 6, 7, 9, 209, 211, 215

Bobcat, 246

Bobolinks: Audubon on, 2; decline of, 2–3; Dickinson poem on, 1–2; in eastern grasslands, 6, 18, 23, 24; hayfields as nesting habitat for, 22; picture of, opposite p. 1; in recently burned prairies, 57

Bobwhites, 16, 215, 255

Bogor Botanical Garden, Java, 238

Bogs, 33, 37, 53

"Booming grounds" (of prairie-chickens), 58–59

Boreal forests: bird declines in managed forests of northern Europe, 138–43; compared with western coniferous forests, 155–56, 157; conservation of birds of, 152–53; crossbills and other nomadic seedeaters in, 143–51; decline of Newfoundland crossbill in, 149–51; fires in, 133–38; Ice Age forest, 132–33; introduction to, 131–32; large-scale harvesting of, 138–40; photograph of, 133; spruce budworm warblers in, 151–52

Bottomland forests, 88–97

Boundary Waters Canoe Area, Minn., 145

Bradley International Airport, Conn., 22

Brazil, 76, 78, 84, 125

Breeding. *See* Courtship and mating behaviors

Breeding Bird Censuses, 100, 102, 105, 107, 185

Breeding Bird Survey (BBS), 4, 5, 29–31, 63, 124, 125, 128, 176, 181, 182, 187, 193

Britain. *See* Great Britain

British Columbia, 165, 168, 171

Brittingham, Margaret, 113

Brown, James, 71

Bryant, Henry, 45

Bucher, E. H., 86

Buckthorn berries, 192, 193

Budworms. *See* Spruce budworms

Buffer areas, 242–43

Bunch grasses, 25

Buntings: Indigo Buntings, 29, 111, 118, 215; Lark Buntings, 62, 64, 69, 71; Lazuli Buntings, 162, 195; Painted Buntings, 29

Burke, Dawn, 115

Burning. *See* Fires

"Bush-hogging," 49

Butcher, Greg, 102

Cabin John Island, Md., 102, 104

Cacao plantations, 123

Cactus, 121; saguaro cactus, 190

Cahokia, 10, 11

Camels, 72, 198

Canada: beavers in, 12, 15; boreal forests in, 131, 132, 134–36, 145–47, 151–53, 155, 181; clearcutting in, 141; crossbills in, 145–47, 149, 181; Golden-crowned Kinglets in, 182; grassland in, 14; Marbled Murrelets in, 168, 170–71; migratory birds in, 124, 194; old-growth forests in, 165; spruce budworm warblers in, 151–52; spruce-fir forests of, 157

Canadian Wildlife Service, 154

Canebrakes, 94–96

Canopy: of cottonwood and willow woodlands, 189–90; formation of, 37; gaps in, 37, 38, 39, 52, 94, 137–38, 252; of longleaf pine woodland, 209, 227

Canyon-bottom woodlands, 195

Cape May Peninsula, N.J., 126

Capercallie, 140, 141, 153

Caribbean Basin, 95

Caribbean Islands, 60, 129

Caribou, 15, 16

Cascade Mountains, 165

Caterpillars, 107

Catesby, Mark, 18, 33

Cathedral Pines, Conn., 246

Cather, Willa, 185

Cats, 110, 112, 117

Cattle, 63–65, 70, 71, 73, 81, 198–202, 212, 237, 242–43

Cedar brakes, 241

Central America, 104, 121. *See also* *specific countries*

Champlain, Samuel de, 7, 8

Chaparral, 192

Chapman, Frank M., 230

Charles M. Russell National Wildlife Refuge, Mont., 62–63

Chasko, Gregory, 111–12

Chats, 270n5; Yellow-breasted Chats, 26–29, 33, 38, 40, 42, 48, 52, 53, 118, 124, 190, 195, 197, 200

Cheat Mountains, 107

Chesapeake Bay region, 8

Chickadees, 131; Black-capped Chickadees, 138; Boreal Chickadees, 135, 138, 153; Carolina Chickadees, 82; Mountain Chickadees, 163

Chicken eggs, 119, 120

Chipmunks, 87, 111, 113

Christmas Bird Counts, 170

Chuck-will's-widows, 82

Citrus groves, 123

Clayoquot Sound, British Columbia, 171

Clearcutting: in Alberta, mixed quaking aspen and white spruce forest, 141–42; as beneficial to birds, 48–50, 117–21, 135, 138, 163, 248, 252; and bird declines, 138–39, 163, 165, 216; in eastern forests, 78; in eastern thickets, 48–50; and Ivory-billed Woodpeckers, 230; in late nineteenth century, 78; and migratory forest birds, 117–21; and nesting, 118–21, 142, 162, 171; in northern European forests, 138–39; in western mountains, 155; and woodpeckers in pine woodland, 222–26. See also Forestry

Clesch, Alfred, 247

Clover, 69

Coca, 124

Cocklebur seeds, 91

Coffee plantations, 121, 123

Coles, J. M., 13

Colorado Desert, Calif., 192

Colorado River Valley, 188, 190, 192, 194, 195–98, 202–4, 237

Condors, 198

Conifers: in boreal forests, 131–41, 152–53; and Clark's Nutcrackers, 177–82; and crossbills and other nomadic seed-eaters, 143–51; in eastern U.S., 37–38, 78–79, 82; and migratory birds, 120, 128; and Pinyon Jays, 172–80; and spruce budworm warblers, 151–52; in western forests, 155–60, 167, 171–81. See also specific conifers

Connecticut College Arboretum, 9, 36, 51, 52, 102, 106

Conrad, H. S., 6

Conservation: and biological diversity, 155, 238–39, 240, 242, 244–45; of boreal forest

birds, 152–53; cooperative management of natural landscapes, 239–44; cowbird control program, 47; of desert streamside birds, 205–6; of eastern grassland birds, 25; forest management and migratory birds, 117–21; and landscape ecology, 238–39; of longleaf pine woodland birds, 226–27; and "lost landscapes," 235–37; of migratory birds of eastern deciduous forest, 128–30; restoration of ecosystems, 237–38; restoration of southwestern floodplains, 200, 202–5, 237; of shrubland birds in eastern thickets, 47, 53; of western forest birds, 182–83; of western prairie birds, 72

Conservation Reserve Program, 22, 71, 72

Coolidge Dam, 197

Copeyon, Carole, 222

Corn, 10

Cornell Laboratory of Ornithology, 108

Coronado National Forest, Ariz., 202

Costa Rica, 121, 122

Coteries (of prairie dogs), 59

Cotton fields, 80, 90

Cottonwoods, 185, 189–90, 195–200, 203–5, 237

Courtship and mating behaviors: of Capercallies, 141; of Marbled Murrelets, 168; of Phainopeplas, 192–93; of Pinyon Jays, 175–76; polygynous birds, 66; of prairie-chickens, 58–59; of Red-cockaded Woodpeckers, 219, 225, 226; and spot-mapping, 100–102

Cowbirds, Brown-headed, 16, 46–47, 68, 98, 111, 113–15, 116, 117, 118, 119, 165, 233, 248

Creepers, Brown, 135, 155, 163, 164, 165, 182, 183, 247

Crop milk, 87

Crosby, Alfred, 12

Crossbills, 143–51, 233; Hispaniolan subspecies of White-winged Crossbill, 150; Newfoundland crossbills, 149–51; Parrot Crossbills, 140, 143, 277n68; Red (or Common) Crossbills, x, 129, 131, 140, 143, 144, 146–47, 148, 149, 150, 155, 159, 181–83, 234, 277n68; Scottish Crossbills, 150; White-Winged Crossbills, 144, 145, 146, 148, 149

Crows, 110, 165, 171, 173; American Crows, 118; Fish Crows, 82

Cryptic species, 146–47

Cuba, 93, 95

Cuckoos, Yellow-billed, 29, 187, 190, 195, 197, 198, 203

Curlews, Long-billed, 64
Cypress trees (baldcypress), 91, 93

Dams, 196, 197, 205
Dawson, W. L., 167
De Soto, Hernando, 10, 12, 209
Deciduous forests: compared with western
    mountain forests, 165; in eastern U.S., 32,
    53, 78–79, 82, 98–99, 102–3, 165; fragmen-
    tation in, compared with boreal forests, 153;
    migratory birds in, 120, 126–30; tropical de-
    ciduous forests, 182. See also Eastern forests;
    and specific types of trees
Declines in populations. See Population de-
    clines
Deep-forest birds. See Forests; Migratory birds
Deer, 8, 39, 75, 106; mule deer, 61; white-tailed
    deer, 106
DeGraaf, Richard, 119
Delmarva Peninsula, Va. and Md., 126
Desert floodplains. See Southwestern flood-
    plains
Devil's Den Preserve, Conn., 115
Diamond, A. W., 124
Dickcissels, 18, 66–68, 70, 235
Dickinson, Emily, 1–2
Disturbance patches, 55, 69–70
Disturbance regimes, 35–39, 56, 134, 158–59,
    211, 214
Disturbances: definition of, 272n1; and forest
    size, 246–48; historical range of variation
    of, 255–56; human interference with envi-
    ronmental disturbances, 233–34; timing
    and pattern of, 235, 240. See also specific dis-
    turbances
Diversity. See Biological diversity
Dogs, 110, 112, 117
Dolan Falls Ranch, Tex., 242
Douglas-fir, 147, 157, 159, 163, 165, 168, 169,
    181
Doves: Bridled Quail Doves, 121; Eared Doves,
    86; Mourning Doves, 86; White-winged
    Doves, 197
Dragonflies, 244
Drought, 56
Dutchess County forest, N.Y., 115
Dwight, Timothy, 39

Early successional forest, 249–52
Eastern forests: birds of southern bottomland
    forests, 88–96; conservation of migratory
    birds of, 128–30; destruction of, 76–78, 88;

endemic species in, 82, 83; extinction of
    Passenger Pigeon, 75–76, 82, 84–88, 95;
    introduction to, 75–76; photographs of,
    89, 92; rebirth of, 78–82; species at risk,
    82–84
Eastern grasslands: and beavers, 12–13, 14,
    15, 21; birds of hay meadows and air fields,
    21–25; conservation of birds of, 25, 249; de-
    cline in bird populations of, 2–6; Dickinson
    poetry on birds of, 1–2; before European
    settlement, 6–12; fires in, 6, 8–9, 13–14, 21;
    habitat creation for birds in, 23–25; habitat
    quality for birds in, 22–23; Hempstead
    Plains, Long Island, 6–7, 19, 21; before In-
    dian agriculture, 12–14; introduction to,
    1–3; management of grassland birds,
    21–25; Montauk Downs, Long Island, 7; In-
    dians in, 7–12; origin of birds of, 17–21;
    photographs of, 3, 7, 9, 14, 21, 23; plants re-
    stricted to, 19–20; Pleistocene steppe and
    savanna, 14–17; sand plain grasslands, 12
Eastern thickets: chats in, 27–28; clearcuts for
    birds in, 48–50; conservation of shrubland
    birds, 47, 53; creation of shrublands by peo-
    ple, 39–42; and decline in acreage of
    cleared farm land, 39–41; decline in shrub-
    land birds, 27–32, 42, 53, 250; history of
    eastern shrubland birds, 32–33; introduc-
    tion to, 27–28; Kirtland's Warbler in, 32,
    42–48; management of shrubland birds,
    48–53; photographs of, 35, 36, 44, 50, 52;
    powerline corridors for birds, 49–53; shrub-
    lands before European clearing, 33–39
Edwards Plateau, 240–43
Egrets, 230
Elephants, 17
Elk, 61, 63
Empidonax traillii extimus, 187
Endangered and threatened species: Black-
    capped Vireos, 241–42; black-footed ferrets,
    63; crossbills, 149–51; Dickcissels, 66, 235;
    eastern grasslands birds, 4; Golden-cheeked
    Warblers, 240–41; Golden-winged War-
    blers, 42; Kirtland's Warbler, 32, 42–48;
    Marbled Murrelets, 170–71; Red-cockaded
    Woodpeckers, 211; Willow Flycatchers, 187;
    Yellow-breasted Chats, 42. See also Extinction
Endemic species, 82, 83, 84, 244
England. See Great Britain
Eremophila alpestris praticola, 17–18
Erosion. See Soil erosion
Erskine, Anthony, 152

Eurasia, 131–33

Europe, 131–33, 138–43, 153, 236

Evergreen shrubs, 34

Ewert, David, 120, 127

Extinction: of Bachman's Warblers, 75, 82, 93–95; of Carolina Parakeets, 74, 75, 82, 90, 91, 94–95; causes of, 232–35; with destruction of eastern forests, 75–76, 78, 82–88; of large mammals, 198; of Passenger Pigeon, 75–76, 82, 84–88, 95; prediction of, from species-area equation, 83, 96; in tropical forests, 84. See also Endangered and threatened species

Exxon Valdez, 170

Fallow farmlands, 22, 25, 28–29, 39, 66

Farmland. See Agriculture

Farnsworth, George, 248

Faulkner, William, 75, 273n1

Fawver, Ben, 107

Ferrets, 61, 63

Fertilizers, 162

Finches: Cassin's Finches, 159; Purple Finches, 148, 149

Finland, 132, 139–40, 142, 143, 144, 153, 238

Fir, 34, 131–32, 135, 136, 151, 153, 166, 180; balsam fir, 151, 157, 234; subalpine fir, 157, 159, 165; white fir, 157

Fire cycles, 134

Firebreaks, 70

Fires: in boreal forests, 135–38, 142; and canebrake, 95; in eastern forests, 94, 246; in eastern grasslands, 6, 8–9, 13–14, 21; and eastern thickets, 36, 37, 38, 42, 45, 47, 53; in Great Plains, 55, 56–59, 64, 66, 69–70; as integral to prairie ecology, 234, 237, 240, 272n1; and Kirtland's Warbler, 42–48; from lightning strikes, 214–15, 236; in longleaf pine woodland, 211–15; in pocosins, 34; and ponderosa pines in Black Hills, 278n14; and prairie birds, 56–59; slash-and-burn agriculture, 8, 39; in western forests, 156, 160–64, 165, 180, 182; in white pine and hemlock forests, 79

Firestorm, 159

Fish, scientific names of, 249

Fish and Wildlife Service, U.S., 187

Fishers, 119

Fleischner, Thomas, 71

Flickers: Gilded Flickers, 190; Northern Flickers, 160

Floaters, 101

Flocking, 87

Flooding, 33, 53, 89, 91, 93–97, 187–89, 196, 199, 203, 205, 234

Floodplains. See Southwestern floodplains

Floyd Bennett Field, Long Island, 22

Flycatchers, 122; Alder Flycatchers, 48, 135; Ash-throated Flycatchers, 190, 203; Brown-crested Flycatchers, 190, 203; Cordilleran Flycatchers, 163, 195; Dusky-capped Flycatchers, 195; Eastern Wood-Pewees, 17, 124; Great Crested Flycatchers, 101; Least Flycatchers, 135, 182; Olive-sided Flycatchers, 160; Pacific-slope Flycatchers, 163; Phainopeplas, 184, 190–94, 233; Pied Flycatchers, 142; Vermilion Flycatchers, 187, 190, 195, 234; Western Wood-Pewees, 160, 163; Willow Flycatchers, 33, 187, 190, 194, 197, 199, 200; Yellow-bellied Flycatchers, 135

Food for birds: acorns and beechnuts, 87, 144, 233; fish, 170; fruit, 190–92; insects, 86, 93, 107, 115, 122–23, 143, 144, 151–52, 160, 193, 217, 225, 233; mistletoe berries, 184, 190–93; pinyon pine seeds, 172–81; seeds, 143–51, 159, 172–83, 215, 233

Forbs, 55, 59, 60

Forbush, Edward Howe, 3, 27, 28–29, 32

Forest Service, U.S., 46

Forestry: and Bachman's Warblers, 94; bird declines in managed forests of northern Europe, 138–43, 153; and boreal forest birds, 138–43, 154–55; and conservation of boreal forest birds, 152–53; and crossbills, 145–46; in Europe, 132; impact of, on western forest birds, 162–66; large-scale harvesting of boreal forest, 136–38; in late nineteenth and early twentieth centuries, 39, 78, 96, 210–11, 230; in longleaf pine woodland, 220–26; and mechanical harvesters, 136–37, 139; and migratory birds, 117–21; types of timber harvesting, 162; in western forests, 162–66. See also Clear-cutting

Forests: "asbestos forests," 37; canopy gaps in, 37, 38, 39, 52, 94, 137–38; cause of declines in birds in, 103–5, 120–25; clearing of, for agriculture, 76–78, 88; coniferous forests, 37–38, 78–79, 82, 120, 127, 131–41, 143–53, 155–60, 167, 171–81; deciduous forests, 32, 53, 78–79, 82, 99, 102–3, 120, 127, 128–29, 153, 165, 182; hardwood forests, 14, 37–38, 88–90, 96, 101, 107, 118–19, 138, 153, 211, 215, 217, 220–22, 227,

Forests (continued)
236; history of New Hampshire forest,
35–37; hostile edges of small forests,
110–16, 165; Ice Age forests, 132–33; management of, and migratory birds, 117–21;
maturation of, 247–48; old-growth forests,
38, 39, 49, 78, 96, 107, 135, 155, 163–67,
169, 182–83, 210, 224, 233; percentage of
land covered with, 80, 81; population
crashes in forest islands, 99–103; population
trends in large forests, 105–8; rain forests,
104, 120–25, 128, 163; "sawtimber" forests,
119; second-growth forests, 78–81, 88, 96,
167, 224; "seedling and sapling" stage of,
42; songbird populations in forests of different sizes, 108–10; source and sink populations of birds in, 116–17, 119; southern bottomland forests, 88–97; as stopover habitat
for migratory birds, 126–28; tropical forests,
76, 78, 84, 104, 105, 120–25, 130; windstorm
destruction of, 34–38, 246–47. See also Boreal forests; Eastern forests; Forestry; Western forests; and specific types of trees
Fort Huachuca, Ariz., 198
Francis Marion National Forest, S.C., 215–16,
219–21, 222–25
Fretwell, Stephen, 66, 67, 68
Frost, Cecil, 211
Frost, Robert, 2
Frothingham, E. H., 42
Fruit-eating birds, 190–92
Fungi, 142, 180, 249

Ganier, Albert, 247
Gates, Edward, 111–12
Genesee River, 39
Gila River, 185, 196, 197
Gila Wilderness Area, N.M., 186
Girdling, 76
Glaciers, 14–15, 132–33, 172, 232
Glades, 7
Gnatcatchers: Black-tailed Gnatcatchers, 190;
Blue-gray Gnatcatchers, 124, 156, 203
Goldfinches, 149
Gophers, 244
Goshawks, Northern, 246
Grand Canyon, 198
Grand Teton National Park, 240
Grasses: blue grama, 62; bluestem, 6, 7, 9,
209, 211, 215; bunch grasses, 25; turf
grasses, 25; wiregrass, 209, 211, 212, 214,
215. See also Grazing; Prairies

Grasslands. See Eastern grasslands; Great
Plains; Prairies
Grayling sand, 44
Grazing, 39, 55, 59–65, 69–72, 81, 198–202,
205–7, 234, 236, 242, 272n1
Grazing lawns, 60
Great Britain, 13, 150
Great Lakes region, 78, 79, 80, 81, 85
Great Plains: bison in, 55, 61, 62, 64–65, 66,
69, 72; conservation of birds of, 72; decline
of Dickcissels in, 66–68; fires and prairie
birds, 56–59, 64, 66, 69–70; grazing of, 55,
59–65, 68, 69–72; impact of grazing on
prairie birds, 71–72; introduction to, 55;
photographs of, 56, 58, 61, 62, 66; prairie
dogs in, 55, 59–64, 68, 72–73; restoration
of prairie mosaic, 68–71
Great Smoky Mountains, 234, 236
Great Smoky Mountains National Park, 38,
107, 112, 248
Greater Antilles, 125
Greater Yellowstone Ecosystem, 240
Green Mountain National Forest, Vt., 49
Greenbrook Sanctuary, N.J., 102
Greenlets, 122
Greenways, 247
Griscom, Ludlow, 3, 29
Grosbeaks: Black-headed Grosbeaks, 163; Blue
Grosbeaks, 29, 197, 215; Evening Gros-
beaks, 148, 149, 157; Pine Grosbeaks, 131,
148, 157; Rose-breasted Grosbeaks, 118, 135,
141
Groth, Jeffrey, 146
Group selection cuts, 49, 253
Grouse: Black Grouse, 140; Capercallie, 140,
141, 153; Sharp-tailed Grouse, 15, 16, 56;
Spruce Grouse, 135, 138, 153
Gruell, George, 160
Guatemala, 122
Gulf Coast, 90, 93, 194
Gulf of Mexico, 93

Habitat fragmentation, 23, 49, 90, 99, 104–5,
108–16, 128–29, 140–42, 153, 156, 165–67,
224–26, 233, 235, 241
Habitat mosaic, 23
Hagan, John, 138
Hahn, Thomas, 144
Hamas, Michael, 127
Haney, Christopher, 247
Hanski, Ilpo, 120
Hardwoods: in eastern U.S., 14, 37–38, 88–90,

96, 101, 236; longleaf pine replaced by, 227; and migratory birds, 107, 117–19; mixed forests of conifers and, 153; removal of, in Texas, 221; in southern bottomland forests, 88–90, 96; transition zone between coniferous forest and, 140; in understory after fires in longleaf pine woodland, 211, 215, 217. See also specific types of trees
Hares, 16
Harper, Roland, 213
Harriers, 4
Harvard Forest, Mass., 35
Hawk Mountain, Pa., 113, 115
Hawks, 85, 142; Ferruginous Hawks, 61, 64; Red-shouldered Hawks, 197
Hay meadows and hayfields, 22, 25, 66, 67, 69
Heath Hens, 19, 20, 230
Heathlands, 12
Hejl, Sallie, 163
Hemlocks, 36, 79, 102, 106, 107, 147, 159, 164, 168, 181
Hempstead Plains, Long Island, 6–7, 21
Herbaceous plants, 149, 199
Herbicides, 49–51, 162
Herbivores, 17, 198
Herbs, 72, 215
Hickory, 10, 106, 118
Highways. See Roads
Hil, Norman P., 29
Hispaniola, 150
Hogs, 224
Holly, gallberry, 34
Holmes, Richard, 107, 125
Honeysuckle, Japanese, 28
Hooper, Robert, 218–19
Hoover Dam, 196
Horn Island, Miss., 127
Horses, 6, 63, 72, 171, 198
Hostile edges of forests, 110–16, 165
Housing developments, 99, 104, 105, 110, 117, 129, 195
Huachuca Mountains, Ariz., 195
Hubbard, Bela, 12–13
Hubbard Brook Experimental Forest, N.H., 107, 115
Hueston Woods, Ohio, 38
Hummingbirds: Black-chinned Hummingbirds, 156; Broad-tailed Hummingbirds, 160, 195; Calliope Hummingbirds, 157, 162, 163; Magnificent Hummingbirds, 195; Rufous Hummingbirds, 157
Hunting: of birds, 85–86, 91; as cause of ex-

tinction, 234; by Indians, 8, 39; by owls, 166–67
Hurricanes, 35–38, 94, 95, 216, 220–24
Hutto, Richard, 162

Ice Age forests, 132–33
Illinois River, 200
Indians: in eastern grasslands, 7–12; and fires in forests, 133–34, 214; flood control by, 203; hunting by, 8, 39; Moundbuilders, 9–12, 90; slash-and-burn agriculture of, 8, 39
Indigo, wild, 6
Inkweed bushes, 194, 202, 203
Insects: defoliating insects, 34, 37, 53, 94, 135, 151, 239; fires and, 135; as food for birds, 87, 91, 107, 115, 122–23, 143, 144, 151–52, 160, 193, 217, 225, 233; introduced insects, 234; scientific names of, 249; in small versus large forests, 116; southern pine beetles and cavity trees, 220–22; in western prairies, 72
Introduced species, 234
Irrigation, 196, 197, 203, 204, 237
Irruptions, 147, 149
Isle Royale National Park, Mich., 135

J. Clark Salyer National Wildlife Refuge, N.D., 58
Jamaica, 104, 124
Java, 238
Jays, 165, 172–73; Blue Jays, 87, 101, 117, 118; Gray Jays, 135, 137, 138, 153, 155, 157; Pinyon Jays, x, 156, 172–78, 181, 183, 233; Siberian Jays, 140–41, 153; Steller's Jays, 171, 172, 176; Western Scrub-Jays, 156, 176
Josselyn, John, 18
Juncos: Dark-eyed Juncos, 135, 162, 248; Yellow-eyed Juncos, 156–57
Junipers, 156, 172–77, 199; Ashe juniper, 240, 241

Kabetogama Peninsula, Minn., 13
Kansas State University, 70
Kennebunk Plains, Maine, 21
Kernen Prairie, Saskatchewan, 57–58
Kestrels, American, 162
Key conifer, 146–47
Keystone species, 59, 194
Kickapoo Cavern State Park, Tex., 244
Killdeer, 200
Kingbirds, Eastern, 215

Kinglets, 131; Golden-crowned Kinglets, 135, 157, 163, 164, 182, 234, 278n31; Ruby-crowned Kinglets, 141, 157, 163, 194

Kisatchie National Forest, La., 210, 218

Kites: Mississippi Kites, 97; Swallow-tailed Kites, 97

Konza Prairie Research Natural Area, Kans., 70, 236

Kulikoff, A., 8

Laguna Dam, 196

Lake Huron, 127

Lake Shelbyville, Ill., 113

Larch, 131

Larks: Dickinson poetry on, 1; Horned Larks, 1, 17–18, 22, 24, 62, 69, 71, 73; Skylarks, 1

Latin America. See Central America; South America; and specific countries

Lee, David, 247

Leopold, Aldo, 69, 88, 177

Lewis and Clark National Forest, Mont., 161

Liatris borealis novae-angliae, 20

Lichens, 142

Lightning strikes, 214–15, 236

Lincoln, Mass., 23–24

Livestock, 63–65, 69–72, 81, 158, 198–202, 207, 212, 237

Llanos, 66–67

Logging industry. See Forestry

Longleaf pine woodland: Bachman's Sparrows in, 211, 215–16, 217, 226–27; conservation of birds of, 226–227; fires in, 211–15; introduction to, 209–11; islands of pine savanna, 224–26; loss of fire and decline of, 211–16; photographs of, 210, 212, 218, 221; Red-cockaded Woodpeckers in, 216–27

Longspurs: Chestnut-collared Longspurs, 64–65, 68, 69; Chestnut-sided Longspurs, 56; Lapland Longspurs, 68, 273n50; McCown's Longspurs, 56, 64, 66, 69, 234; Smith's Longspurs, 68, 273n50

Lorimer, Craig, 251

"Lost landscapes," 235–37

Lostwood National Wildlife Refuge, N.D., 56

Loxia curvirostra percna, 150

Loxia leucoptera megalaga, 150

Lumber industry. See Forestry

Lynx, 75

Madagascar, 84

Magnolia: southern magnolia, 93; sweetbay magnolia, 34

Magpies, Black-billed, 16

Mahogany Creek, Nev., 200

Maize, 10

Mammals: in eastern forests, 75; scientific names of, 249–50; in southwestern floodplain woodlands, 198–200; in spruce parkland in eastern North America, 15, 16; in western forests, 162. See also Grazing; Livestock; and specific mammals

Mammoths, 72, 198

Mangroves, 121, 124, 125, 129, 182

Manomet Bird Observatory, Mass., 30, 138

Maples, 100, 106, 195

Mark Twain National Forest, Mo., 118

Marsh-elder, 10

Martens, 16, 150

Martes americana atrata, 150

Martha's Vineyard, 230

Martins, Purple, 162

Massachusetts Audubon Society, 249

Mast crops, 87, 233

Mastodons, 15, 17

Mating behaviors. See Courtship and mating behaviors

Mayfield, Harold, 3, 45, 46

McDonald, William, 71

Meadowlarks: Eastern Meadowlarks, 3, 4, 18, 22, 24, 28, 69, 70; Western Meadowlarks, 18, 69, 70, 71

Meadows: alpine meadows, 156, 158; beaver meadows, 12–13, 14, 21, 33, 37; hay meadows, 22, 25, 66, 67, 69

Mesquites, 189, 190, 192, 194, 198, 203, 237

Metapopulations, 247, 249

Mexico, 10, 104, 121, 122, 124, 128, 172, 182, 194, 195, 203

Mice, 113, 120

Michigan Department of Natural Resources, 46

Migratory birds: conservation of, in eastern deciduous forest, 128–30; and cowbird parasitism, 113–15, 116, 117; decline of, in small forests, 4, 104–16; and forest management, 116–21; and hostile edges of small forests, 110–16; loss of winter habitat as cause for population declines, 120–25, 235; and nest predation, 110–13, 116, 117, 119–21; population changes based on interplay of summer and winter habitat changes, 125–26; population trends in large forests for neotropical migrants, 105–8; songbird populations in forests of different sizes, 108–10; of south-

western floodplains, 194; stopover habitat for, 126–28, 194
Mining, 24–25
Mississippi River and Mississippi River Valley, 9–12, 90–94
Mississippian Period, 10
Missouri Department of Conservation, 120
Missouri Ozark Ecosystem Project, Mo., 120–21
Missouri River, 10
Mist nets, 123
Mistletoe, desert, 184, 190–93
Mixed-grass prairie, 56, 63, 66, 68, 69, 71, 72
Mockingbirds, Northern, 191
Montauk Downs, Long Island, 7
Moose, 75
Moose-spruce biome, 132
Moosehorn National Wildlife Refuge, Maine, 49
Morton, Thomas, 19
Moundbuilders, 9–12, 90
Mount Baker, Wash., 167
Mount Rainier National Park, Wash., 159
Mountain forests of West. See Western forests
Mountain lions, 75
Murrelets: Kittlitz's Murrelets, 168; Long-billed Murrelets, 278n47; Marbled Murrelets, x, 167–71, 183, 278n47

Nantucket Island, Mass., 12
Nassau County Community College, Long Island, 6
Nassau County nature preserve, Long Island, 6
National Audubon Society, 100, 229
National parks, ix–x, 13, 38, 60–62, 85, 107, 112, 120–23, 129, 135, 157–59, 171, 236, 239–40
Native Americans. See Indians
Natural Chimneys, Va., 16, 17
Natural disturbances, 12–14, 25, 34–39, 55, 69–72, 94–96, 133–36, 151, 155–56, 158–60, 187–96, 198, 211–16, 220–21, 227, 237
Nature Conservancy, xi, 6, 70, 115, 240–43, 246
Nature preserves, 6, 115, 129, 142, 227, 229, 238–44
Naval stores, 210
Nesting: artificial nests for research on, 111–13, 119–20, 171; of Bachman's Sparrows, 215–16; of Bachman's Warblers, 94; cavity-nesting birds, 142, 160, 190, 203, 208, 217–23; and clearcutting, 119–21, 142, 162,

170; of crossbills, 144, 147; of forest birds, 81; at forest edge, 111; of Golden-cheeked Warblers, 241; hayfields for, 22; of Ivory-billed Woodpeckers, 91–93; of Kirtland's Warblers, 43, 44, 46–47; of Marbled Murrelets, 167–72; of neotropical migratory birds, 110, 120; of Ovenbirds, 235; of owls, 166, 167; of Passenger Pigeons, 85, 87–88; of Phainopeplas, 192–93; of Pinyon Jays, 176; on powerline corridors, 52, 111–12; of prairie-chickens, 59; and predators, 110–13, 116, 117, 119–21, 165, 170, 217, 233, 248; of Red-cockaded Woodpeckers, 217–23; in small forests, 112–13; in southwestern floodplains, 185, 190
New Brunswick forests, 151
New England. See Eastern grasslands
New Paris No. 4, Pa., 15, 16
Newfoundland, 150
Newfoundland Wildlife Service, 150
Nitrogen, 57, 61
Nol, Erica, 115
Nomadic behavior, 66–67, 86–88, 91, 143–45, 147–49, 150–52, 175, 181–83, 233, 239
North Carolina Sandhills, 219
North Carolina State University, 222
Northern boreal forests, 131–38
Norway, 132, 153
Nutcrackers, 173; Clark's Nutcrackers, x, 159, 162, 178–81, 183, 233
Nuthatches: Brown-headed Nuthatches, 82, 215, 227; Pygmy Nuthatches, 156–57, 163, 182; Red-breasted Nuthatches, 17, 135, 141, 147–49, 163, 234
Nuts, 87–88, 233
Nuttall, Thomas, 32

Oaks: in eastern U.S., 7, 10, 12, 19, 79, 87–88, 93, 102, 106; Ivory-billed Woodpeckers in, 228–29; Lacey oak, 240; in Missouri, 118; scrub-oak, 19; in southwestern floodplains, 192, 193, 195; in Texas Hill Country, 241–42; in understory after fires in longleaf pine woodland, 211, 214, 215; Virginia live oak, 240; water oaks, 215; in western mountains, 157
Oat fields, 66
Ocala National Forest, Fla., 219
Ohio River and Ohio River Valley, 93, 236
Ohio Valley, 38, 76, 78, 99
Ohmart, Robert, 202
Oil spills, 170
Okefenokee National Wildlife Refuge, Ga., 210

Old-growth forests, 38, 39, 49, 78, 96, 107, 135, 163–70, 182–83, 210, 224, 233

Ontario, 12, 145, 152, 153

Opossums, 110

Orinoco River, Venezuela, 67

Orme, B. J., 13

Osceola National Forest, Fla., 218–19

Ouachita Mountains, Ark., 215

Ovenbirds, 48, 103, 106, 107, 109, 113, 114, 115, 118–19, 120, 182, 235, 253

Overstory, 28, 34, 57, 142, 166, 227, 242

Owls, 16, 183; Barred Owls, 81; Boreal Owls, 157, 166; Burrowing Owls, 61, 69; Elf Owls, 190; Flammulated Owls, 166; Great Gray Owls, 166; Spotted Owls, x, 166–67, 182–83, 233

Oxen, 78

Ozark Mountains, 104, 215

Pachaug State Forest, Conn., 50

Pacific Northwest, 159, 162, 166, 181

Palo verde, 192

Parakeets, Carolina, 74, 75, 82, 88, 90, 93, 94, 95, 230

Parasitism: cowbirds, 46–47, 68, 111, 113–15, 116, 117, 165, 233; desert mistletoe, 190–92

Parks. *See* National parks

Partners in Flight Program, 129–30, 246

Parulas, Northern, 96, 103, 121, 123

Passerculus sandwichensis princeps, 19

Pastures, 79, 129, 200. *See also* Cattle; Grazing; Livestock

Patuxent Wildlife Research, Md., 108

Peccaries, 15, 16

Pecos River Valley, 197, 198

Pewees: Eastern Wood-Pewees, 17, 124; Greater Antillean Pewee, 48; Western Wood-Pewees, 160, 163

Phainopeplas, 184, 190–94, 233

Philbrick, Margarett, 110

Philippines, 84

Pigeons: Passenger Pigeons, x, 75–76, 82, 84–88, 95, 144, 233; Scaly-naped Pigeons, 121

Pimm, Stuart, 84

Pine barrens, 19

Pineo Pond, Maine, 9

Pines: in boreal forests, 131, 132, 138, 139, 141, 146, 149; bristlecone pine, 157; and Clark's Nutcrackers, 178–81; in eastern U.S., 34, 42–45, 47, 48, 78, 79, 82, 91; jack pine,

42–46, 134, 137, 159; Jeffrey pine, 178; and Kirtland's Warbler, 42–48; limber pine, 157, 178, 180; loblolly pine, 209, 211, 216, 221, 222; lodgepole pine, 157, 158–59, 181, 236; longleaf pine, 209–27, 244; and migratory birds, 118, 125, 127, 128; and Pinyon Jays, 172–80; pinyon pine, 156, 172–78; pond pine, 34; ponderosa pine, 147, 156, 158, 159, 160, 163, 176, 181, 234, 278n14; red pine, 45, 137; shortleaf pine, 209, 216, 220; slash pine, 211; in southwestern floodplains, 199; table mountain pine, 147; white pine, 45, 79, 107, 137; whitebark pine, 157, 178–79, 181

Pipits, Sprague's, 64

Pisgah Forest, N.H., 35–37

Plains: Hempstead Plains, Long Island, 6–7, 21; Kennebunk Plains, Maine, 21. *See also* Great Plains

Plants, scientific names of, 247–49

Pleistocene, 14–17, 132, 172, 198, 232

Plovers, Mountain, 54, 62–63, 64, 69, 234

Plums, 10

Pocosins, 34, 37, 53

Polygynous, 66

Poplar, balsam, 141

Population declines: causes of, 103–5, 120–25, 232–35; and clearcutting, 138–39, 163, 165, 216; of Dickcissels, 66–68; in eastern grasslands, 2–6, 249; and "lost landscapes," 235–37; of migratory birds, 4, 104–16, 120–25, 235; of Newfoundland Crossbill, 149–51; in northern European managed forests, 138–43, 153; of shrubland birds, 27–32, 42, 53, 249

Population increases, in Clark's Nutcrackers, 181; in generalists, 140; in neotropical migrants, 106–7, 128; in Red-cockaded Woodpeckers, 224; in shrubland birds, 29

Population sinks, 116–17, 119, 275n52

Population sources, 116–17, 248

Powerline corridors, 49–53, 111–12, 117

Prairie birds. *See* Great Plains

Prairie-chickens, 58–59; Greater Prairie-Chickens, 19, 20, 68; Heath Hens, 19, 20

Prairie dogs, 55, 59–64, 69, 72, 73, 234, 236

Prairies: bluestem prairie, 6, 7, 9; fires in, 56–59; mixed-grass prairies, 56, 63, 66, 68, 69, 72; shortgrass prairies, 55, 59, 62, 63, 68, 69, 72; tallgrass prairies, 6, 25, 55, 56, 66, 68, 69, 72, 200, 234. *See also* Great Plains

Predators, 110–13, 116, 117, 119–21, 150, 165, 171, 217, 233
Prince of Wales Island, Alaska, 163
Prince William Sound, Alaska, 168

Quabbin Reservation, Mass., 12
Quail: Japanese Quail, 112–13; Mountain Quail, 162
Quail bush, 194, 202–3
Quail eggs, 119–20, 121–13
Quaker Run Valley, N.Y., 105
Quetico Provincial Park, Ontario, 145
Quinnipiac River, 7

Rabbits, 42, 245
Raccoons, 85, 110, 111, 112, 117
Rain forests, 104, 121–25, 128, 163
Ranches, 242–43. *See also* Cattle
Raphael, Martin, 165
Rats: kangaroo rats, 63; woodrats, 166
Rattlesnakes, diamondback, 16
Ravens, 165, 171
Redbay, 34
Redcedar, 159
Redpolls, Common, 148, 149
Redstarts, 140, 142; American Redstarts, 81, 102, 103, 106, 119, 121–26, 135, 182; Painted Redstarts, 195
Redwoods, 159, 168, 171
Refugia, 82, 87
Reinikainen, A., 146
Remsen, J. V., 94
Reptiles, scientific names of, 249
Resin wells, 217, 222
Rhinos, 72
Rhododendron, 96
Rice fields, 67, 68, 90
Richter, Conrad, 99
Rio Grande Valley, 194, 195, 197, 203
Riparian woodlands, 194, 197–98, 202–5
Roaches, 217, 225
Roadrunners, 185
Roads, 117, 195, 212, 215
Robbins, Chandler, 108
Robins: American Robin, 1, 135, 190, 200; European Robins, 140; poem on, 1
Robinson, S. K., 115
Rockrose, bushy, 20
Rocky Mountain National Park, Colo., 157
Rocky Mountain Plateau, 62
Rocky Mountains, 146, 155–62, 166, 195, 278n28

Rölvaag, Ole Edvart, 55
Roosevelt (Theodore) National Park, N.D., 60, 61
Rose Lake Wildlife Research Area, Mich., 111
Rosenberg, Kenneth, 165
Runkle, James, 38
Runoff, 49
Russell National Wildlife Refuge, Mont., 62–63
Russia, 135, 142

Saab, Victoria, 199
Sable Island, 19
Saco River, 14
Sacramento Valley, 197
Sage, 199
Sagebrush, 156
St. Croix River valley, 127
St. Johns River, 91
St. Marks National Wildlife Refuge, Fla., 214
Salamanders, woodland, 49
Salt marshes, 200, 237
Salt River, 197
Saltbush, 194
Saltcedar, 196–97, 199, 202–3, 234, 237
Salyer (J. Clark) National Wildlife Refuge, N.D., 58
Sam Houston National Forest, Tex., 220
San Josecito Cave, Mexico, 172
San Pedro River, 200, 201, 206, 236
San Xavier, Ariz., 198
Sand bars, 196
Sand plain grasslands, 12
Sandblast Cave, 198
Sandpipers, Upland, x, 3–4, 6, 7, 16, 18, 22, 24, 58, 62, 70, 233, 235, 249
Santee Swamp, S.C., 230
Sapsuckers: Red-breasted Sapsuckers, 163; Red-naped Sapsuckers, 162; Williamson's Sapsuckers, 57, 156–57
Saunders, Aretas, 105
Savannas, 14–17, 60, 200, 209, 211, 215–16, 224–27, 234, 244
"Sawtimber" forests, 119
Scandinavia, 131–32, 139–42, 153, 236, 238
Schafer, Christopher, 49
Scientific names: birds, 251–56; fish, 249; fungi, 249; insects, 249; mammals, 249–50; plants, 247–49; reptiles, 249
Scotland, 150
Second-growth forests, 78–81, 88, 96, 167, 224

Sedges, 43
Seed-eating birds, 143–51, 159, 172–83, 233
Shawnee National Forest, Ill., 115
Sheep, 6, 12, 69, 237
Shelterbelts, 128
Sherry, Thomas, 125
Shortgrass prairies, 55, 59, 62–63, 68–72
Shrikes, Loggerhead, 4, 215
Shrubland birds. See Eastern thickets
Siberia, 131, 168, 236
Sierra de los Tuxtlas, Mexico, 121
Sierra Nevada, Calif., 178–81
Silt banks, 196
Simons, Theodore, 248
Singer Tract, La., 229, 230
Sink populations, 116–17, 119
Siskins, Pine, 148, 149, 159
Skunks, striped, 112
Slash-and-burn agriculture, 8, 39
Sloths, 17, 172
Smith, John, 7
Snags, 142, 143, 153, 158, 162, 164, 203, 217, 235, 237
Snakes: indigo snakes, 244; rat snakes, 217
Snowberry bushes, 234
Snowy Range, Wyo., 165
Soil erosion, 49, 71, 72, 199
Soil of floodplains, 187–89, 202
Sonoran Desert, 190, 191, 203
Sorghum fields, 67, 68
Source populations, 116–17
South America, 104. See also specific countries
Southwestern floodplains: birds of, 189–94; conservation of birds in desert streamsides, 205–6; grazing in, 198–202, 205; introduction to, 185, 187; landscape of, 187–89; loss of woodlands along Colorado River, 195–97; in mountain canyons, 195; Phainopeplas in, 184, 190–94; photographs of, 186, 201, 205–7; restoration of, 200, 202–5, 237; terraces of, 189–90; threats to riparian woodlands, 197–98; winter residents and migrants in transit, 194
Soybeans, 90, 96
Sparrows, 149; Bachman's Sparrows, 40, 82, 209, 211, 215–17, 226–27; Baird's Sparrows, 56, 58, 64, 69, 73; Black-throated Sparrows, 185; Cassin's Sparrows, 64; Chipping Sparrows, 48, 160, 215; Clay-colored Sparrows, 58, 69; Eastern Henslow's Sparrows, 19; Field Sparrows, 28, 29, 33, 48, 51, 58, 111;

Fox Sparrows, 162; Grasshopper Sparrows, 3, 4, 6, 18, 22–24, 28, 57, 69, 70, 71, 233, 238, 249, 272n11; Henslow's Sparrows, 4, 19, 22, 23, 24, 57, 69, 70, 260n11; Ipswich Sparrows, 19; Lark Sparrows, 18, 160; LeConte's Sparrows, 68; Lincoln's Sparrows, 138, 200; Sage Sparrows, 194, 203; Savannah Sparrows, 19, 23, 24, 58, 71, 162, 200; Song Sparrows, 111, 138, 163, 200; Vesper Sparrows, 4, 6, 7, 24, 162; White-crowned Sparrows, 194, 200; White-throated Sparrows, 118, 135, 153
Species-area equation, 84, 273–74n24
Spot-mapping, 100–102
Spruce: black spruce, 134, 144, 150, 159; blue spruce, 157; in boreal forests, 130–32, 135, 136, 139, 140, 142, 153; and crossbills, 146, 149; in eastern U.S., 14–17, 20, 34; Engelmann spruce, 157, 159, 165; red spruce, 107, 144; Sitka spruce, 159, 164, 168, 181; and spruce budworm warblers, 151–52; in western mountains, 166, 180; white spruce, 130, 141, 144, 157
Spruce budworms, 34, 135, 143, 144, 151–52, 233
Squirrels, 143, 173; flying, 166; ground, 17; red, 16, 150
Stopover habitat, 126–28, 194, 253–55
Strip mines, 24–25
Sugar cane fields, 67, 90
Sugeno, Dave, 110
Sunflower, 10
Superflights, 149
Swallows, Violet-green, 163
Swamps, 76, 88–96, 121, 124, 127, 129, 230
Sweden, 132, 142, 153
Sweetfern, 43
Sweetgums, 92–93, 228–29
Sycamores, 192, 193, 195

Taiga, 132
Tallahatchie River, 75
Tallgrass prairies, 6, 25, 55, 56, 66, 68, 69, 72, 200, 234
Tamarack, 144, 146
Tanagers, 99; Scarlet Tanagers, 102, 103, 108, 109–10, 118, 248; Summer Tanagers, 190, 197, 200, 203; Western Tanagers, 157
Tanner, James, 91, 229–30, 237, 242
Temple, Stanley, 113
Termites, 225

Texas Hill Country, 240–44
Theodore Roosevelt National Park, N.D., 60, 61
Thickets. *See* Eastern thickets
Thistle seeds, 91
Thompson, Frank, 118
Thoreau, Henry, 75, 131
Thrashers: Brown Thrashers, 28–29, 33, 37, 38, 58; Crissal Thrashers, 190, 203; Sage Thrashers, 190
Threatened species. *See* Endangered and threatened species
Thrushes, 99, 129; Bicknell's Thrushes, 252; Song Thrushes, 140; Swainson's Thrushes, 163, 234; Varied Thrushes, 154, 163, 164; Wood Thrushes, 1, 100, 102, 106, 109, 113, 115–17, 124, 215, 247–48, 252–54, 275n52
Thunderstorms, 37, 199
Tikal National Park, Guatemala, 122
Timber harvesting, 117–20, 133, 136–40, 152–53, 162–65, 195–96, 219
Timber industry. *See* Forestry
Timothy, 69
Tionestra Scenic and Research Natural Areas, 247
Titi, 34
Titmouse, Plain, 156
Tits: Crested Tits, 140, 142; Great Tits, 140; Siberian Tits, 140, 141, 142, 153; Willow Tits, 142
Tobacco fields, 80
Toiyabe Mountains, Nev., 195
Tomback, Diana, 180
Törmävaara, Finland, 140, 238
Tornadoes, 34, 35, 37, 38, 94, 246–47
Tortoises, 244
Towhees: Eastern Towhees, 29–32, 33, 34, 49; Spotted Towhees, 31, 163
Trani, Margaret, 249
Tree canopy. *See* Canopy
Treecreepers, 142
Trees. *See* Boreal forests; Eastern forests; Forests; Western forests; *and specific types of trees*
Trogons, Elegant, 195
Tropical forests, 76, 78, 84, 104, 105, 120–25, 129
Trumbull, B., 8
Tupelo, water, 93
Turf grasses, 25
Turkeys, 75; Wild Turkeys, 87
Turtles, 60, 244
Tympanuchus cupido cupido, 19

Understory: in eastern forests, 95; in eastern thickets, 28; in longleaf pine woodland, 209, 211, 214, 217, 227; and migratory birds, 106, 107, 118, 123; in older vs. younger forests, 138; and Ovenbirds, 235; in southwestern floodplains, 199, 200; in western forests, 158–60
Unicoi Mountains, 247–48
University of Delaware, 115

Vancouver Island, 171
Veery, 106, 107, 119
Venezuela, 66, 67, 235
Verde River, 185
Verdins, 187, 190, 203
Vernal pools, 246
Verrazano, Giovanni da, 7
Vines, 28, 51
Violets, birdfoot, 6
Vireos, 99, 129, 153; Bell's Vireos, 34, 58, 197, 200; Black-capped Vireos, 234, 241–44; Gray Vireos, 156; Philadelphia Vireos, 138; Red-eyed Vireos, 48, 100, 102, 103, 106–9, 118, 135, 215; Warbling Vireos, 195; White-eyed Vireos, 34, 38, 40, 51–53, 234, 253; Yellow-throated Vireos, 102, 103, 110
Virgin Islands National Park, 120, 123
Voyageurs National Park, Minn., 13
Vuotso region, Finland, 140

Wade Tract, Ga., 210, 220, 236
Walkinshaw, Lawrence, 46
Wall, Stephen Vander, 180
Walnuts, 10
Walters, Jeffrey, 219
Warblers, 99, 129, 153; Bachman's Warblers, 75, 82, 88, 93–95, 234; Bay-breasted Warblers, 136, 138, 151, 152; Black-and-white Warblers, 48, 103, 118, 121–23, 135, 138, 253; Black-throated Blue Warblers, 108, 117, 125–26; Black-throated Green Warblers, 102, 103, 106–7, 141, 152, 248; Blackburnian Warblers, 34, 106, 136, 152, 153, 248; Blue-winged Warblers, 33, 34, 40, 49–52, 124, 252; Canada Warblers, 102, 103, 108, 135; Cape May Warblers, x, 136, 151, 152; Cerulean Warblers, x, 106, 108, 110, 124–25, 233, 235, 238, 252; Chestnut-sided Warblers, 29, 32–34, 37, 38, 40, 48–53, 118, 124, 135, 138, 234, 245, 248, 252; Golden-cheeked Warblers, 240–43; Golden-winged Warblers, x, 29, 30, 42, 48, 52, 234; Hooded

Warblers (*continued*)
  Warblers, 98, 102, 103, 106, 116, 122, 215,
    252–53; Kentucky Warblers, 102, 103,
    118, 122, 124, 253; Kirtland's Warbler, 32,
    42–48, 125, 234; Lucy's Warblers, 190, 197;
    MacGillivray's Warblers, 157; Magnolia War-
    blers, 34, 48, 106, 152; Mourning Warblers,
    34, 48, 49, 135, 138, 153; Nashville War-
    blers, 29, 34; Olive-capped Warblers, 48;
    Olive Warblers, 156–57; Orange-crowned
    Warblers, 163, 194, 203; Palm Warblers, 48;
    Pine Warblers, 118, 215; Prairie Warblers,
    29, 33, 34, 48–52, 118, 121, 215; Prothono-
    tary Warblers, 95–97; Red-faced Warblers,
    163; Swainson's Warblers, 82, 96–97; Ten-
    nessee Warblers, 141, 151, 152; Townsend's
    Warblers, 163, 164; Willow Warblers, 140;
    Wilson's Warblers, 157, 162, 163; Worm-eat-
    ing Warblers, 82, 115, 118, 253, 275n37; Yel-
    low Warblers, 190, 195, 200; Yellow-rumped
    Warblers, 152, 194; Yellow-throated War-
    blers, 82
Waterthrushes, Northern, 121
Waxwings: Bohemian Waxwings, 131; Cedar
  Waxwings, 162, 190
Wells, pumping of, 198
West Indies, 104, 121, 123
West Virginia strip mines, 24–25
Western forests: Clark's Nutcrackers in,
  177–81; compared with boreal forests,
  155–56, 157; conservation of birds, 182–83;
  fires, 146, 160–64, 165, 180; forest types by
  elevational belts, 156–58; logging, 162–66;
  Marbled Murrelets in, 167–71; owls of old-
  growth forest, 166–67; photographs of, 157,
  161, 169, 173; Pinyon Jays in, 172–77;
  pinyon-juniper woodland, 172–77; Red
  Crossbills in, 181–82
Western Hemisphere Shorebird Reserve Net-
  work, 246
Westover Air Reserve Base, Mass., 22, 23
Wheat fields, 66
Whitcomb, R. F., 5
White Memorial Foundation, Conn., 107
White Mountain National Forest, N.H., 115,
  118–20
White Mountains, 14, 117, 125–26
White pine blister rust, 180
Wilbert, Mark, 247
Wilcove, David, 107, 112–13
Wildebeest, 60

Wildfires. *See* Fires
Wildflowers, 49
Wildlife refuges, 129
Williams, A. B., 100
Williams, Roger, 8
Willows, 127; desert willows, 192, 194,
  195–200, 203; Goodding willows, 189–90
Wilson, Alexander, 18, 27, 32, 85, 90
Wind Cave National Park, S.D., 60–61, 62
Windstorms, 34–38, 53, 94, 95, 135, 234, 246,
  248
Winne, Chris, 9
Winter habitats, 42, 47–48, 67–68, 95, 104,
  120–26, 128–29, 182, 194
Winthrop, John, 14
Wiregrass, 209, 211, 212, 214, 215
Wolfberries, 192, 202
Wolverines, 75
Wolves, 75
Wood, William, 18, 19
Woodcocks, American, 255
Woodcreepers, 122
Woodpeckers, 142, 151, 216; Acorn Woodpeck-
  ers, 165; Black-backed Woodpeckers, 135,
  136, 155, 160–63, 182; Black Woodpeckers,
  140, 142, 153; Downy Woodpeckers, 101;
  Gila Woodpeckers, 190, 191, 203; Great
  Spotted Woodpeckers, 142; Hairy Wood-
  peckers, 135, 160; Ivory-billed Woodpeckers,
  x, 75, 81, 82, 88, 91–94, 228–33, 239, 242,
  273n1; Ladder-backed Woodpeckers, 203;
  Pileated Woodpeckers, 81, 93, 135, 138, 153,
  165, 273n1; Red-bellied Woodpeckers, 17,
  87; Red-cockaded Woodpeckers, 75, 82,
  208, 209, 211, 215–27, 244; Red-headed
  Woodpeckers, 16; Strickland's Woodpeckers, 195; Three-toed Woodpeckers, 131, 135,
  138, 140, 142, 153, 160, 162, 163
Woodrats, 166
World Resources Institute, 234
Wrens: Cactus Wrens, 185; House Wrens, 160;
  Rock Wrens, 156, 163; Sedge Wrens, 70;
  Winter Wrens, 163, 165, 183, 247–48

Yellowstone National Park, ix–x, 85, 158–59,
  236, 239–40
Yellowthroats, 135, 199, 200
Yucatan Peninsula, 104, 122

Zebra, 60
Zuckerberg, Benjamin, 49